2 Springer Series in Chemical Physics
Edited by Robert Gomer

Springer Series in Chemical Physics

Editors: V. I. Goldanskii   R. Gomer   F. P. Schäfer   J. P. Toennies

M. A. Van Hove   S. Y. Tong

Surface Crystallography by LEED

Theory, Computation and Structural Results

With 19 Figures

Springer-Verlag Berlin Heidelberg New York 1979

Michel A. Van Hove, Ph.D.

Materials and Molecular Research Division, Lawrence Berkeley Laboratory
and Department of Chemistry, University of California, Berkeley, CA 94720, USA

Shuk Yin Tong, Ph.D.

Surface Studies Laboratory, Department of Physics
University of Wisconsin-Milwaukee, Milwaukee, WI 53201, USA

Series Editors:

Professor Vitalii I. Goldanskii

Institute of Chemical Physics
Academy of Sciences
Vorobyevskoye Chaussee 2-b
Moscow V-334, USSR

Professor Robert Gomer

The James Franck Institute
The University of Chicago
5640 Ellis Avenue
Chicago, IL 60637, USA

Professor Dr. Fritz Peter Schäfer

Max-Planck-Institut für
Biophysikalische Chemie
D-3400 Göttingen-Nikolausberg
Fed. Rep. of Germany

Professor Dr. J. Peter Toennies

Max-Planck-Institut für Strömungsforschung
Böttingerstraße 6–8
D-3400 Göttingen
Fed. Rep. of Germany

ISBN-13:978-3-642-67197-5   e-ISBN-13:978-3-642-67195-1
DOI: 10.1007/978-3-642-67195-1

Library of Congress Cataloging in Publication Data. Van Hove, Michel André, 1947-. Surface crystallography by LEED. (Springer series in chemical physics; v. 2) Bibliography: p. Includes index. 1. Crystallography. 2. Surface chemistry. 3. Electrons—Diffraction. I. Tong, David Shuk Yin, 1942-. joint author. II. Title. III. Series. QD921.V33 548   78-20806

2153/3130-543210

Preface

Surface science has experienced an impressive growth in the last two decades. The attention has focussed mainly on single-crystal surfaces with, on the atomic scale, relatively simple and well-defined structures (for example, clean surfaces and such surfaces with limited amounts of additional foreign atoms and molecules). One of the most fundamental types of information needed about solid surfaces concerns the relative atomic positions. The geometrical arrangement of surface atoms influences most physical and chemical properties of surfaces, the list of which is long and includes a number of important technological applications: electronic surface states, contact potentials, work functions, oxidation, heterogeneous catalysis, friction, adhesion, crystal growth etc.

Surface crystallography - the determination of relative atomic positions at surfaces - has found a successful tool in Low-Energy Electron Diffraction (LEED): this technique has now determined the atomic positions for nearly a hundred surfaces, whether in the clean state or with additional foreign atoms or molecules. The main aim of this book is to publish a set of computer programs that has been specifically designed for and extensively used in surface crystallography by LEED. These programs are based on the dynamical (i.e. multiple-scattering) theory of LEED. They include a number of features reducing the computing time and computer-core requirements. They can be used as they stand for a selected set of surfaces, or they can be slightly modified to handle an unlimited number of other types of surfaces. The computer programs have been conceived in a building-block form to allow the selection in any given situation of the most efficient combination of several theoretical methods. To that end extensive information is supplied about the underlying theory, about the structure of the programs and about the choice of suitable input parameters. The intention is to enable the user, who might be an experimentalist or a graduate student, to perform independently a surface structural analysis by LEED.

Since the LEED theory describes the phenomenon of propagation of slow electrons (energy in the range 0-400 eV) through ordered arrays of atoms, this description is valid for a variety of other modern surface analysis techniques that also involve the propagation of slow electrons through crystalline lattices: thus the methods and computer programs described in this book would find ready applications to, for example, ultraviolet photoemission spectroscopy, Auger electron spectroscopy and high-resolution inelastic electron energy loss spectroscopy.

To give the reader a better feeling for the LEED process, which is not as complicated as might appear at first sight, we supplement the basic theory with a more phenomenological description of the diffraction of electrons at surfaces. Both to illustrate the capabilities of surface crystallography by LEED and to provide a comprehensive and organized review of its results, a chapter is devoted to describing and referencing all surface structures obtained in this way (and known to the authors).

In writing this book we are mainly indebted to John Pendry, who introduced one of us (MAVH) to the field of LEED and whose influence is felt in particular in this book. We also wish to thank a number of colleagues for their critical and constructive comments arising from the actual use of our programs: we should mention in particular B.J. Mrstik, N.L. Stoner, P.R. Watson, A. Ignatiev, B.W. Lee and S.L. Cunningham. We gratefully acknowledge essential help on the part of Sandie Tong (for the figures) and Joyce Miezin (for the typing). Special thanks to S.-W. Wang are also appropriate. One of us (MAVH) is moreover indebted for hospitality and the availability of computational facilities during the preparation of this book to the groups of Prof. W.H. Weinberg at the California Institute of Technology and of Prof. G. Ertl and Prof. H. Jagodzinski at the University of Munich (Sonderforschungsbereich 128). S.Y. Tong acknowledges the hospitality of the Department of Chemistry and the Lawrence Berkeley Laboratory, University of California, Berkeley, where the manuscript received its finishing touches. And last but not least, our thanks go to Prof. R. Gomer and Dr. H. Lotsch (Springer-Verlag) for making the publication of this book a reality.

M.A. Van Hove

Munich and Berkeley, November 1978   S.Y. Tong

Contents

1. Introduction

1.1 LEED as a Tool for Surface Studies

The interference phenomenon involved in the scattering of low-energy electrons from a solid is among the first successful demonstrations of the predictions of modern quantum mechanics. The pioneering experiments were performed by DAVISSON and GERMER in 1925-27 [1]. From the very start, they recognized the potential of using this scattering from the surface region to study geometric positions of surface atoms. Incident electrons in the energy range 10-400 eV penetrate not more than 5-10 Å into the surface region. The wave interference basic to LEED contains an abundance of information about the geometrical configuration of atoms at surfaces.

Although some early enthusiasts had hoped for quick results, successful extraction of surface structural information by LEED did not come until 1971. Part of this is because in order to realize the potential of LEED, it is first necessary to obtain experimentally-controlled surfaces whose chemical composition could be reproduced and maintained during the time required to take the LEED intensity data. This requirement implied the use of ultra-high vacuum ($\sim 10^{-10}$ Torr) equipment, whose unavailability until about 1960 partially explains the state of dormancy of the art before then. The second difficulty in the development of the LEED technique is a theoretical one. It was recognized early that the scattering of low-energy electrons from atomic centers is many orders of magnitude stronger than that encountered in X-ray scattering. As a result, many more physical processes are involved, contributing to rather complex intensity-voltage (IV) LEED spectra. Thus, simple ideas borrowed from X-ray diffraction theory proved largely fruitless. The only important exception is in the extraction of unit cell information from the periodicities of the two-dimensional diffraction patterns. The realization that multiple scattering of LEED electrons was responsible for the difficulties encountered in the theoretical interpretation resulted in the development of "dynamical" theories. The application of these

theories, initially to metal surfaces, established that the large cross-
section of the electron-atom scattering has to be taken into account not
only in terms of multiple scattering but also in terms of the angular dependence of the atomic scattering amplitude.

The major processes that are involved in the electron scattering by a
solid surface have now been identified. The main ingredients in LEED theory may be summarized as follows: (i) calculation of energy-dependent electron scattering amplitudes from surface atoms, (ii) summation of intraplanar
and interplanar multiple scattering of electrons from surface layers, (iii)
inclusion of inelastic damping of the incident electrons by bulk and surface
plasmons and single-particle excitations, (iv) inclusion of electron-phonon
scattering which contributes to an effective Debye-Waller factor.

The first structural results determined by LEED were obtained on simple
metals in 1971. In the years that followed, the list extended to include
surface structures of transition metals, semiconductors, layer compounds,
chemisorbed atomic and molecular overlayers on metal surfaces, "underlayers"
and incommensurate overlayer-substrate lattice structures. By early 1978,
results for over 90 surface structures were reported (see Chap.12). The
results are impressive both in terms of the variety of materials studied
and the (relative) complexity of some of the structures analyzed. The trend
.is moving towards studying increasingly complex surfaces on materials
having practical and chemical applications. Substances that are catalyti-
cally active seem to receive particular attention. The emphasis of the
theoretical development is towards more efficient treatments of multiple
scattering processes for more atoms per unit cell and smaller interlayer
spacings.

It is perhaps important to recognize that surface crystallography by
LEED will never be as simple as its highly successful counterpart, bulk
crystallography by X-ray diffraction. This is because X-rays interact only
weakly with the crystal matter (allowing a single-scattering description
with simple atomic scattering factors), while the surface sensitivity in
LEED results from strong interactions of the electrons with surface atoms
of a solid. To provide a useful theoretical interpretation of LEED data,
one would necessarily use a somewhat sophisticated description of this
strong interaction and include many important multiple scattering events.
In other words, the LEED problem by its very nature of being surface sensi-
tive, requires a rather complex theoretical description. However, this complexity of LEED also has an important consequence. The problems of strong

electron-ion-core scattering and multiple scattering are shared by every
surface technique that involves the passage of slow electrons through atomic layers. The development of an accurate and effective treatment of strong
electron-ion-core scattering and the efficient inclusion of multiple scattering processes are both important goals in the application of current theoretical methods to solid surfaces. The by-products of the difficulties
faced in LEED are the formulation, out of necessity, of a number of very
efficient methods for treating electron scattering by periodic two-dimensional layers. These theoretical techniques are useful not only in LEED but also
in the interpretation of other surface probes. They also share many common
features with layer and surface calculations of electron energy bands, surface states and surface bonding.

Since the knowledge of exact geometric configurations at a surface is a
prerequisite for quantitative studies of many surface problems, it is felt
that the computational techniques developed for surface structural determination by LEED should be made widely available. An important purpose of
this book is to provide and explain the use of a set of computational programs that have been designed and used for many surface structural determinations. Chapter 2 introduces the basic theoretical ideas on which the
computations are based, while Chaps. 3-9 discuss the various techniques,
formulas and ingredients used and their implementation in the programs. The
main program is discussed and listed in Chap. 10, while annotated subroutine listings appear in Chap. 11. Results of surface structure determinations by LEED are given in Chap. 12. A plotting program is given in App. C.

1.2 Purpose of the Computational Programs

The computer programs included in this book basically serve the purpose of
computing elastic LEED intensities. They are primarily designed to be used
in surface crystallography, i.e. in the search for the atomic positions at
the surface of a well-ordered crystal, through the comparison of computed
and measured LEED intensities. In addition, these programs may be used in
the closer study of several physical quantities related to the surface
under consideration (such as the electron-atom scattering potential, the
surface potential step, the electronic mean free path and atomic thermal
vibration amplitudes at the surface).

Many of the subroutines are also useful for or adaptable to the calculation of the final (scattering) state in angular-resolved ultraviolet photoemission spectroscopy and angular-resolved Auger electron spectroscopy.

1.3 Physical Processes Included in the Programs

The electron scattering by individual atoms of the crystal surface under consideration is taken into account by assuming spherically symmetrical interaction potentials surrounded by regions of constant potential (the "muffin-tin model"). Electron scattering by the potential is described by partial-wave phase shifts (cf. Sect.3.4). The effect of surface atomic thermal vibrations is included through temperature corrections to the phase shifts. These temperature corrections incorporate a Debye-Waller factor correction to the atomic scattering amplitude. The correlation of the vibrations of neighboring atoms is neglected.

The effect of inelastic damping, which removes electrons from the flux of elastically scattered electrons, is incorporated through an imaginary part of the constant potential in the muffin-tin model. Multiple scattering of the electrons by the lattice is included through various schemes; most of these are perturbation treatments, all of which converge to the results of "exact" computation schemes.

No surface potential step is included in the programs, but such a step can easily be included as an additional scattering layer.

The underlying theory is non-magnetic and non-relativistic (except insofar as the user may, for example, choose a relativistic atomic potential).

1.4 Capabilities of the Programs

The dynamical LEED programs can be used to calculate the diffraction of electrons by many ordered crystal surfaces with or without superstructures. This includes clean surfaces, reconstructed surfaces with or without foreign atoms (including interstitial or substitutional insertion), overlayers (from submonolayers to multilayers), cases of coadsorption and molecular adsorption. This flexibility is mainly due to the "building-block" nature of the programs: all one needs to do is to choose the right combi-

nation of subroutines and assemble an appropriate program for a given surface structure.

Features of the program that reduce the computational effort include lattice sums done over superlattice-induced sublattices and angular sectors
(Chap.5), the use of symmetries in reciprocal space (Chap.4) and the energy dependence of the number of plane waves used (Chap.3).

1.5 Size, Speed and Limitations of the Programs

It is not possible to give general rules concerning the LEED program sizes
and time requirements, because these depend on too many factors. Only guidelines can be given in this regard.

Situations that involve large two-dimensional surface unit cells ($\gtrsim 50\overset{\circ}{A}^2$)
are unfavorable because of the resulting large number of plane waves that
must be considered, giving rise to matrices with large dimensions. Similarly, large numbers of phase shifts ($\gtrsim 10$) are unfavorable. Also, high energies ($E \gtrsim 400eV$) are problematic, because they require many plane waves
and many phase shifts. Low symmetry increases matrix dimensions as well.

When many ($\gtrsim 5$) closely-spaced atomic layers have to be treated by the
Beeby matrix inversion and/or Reverse Scattering Perturbation methods, many
large propagation matrices G^{ij} (see Chaps.6,7) must be generated, stored
and used. Furthermore, at very low electron energies ($\gtrsim 15eV$), where the
electron damping is small in many materials, the convergence of the lattice
sums and the multiple-scattering perturbation expansions become poor.

Since surface crystallography by LEED implies a search for the correct
structure by trial and error, this search can involve fair numbers of computations for different structural guesses. Although many time-saving features are incorporated in the programs to deal with this situation, complicated structures can give rise to a large computational effort.

The set of structures described in Chap. 12 gives a good idea of what
kind of surfaces can in practice be analyzed with the current LEED methods.

1.6 Relation to Other LEED Programs

Dynamical LEED programs are available from a book by PENDRY [2]; but these
programs apply, as they stand, only to a rather restricted class of sur-

faces. They do not accept closely-spaced atomic layers and do not exploit symmetries in reciprocal space. They do, however, include, in addition to Layer Doubling and Renormalized Forward Scattering, the Bloch wave method (which is relatively more time-consuming), and the Kambe lattice summation, which must be used in the presence of very small damping.

Symmetrized programs have been published by RUNDGREN and SALWEN [3]. There the symmetrization involves less user intervention than in the programs presented in this book. However, as with PENDRY's programs, only a limited class of surfaces can be analyzed.

None of the above published programs uses the Combined Space method available in this book (see Chaps.6,7), which removes many limitations on the kind of surface the calculations can handle.

1.7 Some Computational Considerations

Some subroutines used in our programs were originally developed by PENDRY, although in many cases extensive revisions have been made. The incorporation of the Combined Space method involved merging with programs developed by TONG on the basis of the Beeby formalism (cf. Chap.6). This resulted in the mixing of two differing conventions, e.g. two equivalent but different sets of Clebsch-Gordon coefficients are required in the Combined Space method for generating layer diffraction matrices (not counting a third set for the generation of temperature-dependent phase shifts).

The programs are written in a Fortran that is standard enough to be accepted by compilers on most machines: usually trivial adaptations are sufficient. Although most of the supplied program components have been extensively used in many computations, we cannot guarantee that they are error-free: feedback from users would be appreciated in this respect.

The authors intend to make available magnetic tapes containing an up-to-date version of the programs listed in this book, including additional main programs for other surface geometries, new subroutines, a phase shift program and a series of atomic potentials and phase shifts. This may also be useful for avoiding errors in manual punching from the listings presented here. To obtain the tapes, users should address their requests to S. Y. Tong.

2. The Physics of LEED

Low-energy electron diffraction is termed a "dynamical" process to express the fact that non-geometrical parameters play an important role in its description: these parameters (mainly atomic scattering amplitudes, inner potential and inelastic electron damping), together with multiple scattering of the electrons through the crystalline lattice, complicate the understanding of the diffraction process. For example, the analogy with X-ray diffraction (with its simple Bragg reflection conditions, etc.) is obscured by the dynamical character of the scattering. This is readily seen in the failure of "kinematic" theory (single-scattering theory) to explain LEED intensity vs. energy curves (IV-curves). Background information about the development and application of LEED, including experimental questions, will be found in Refs. [2,4-9].

While IV-curves usually do have a complicated appearence, it is nevertheless possible to understand how they come about in a simple way. This section will attempt to show this by introducing one by one the various complicating factors of LEED, starting from the familiar situation of X-ray diffraction.

2.1 A Simple Description of the LEED Process:
Clean Crystals and Bragg Reflections in One Dimension

Let us first consider a one-dimensional space, one half of which (x>0) is occupied by a crystal composed of a semi-infinite row of identical equally-spaced "atoms" (this represents a clean, ideal surface). An undamped quantum-mechanical wave exp(ikx), where k is the wavevector, is incident on the crystal from x = -∞. This wave is scattered (towards x = -∞ and x = +∞) by each atom, as described by complex reflection and transmission coefficients r and t.

The weak-scattering X-ray limit is characterized by a very small value of |r| and essentially total transmission |t| = 1 [in that limit the phase

arg(t) of the transmission coefficient must vanish for wavefunction conti-
nuity]. Our one-dimensional crystal reflects a part of the wave exp(ikx)
into a wave given by the following expression (a: lattice parameter, x = 0
at first atom):

$$\sum_{j=0}^{\infty} re^{ik2aj} e^{-ikx} = \frac{r}{1 - e^{ik2a}} e^{-ikx} \quad . \tag{1}$$

Generally the amplitude of the reflected wave is small, but strong reflection takes place when the Bragg condition k2a = n2π (n integer) is satisfied. An illustration of the energy-dependence of this reflection amplitude
(squared) is given in Fig. 2.1a (energy E and wavevector k being related by
E = (Ħk)^2/2m): infinitely sharp Bragg reflection peaks occur.

Moreover, the reflection amplitude is infinite at each Bragg condition,
clearly violating current conservation. Current conservation is actually
always violated because of the assumption of total transmission (|t| = 1).
For an improved physical description we must accept that |t| < 1. Equation
(1) is easily rewritten for this case, if we recognize that rexp(ik2aj) represents the round-trip amplitude factor for traveling from the crystal
surface to atom j + 1 and back: this quantity should be replaced by
rt^{2j}exp(ik2aj) because the wave is transmitted twice through each interven-
ing atom. We then have, instead of (1):

$$\sum_{j=0}^{\infty} rt^{2j}e^{ik2aj} e^{-ikx} = \frac{r}{1 - t^2 e^{ik2a}} e^{-ikx} \quad . \tag{2}$$

The new reflection amplitude of (2) still has reflection maxima at the same
energies, but they are not of infinite height, since |t| < 1. Nor are these
peaks infinitely sharp any more. Assuming that t is real [arg(t) = 0], the
resulting energy dependence of the reflection amplitude has a behavior as
shown in Fig. 2.1b.

Current conservation has not however been introduced quite correctly in
(2). At each of the transmissions mentioned above, the lost current goes
into a new reflected wave, which in turn can reflect, etc.: in short, multiple reflections take place, at each of which current conservation should
be respected. In fact, current conservation will only be satisfied if all
multiple reflections (to infinite order) are included. This cannot be con-

---

Ħ = h/2π (normalized Planck's constant)

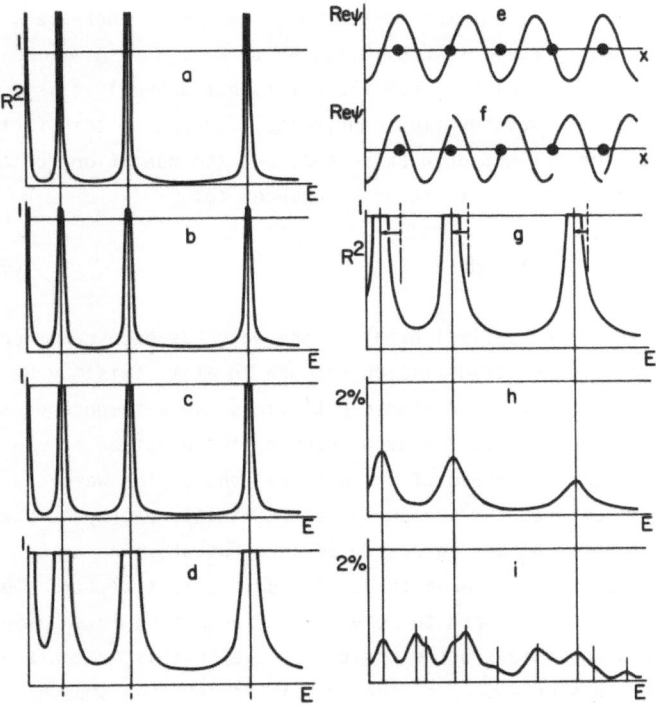

Fig. 2.1 (a)-(d) and (g)-(i) schematically show reflected intensities (amplitudes squared) as a function of energy E as dynamical effects are gradually included (cf. text); vertical bars in (g) line up with those in (d); (e) and (f) show a wave propagating through a lattice with zero (e) and non-zero (f) atomic transmission phases (neglecting amplitude damping)

veniently expressed in a relation like (2). The result is given by the Bloch wave theory, which treats the multiple scattering self-consistently: the result is that multiple scattering leaves the reflection peaks essentially where they are and does not appreciably change their widths (cf. Fig. 2.1c). But the peaks are now flat-topped, with a height of 1: the multiple reflections conspire exactly to produce total reflection in a neighborhood of each Bragg condition. This merely corresponds to the familiar band gaps of a crystalline material: at electron energies within a band gap no propagation can take place, therefore total reflection must occur (whether in the bulk or at the surface). Within the bands an electron can penetrate a crystal infinitely far (if we ignore inelastic scatterings), and so there is only partial reflection.

The peak widths must increase as the scattering strength is increased, corresponding to wider band gaps: cf. Fig. 2.1d, Sect. 2.2.

We have assumed a real transmission coefficient t, but actually t is normally complex (to satisfy wavefunction continuity). Inserting this fact in (2) does not affect the reflection peak shapes. But the condition for a reflection maximum (constructive interference) changes to:

$$k2a + 2 \arg(t) = n2\pi \quad \text{(n integer)} \quad . \tag{3}$$

What has happened is that the "optical path" of the wave has been modified by a phase shift $\arg(t)$ at each transmission through an atom. This may be understood, in the case of attractive atomic potentials, as a temporary speeding up of the electron through the atom, causing the electron to emerge behind the atom somewhat ahead of its original phase (the wavelength is momentarily shortened through an increased kinetic energy, thereby making the phase of the scattered electron advance faster). The effect is illustrated in Fig. 2.1e,f: for an attractive atomic potential, the electron wave effectively gets telescoped to some extent, resulting in an effective overall wavelength reduction in the crystal (the effect is dramatized here by making the wave discontinuous at the nuclei, rather than smoothly varying in wavelength). To keep the Bragg condition satisfied, this shortened wavelength must be increased through a decrease of the energy, and so Bragg peaks are normally found at energies substantially below the kinematic prediction: see Fig. 2.1g. Note that the phase of the reflection coefficient r has no effect on this result, apart from a relatively small indirect effect through multiple scattering.

The physical origin of this "transmission phase effect" can be split into two components. The first is the familiar inner potential, whose effect can be more directly described as a rigid shift of the energy axis (insofar as the small, gradual energy dependence of the inner potential is ignored) by about 5 to 15eV, depending on the material. The second is due to the multiple scattering of an electron within a single atomic core, i.e. the diffraction of the electron through the complicated "electrostatic lens" which an atom is. Here strong resonance effects can and do occur (just as with a particle moving over a square potential well), giving rise to substantial transmission phase shifts. The resonance effects are clearly energy-dependent, and so therefore are the transmission phase shifts. As a result, peaks in IV-curves occur not only well below their kinematically expected energies (due to the inner potential) but also scatter about their inner-potential-corrected positions (due to intra-atomic resonances); this scatter is mostly towards even lower energies. In dynamical LEED calculations the effects of intra-atomic multiple scattering are contained in the partial-wave phase shifts.

The effect of the intra-atomic resonances is one of the main causes of the difficulty that has been experienced in LEED theory with perturbation methods based on the number of scatterings undergone by an electron. Figure 2.1e,f shows the problem clearly: without correction for the transmission phase shift, the wave exp(ikx) of Fig. 2.1e is, after a few lattice spacings, a very poor approximation to the actual wave, which is more like that of Fig. 2.1f. The lowest order of a simple perturbation theory includes only reflection and no transmission phase factor: hence it produces peaks at wrong energies (even after an inner-potential correction). The next orders of perturbation must both destroy those peaks and create new peaks at new energies, since that is what is involved in shifting a peak: hardly a perturbation.

2.2 Peak Width and Electron Penetration Depth

We wish to illustrate the close relationship between the peak widths in IV-curves and the electron penetration depth into the crystal surface.

If we consider a wave that satisfies the Bragg conditions (generalized to take transmission phases into account), this wave must decay into the surface, since it is eventually totally reflected (we still neglect damping). The decay length can be easily estimated. With the same assumptions as in the preceding section, we may say that each atom reflects a fraction |r| of the amplitude incident on it, so that N atoms, where N|r| = 1, will reflect almost all the amplitude. The penetration depth is therefore about Na = a/|r|, where a is the interatomic distance.

To obtain the peak width, we use the fact that an array of N equally-spaced wave sources (corresponding to our N participating surface scatterers) produces diffraction maxima that have a width $2\Delta k \sim \pi/(2Na)$ in reciprocal space. This results from the fact that a pair of sources produces peaks $\Delta k = \pi/a$ apart; a chain of N (strictly, N + 1) such sources sharpens those peaks and produces N - 1 new smaller intermediate peaks, separated by $\Delta k = \pi/(Na)$; the peak width $2\Delta k$ is about equal to half that separation, which gives the above result. The corresponding energy width is

$$2\Delta E = 2 \frac{\hbar^2}{m} k\Delta k \sim \frac{\hbar^2}{m} k \frac{\pi}{2} \frac{1}{Na} = \frac{\pi\hbar^2}{2m} k \frac{|r|}{a} \quad . \tag{4}$$

This result agrees, not surprisingly, with the band gap width obtained in the first-order perturbation treatment of band structure theory.

The link between penetration depth and peak width, as given by (4), is particularly useful because it can be immediately used also to describe the effect of inelastic damping. The electron mean free path λ also acts to reduce the number of participating scatterers, and the above reasoning can be repeated for this case. The penetration depth λ is simply Na in (4), and so

$$2\Delta E \sim \frac{\pi \hbar^2}{2m} \frac{k}{\lambda} \sim 2V_{0i} \quad . \tag{5}$$

Here we have introduced the quantity V_{0i}. If the electron energy is made complex by an imaginary part $-iV_{0i}$ ($V_{0i}>0$), sharp peaks are smoothed out to a width $2V_{0i}$. In other words, we may simulate the mean free path by an imaginary component of the energy. A more accurate comparison of λ and V_{0i} (obtained by directly relating λ via the imaginary part of k with V_{0i} for small V_{0i}) gives the following relationship:

$$V_{0i} \sim \frac{\hbar^2}{m} \frac{k}{\lambda} \quad . \tag{6}$$

Experimentally observed peak widths are usually governed by inelastic effects (not counting peak overlaps): these peak widths vary from ~3eV at low electron energies, to ~15eV at high energies. But at very low energies (E $\lesssim$ 10eV), strong elastic scattering can open up wide band gaps that generate peak widths (~5eV) in excess of the inelastic peak width $2V_{0i}$: there the electrons stand a higher chance of being totally reflected to the vacuum than of being absorbed (reflectivities larger than 50% have been measured at low energies). At higher energies electrons are much more likely to be absorbed than elastically reflected (reflectivities of the order of 1% or lower are the rule at energies of 50eV or more): see Fig. 2.1h. A further effect of damping is to smear out the sharp-edged flat tops of reflection maxima mentioned in the previous section: to obtain flat tops infinite penetration must be possible.

We stress the observation made in this section that peak widths can be described in terms of penetration depth for both elastic and inelastic scattering: the same mechanism operates in both situations.

A further observation can be made: peak shapes (not counting overlapping peaks) are determined primarily by the total penetration distance rather than by local variations of, say, the inelastic effects. It is pointless to make the quantity V_{0i} dependent on position within a unit cell, as such dependence is essentially averaged out in the diffraction process. It may at most be useful to make V_{0i} layer-dependent (thinking in three dimensions), so that electron path lengths parallel to the crystal surface can be different at different depths.

2.3 Three-Dimensional Effects

The major new feature introduced by three-dimensional, in contrast to one-dimensional, crystals is the increase in the number of beams in which the elastically diffracted electrons can travel. This is a result of the periodicities present in more than one dimension. In X-ray diffraction the three-dimensional periodicity of crystals gives rise to a set of beams characterized by *three*-dimensional reciprocal lattice vectors $\underline{g}$. At surfaces only a two-dimensional periodicity exists and so one obtains a set of beams characterized by *two*-dimensional reciprocal lattice vectors $\underline{g}$ (observed as spots on the LEED screen). These beams are sharply defined (if the surface is well ordered) and their positions are uniquely determined by the two-dimensional periodicity and the wavelength. The scattering mechanism (whether kinematic or multiple-scattering) does not affect the positions of beams, because these positions are determined solely by the *relative* phase of reflected waves emanating from scattering atoms equivalent under the periodic translations of the surface: this relative phase depends on the unit cell dimensions, but not on the scattering mechanism. The scattering mechanism, no matter how complicated, influences only the *absolute* phase (as well as the intensity) of the reflected waves, a quantity that disappears when relative phases are considered. Therefore LEED spots always have positions that can be determined kinematically.

Each layer in the clean three-dimensional crystal surface diffracts a beam $\underline{g}$ into a beam $\underline{g}'$ with a reflection coefficient $r_{\underline{g}'\underline{g}}$, or a transmission coefficient $t_{\underline{g}'\underline{g}}$. Each beam has its particular wavevector $\underline{k}(\underline{g})$; $k_{\perp}(\underline{g})$ will denote the component of $\underline{k}(\underline{g})$ perpendicular to the layers, and the letter a stands for the layer separation.

As in one dimension, we can generalize the familiar Bragg reflection conditions used in X-ray diffraction by including transmission phases induced by the strong potentials in the atomic layers:

$$k_\perp(\underline{g}).a + \arg\left(t_{\underline{g}\underline{g}}\right) + k_\perp(\underline{g}').\,a + \arg\left(t_{\underline{g}'\underline{g}'}\right) = n.2\pi \quad (\text{n integer}) \quad (7)$$

gives conditions for maximum reflection from beam $\underline{g}$ into beam $\underline{g}'$. This equation is rigorous for kinematic (i.e. single) scattering. In the presence of multiple scattering, it holds inasmuch as, in diffraction off a single atomic layer, the zero-angle forward-scattered beam is much stronger than all other diffracted beams. Reflection maxima are substantially shifted from the weak-scattering positions by amounts that are very much energy-dependent as a consequence of resonances.

In the limit of weak multiple scattering, the various reflected beams have only intensity maxima corresponding to (7), $\underline{g}$ being the incident beam, i.e., $\underline{g} = \underline{0}$. Strong multiple scattering can however generate additional maxima, that have the same general appearance and often intensities as the single-scattering ones. This is easily visualized in terms of intermediate beams. The incident beam $\underline{0}$ can scatter into an intermediate beam $\underline{g}_1$, which in turn can scatter into an emerging beam $\underline{g}'$; more successive intermediate beams are of course possible, but they tend to be increasingly weaker because of the more scatterings involved. Either of the two scatterings $\underline{0}$ to $\underline{g}_1$ and $\underline{g}_1$ to $\underline{g}'$ will at particular energies satisfy the condition (7), thereby generating peaks in the emerging beam $\underline{g}'$ at energies different from those predicted by the single-scattering conditions (from $\underline{0}$ to $\underline{g}'$). We may therefore say that a peak occurs whenever, in a chain of scatterings, one of the scatterings satisfies the "Bragg condition" (7). Thus is explained the often abundant set of "multiple-scattering" peaks that is so characteristic of LEED IV-curves, and that are responsible for part of the difficulties encountered in treating these IV-curves with kinematic (single-scattering) methods of data averaging and data reduction.

2.4 Overlayer Effects

So far, we have only analyzed "clean" surfaces. The additional effects resulting from a more complicated termination of the bulk structure at the surface can be illustrated with the case of an overlayer adsorbed on a clean substrate.

We consider an overlayer of mono-atomic thickness adsorbed at a distance d from the clean substrate. The total electron reflection coefficient R can be regarded as being composed of interfering reflections from the overlayer and the substrate. In one dimension especially, this is a simple situation: the substrate reflection coefficient R_s and the overlayer reflection and transmission coefficients r and t then combine to give

$$R = r + e^{2ikd} \, t \, R_s (1 - e^{2ikd} \, r \, R_s)^{-1} t \quad . \tag{8}$$

An analogous matrix equation holds for three dimensions. The factor $(1 - e^{2ikd} \, r \, R_s)^{-1}$ describes multiple scattering between substrate and over-layer, as proved by a geometric expansion. This factor usually plays no significant role in practice because of absorption and the not overly large values of $|r|$ and $|R_s|$ (typically 0.1 and 0.5 to 0.1, respectively). Neglecting this multiple scattering factor, (8) has interference maxima between overlayer and substrate when

$$2kd + 2 \arg(t) + \arg(R_s) - \arg(r) = n2\pi \quad (n \text{ integer}) \quad . \tag{9}$$

Again, we have energy-dependent shifts of the maxima away from simple geometric Bragg-like conditions. These maxima are superimposed on the strong energy dependence of the substrate reflection R_s (a complex number) in such a way that only by including the relevant scattering phase shifts can one reproduce experiment faithfully. This is true with one- as well as with three-dimensional crystals. Interference also produces intensity minima in addition to maxima, and total destruction can easily occur.

This interference between overlayer and substrate is obviously strongly geometry-dependent and explains the sensitivity of IV-curves to the surface layer position. The maxima due to this interference generally have a larger width than those due to the substrate alone, since the width is inversely proportional to the number of interfering components. But the general appearance of the IV-curves remains essentially the same.

With more complicated surface structures the above considerations still apply, but it is clear that it becomes rather difficult to predict the result of all the interference conditions, especially in three dimensions, without doing the actual calculations. Thus, IV-curves become difficult to

understand directly: the basic mechanisms of the diffraction are simple,
but the multitude of simultaneous events obscures that simplicity.

2.5 Elements of a LEED Theory

The preceding sections have exhibited the basic ingredients which an accurate LEED theory should contain. These are the properties of the scattering
of electrons by single atoms (including, in particular, transmission, i.e.
forward scattering), multiple scattering, the inner potential and the damping of waves due to inelastic processes. In practice it is desirable to add
the effects of thermal vibrations of the crystal atoms. We shall briefly
discuss these topics in this section so as to establish the connection with
the full theory as implemented in the computer programs presented in this
book. More detailed discussions can be found in the literature [2].

The potential responsible for the scattering of the LEED electrons is
complicated: it should be considered non-local, it should include electro-
static as well as exchange and correlation effects and it should be treated
in a self-consistent manner. To generate and use such a potential would be
a vast task and so a strongly simplified model is used, which nevertheless
gives surprisingly good results. One chooses a local (one-electron) approximation cast in the form of the "muffin-tin model". Here the potential is
split in two parts: within non-overlapping, usually touching, spheres cen-
tered on each nucleus (the "ion cores") a spherically symmetrical potential
is assumed, while in the region left between these spheres the potential is
taken to be constant. The advantage of the muffin-tin potential is that the
scattering by single atoms is made relatively simple (described by just a
set of phase shifts that depend on chemical species, angular momentum and
electron energy), and the electron propagation between atoms is made even
simpler (free-space wavefunctions can be used). The generation and use of
the spherically symmetrical part of the potential is discussed further in
Sect. 3.4.

The constant part of the potential in the interstitial regions, V_0, is va-
riously called the "muffin-tin constant" (or "muffin-tin zero") or also the
inner potential, although these designations do not really present the same
physical quantities. However, since no consensus exists as to the proper
use (or even exact meaning) of these terms, and since their exact relation-

ship is not needed in LEED, we shall neglect the distinction between them and use the designation "muffin-tin constant" (or "muffin-tin zero") for V_0. The values of V_0 found in LEED (which already depend on the particular calculations) cannot be compared with values known from other sources.

Mainly because of the energy dependence of the exchange potential, the muffin-tin constant is slightly energy-dependent, but this is usually neglected.

The form of the transition of the potential through the surface from the bulk muffin-tin constant to the vacuum zero-potential level is relatively poorly known. This potential step has basically three effects: (1) refraction of any plane wave passing the potential step occurs, because the momentum parallel to the surface is conserved, while the momentum perpendicular to the surface is changed; (2) reflection by the step occurs, but this is of importance only when the electron energy is less than a few times the step height; (3) if the potential step has structure parallel to the surface (due to the underlying atomic lattice), additional diffraction can occur in a way similar to that due to any atomic layer. The first of these effects (refraction) is included in our programs; the second one (reflection) is neglected, as its influence is felt only at low energies, below about 40eV; there is no evidence that the third effect (diffraction) needs to be considered.

To represent the loss of electron flux due to inelastic processes, the potential is given an additional imaginary part V_{0i}, cf. (5,6). This quantity in effect lumps together many complicated energy-loss processes, such as single-electron and plasmon excitations. Its calculation is avoided in practice and, since its value is not very critical to the LEED problem, good guesses are easily obtained (V_{0i} is made energy-dependent for some materials): see Sect. 10.1. The effect of the complex nature of the potential is to make the electron wavevector $\underline{k}$ complex; many numerical operations must therefore be performed in the complex plane. For example, spherical harmonics $Y_{\ell m}(\theta,\phi)$ with complex $\cos\theta$ must be used. For these the following definition is chosen in our programs:

$$Y_{\ell|m|}(\theta,\phi) = (-1)^{|m|}\left[\frac{(2\ell+1)}{4\pi}\frac{(\ell-|m|)!}{(\ell+|m|)!}\right]^{1/2}P_\ell^{|m|}(\cos\theta)\exp(i|m|\phi) \quad .$$

$$Y_{\ell-|m|}(\theta,\phi) = \left[\frac{(2\ell+1)}{4\pi}\frac{(\ell-|m|)!}{(\ell+|m|)!}\right]^{1/2}P_\ell^{|m|}(\cos\theta)\exp(-i|m|\phi) \quad .$$

$$(10)$$

Complex conjugation must be done through the relation

$$Y^{*}_{\ell m}(\theta,\phi) = (-1)^{m} Y_{\ell -m}(\theta,\phi) \tag{11}$$

With the muffin-tin potential it is natural to express the wavefunction of the LEED electron as linear combinations of either plane waves or spherical waves. Both types of elementary waves have advantages in the present problem and therefore both are used side by side. The spherical waves are convenient when describing the scattering by the spherically-symmetrical ion cores and the multiple scattering between atoms in individual layers parallel to the surface. On the other hand, plane waves are the more convenient choice to describe the wavefunction between successive atomic layers, since the diffraction by each layer produces a discrete set of "beams", each of which can be represented by a plane wave. These considerations are responsible for the following basic order of operations in a LEED calculation: (1) computation of the single-atom scattering amplitudes (in angular-momentum space); (2) computation of all scattering within a single layer of atoms (in angular-momentum space); (3) computation of all scattering between the various atomic layers composing the crystal surface under consideration (in linear-momentum space).

To treat multiple scattering exactly, i.e., to infinite order in the number of scatterings, can lead to prohibitively large computational efforts: in this respect the determining factors are mainly the degree of complexity of the surface geometry (complicated surfaces can lead to large time and core requirements) and the strength of the atomic scattering (strong atomic scattering induces strong multiple scattering). A reduction of the computational effort has been achieved through the introduction of suitable approximations or convergent iterative perturbation expansions. Thus one can approximate the semi-infinite material by a slab of finite thickness. The finite mean free path justifies this choice, which is used in the methods of computation described in this book; usually (as in Layer Doubling, see Chap. 8, and Renormalized Forward Scattering, see Chap.9) the thickness of the slab is automatically increased during the computation until convergence of the LEED reflectivities occurs. An efficient perturbation expansion was developed in the Renormalized Forward Scattering (RFS) method. This technique expands the total surface reflectivity in terms of the number of backward scatterings (because these are usually relatively weak); but it takes into account the fact that forward scattering through atoms (or layers of atoms) is not to be treated as a weak process: as discussed in Sect. 2.1 the forward scattering (transmission) phase must be included to produce a good description of the propagation of electrons through the crystal. While the

RFS method implements this perturbation expansion in the linear-momentum space representation to handle the multiple scattering *between* atomic layers, the Reverse Scattering Perturbation method does the same in the angular-momentum space representation (cf. Chap.7) to handle the multiple scattering *within* atomic layers having more than one atom per unit cell.

A topic not touched upon in the preceding sections concerns the effect on LEED of disorder in the crystal lattice. There are many types of disorder: domains (identically structured but mutually mismatched patches of surface, particularly frequent in overlayers with a superlattice), terraces separated by steps of atomic heights (really a form of domains), impurities, point defects, bulk defects terminating at the surface, randomness in the occupation of adsorption sites for overlayers with less than a monolayer coverage, thermal vibrations of the atoms about their stable lattice sites, and many others. The diffraction process has a filtering effect that removes sensitivity to a number of types of disorder, focusing on the perfectly periodic component of the surface structure. In general, the effect of disorder is an overall reduction in the intensity of diffracted beams (while incoherent scattering in all other directions increases). This seems to be the prime cause of the frequent large mismatch that occurs between the experimental and calculated intensity scales: the relation between the intensity scales is therefore often ignored and the intensities are then plotted in arbitrary units. In the case of domains we assume that the area of each of them is appreciably larger than the so-called coherence area of the beam (an area of the order of that of a circle with a radius of a few hundred Ångstroms), so that reflected contributions from different domains do not interfere, but simply add their incoherent intensities. In general, therefore, averaging of the intensities reflected from the different possible domains should be carried out. Because the effects on LEED intensities of the thermal motions of atoms depend strongly (and non-uniformly through multiple scattering) on the energy, the theory includes a correction for these effects. In X-ray diffraction Debye-Waller factors (that reduce the atomic scattering amplitudes) have proved adequate for this purpose. In the case of LEED the situation is complicated by multiple scattering: not only must a correction be applied at each scattering, but in addition the correlation between the thermal motions of nearby atoms should be taken into account, since the path length between nearby atoms is involved. Such correlation effects are disregarded in LEED theory, while the correction at each scattering is included through a Debye-Waller factor applied to each atom's scattering amplitude (cf. Sect.3.5). Other forms of disorder are neglected.

3. Basic Aspects of the Programs

3.1 Units and Geometrical Conventions

The LEED programs assume the input to be expressed in Ångstroms for lengths, electron-volts for energies (except in the phase shift input, where energies are in hartrees), degrees for angles (radians for phase shifts) and kelvins for temperatures. Internally these programs work in atomic units ($\hbar=e=m=1$), i.e. in Bohr radii for lengths and hartrees (double rydbergs) for energies, and in radians for angles and kelvins for temperatures. Output quantities have mixed units: see Chap. 10.

In inputting phase shifts, the main program reads a list of phase shifts with energies in hartrees (cf. Sect.3.4). From this list, a linear interpolation is used to evaluate the phase shifts at the particular electron energies of the calculation. The phase shifts in this input list, for the purpose of interpolation, must be continuous functions of energy. Thus, for phase shifts generated modulo π and energies in rydbergs (as appears in the output of many phase shift programs), the discontinuities must be manually removed by the addition or subtraction of appropriate multiples of π and the conversion from rydbergs to hartrees must be carried out. The number of energies at which phase shifts are given in the input list is rather arbitrary, but the lowest energy in the list must be below the lowest energy of the LEED calculation. This is because our interpolation scheme is not designed to go to lower values of the energy (the interpolation scheme may be used to go to higher energies, i.e. to extrapolate linearly above the upper limit of the list of phase shifts, although the accuracy obviously progressively gets worse).

For the LEED programs, a Cartesian coordinate system is assumed, with the x-axis pointing perpendicularly *into* the crystal surface, the y- and z-axes being parallel to the surface: this choice produces a left-handed coordinate system, but this is of no consequence. The crystal lattice parallel to the surface for different layers is described by pairs of

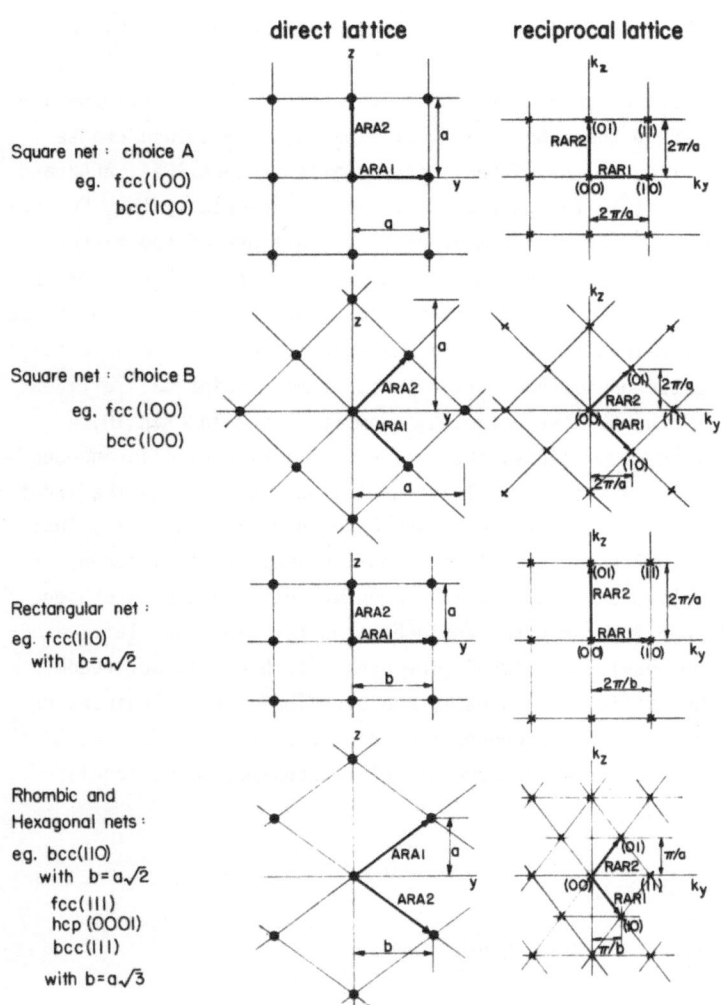

direct lattice **reciprocal lattice**

Square net : choice A
eg. fcc(100)
bcc(100)

Square net : choice B
eg. fcc(100)
bcc(100)

Rectangular net :
eg. fcc(110)
 with b=a√2

Rhombic and
Hexagonal nets :
eg. bcc(110)
 with b=a√2
 fcc(111)
 hcp(0001)
 bcc(111)
 with b=a√3

Fig. 3.1 Preferred direct and reciprocal lattice orientations, at left and at right, respectively, for common surfaces (other acceptable choices exist). If superlattices are condidered, these diagrams apply to them as well, substituting ARB1, ARB2, RBR1, RBR2 for ARA1, ARA2, RAR1, RAR2. The labels (0,0), (1,0), (0,1), ... identify beams (plane waves, spots) in standard LEED notation

two-dimensional basis vectors spanning the unit cells: in the program these basic vectors are called ARA1, ARA2 for substrate layers and ARB1, ARB2 for overlayers or other top atomic layers, which often have a superlattice, assumed to be simply related to the substrate lattice. The corresponding re-

ciprocal-lattice basis vectors are called RAR1, RAR2 and RBR1, RBR2, respectively.

Because of our particular use of symmetries (namely because of some assumptions that speed up the computation, cf. Chap.4), one should choose
certain preferred orientations of the lattice basis vectors ARA1, ARA2 and
ARB1, ARB2 with respect to the (x,y,z) Cartesian coordinate system. For example, a system with mirror planes must have at least one of the mirror
planes oriented at 0°, $\pm45^o$ or 90° to the (xy)-plane. The choice of orientation is unrestricted when no symmetry is used, but it is safest to choose
the preferred orientations there as well. Examples of acceptable orientations for the more common surface lattices with their reciprocal lattices
are represented in Fig. 3.1 (see also the examples given in Chap.10).

To speed up lattice summations, the lattice symmetry (which is independent of the wavefield symmetry) is always exploited. To this end the input
number IDEG must be given one of the values 2, 3, 4 or 6, indicating that
the lattice has an IDEG-fold axis of rotational symmetry. If different layers have different symmetry, the lowest common symmetry should be chosen.
In the case of IDEG = 3, 4 or 6 the angle between the two input lattice basis vectors must be equal to 2π/IDEG. (The use of IDEG = 6 is not recommended, because the lattice sum is performed over IDEG parallelepipeds ra-
diating from the origin: this produces a 6-pointed star for IDEG = 6;
IDEG = 3 produces a hexagon which more closely approximates the ideal
circle.)

3.2 Layers, Subplanes and Plane Waves

Any crystal surface is imagined to consist of a stack of atomic layers,
each parallel to the surface and therefore extending to $\pm\infty$ in the y- and
z-directions. A layer may have a Bravais lattice, i.e. have one atom per
unit cell. We call this a "Bravais-lattice layer". However, for reasons to
be explained later in this section, when layers have a vanishing or small
mutual spacing (measured perpendicular to the surface), it is more efficient
to treat the whole group of closely-spaced Bravais-lattice layers as one
single layer, called a "composite layer". A composite layer can always be

subdivided into Bravais-lattice components which we now call "subplanes". This means that for layers having more than one atom per unit cell, each atom in the unit cell belongs to a different subplane.

The electron wavefield between layers (whether of Bravais-lattice or composite type) is expanded in terms of plane waves (often loosely called "beams"). Each of these plane waves corresponds to a reciprocal lattice vector $\underline{g}$ of the two-dimensional surface structure and its wavevector $\underline{k}_{\underline{g}}^{\pm}$ is given by

$$k_{\underline{g}x}^{\pm} = \pm \sqrt{2(E - V_0 - iV_{0i}) - k_{\underline{g}y}^2 - k_{\underline{g}z}^2} \quad ,$$

$$k_{\underline{g}y} = k_{0y}^{v} + g_y \quad , \tag{12}$$

$$k_{\underline{g}z} = k_{0z}^{v} + g_z \quad ,$$

where E is the electron energy, V_0 is the constant potential level between muffin-tin spheres ($V_0 < 0$), V_{0i} is the imaginary part of the potential ($V_{0i} < 0$), $\underline{k}_0^{v}$ is the incident wave vector in vacuum and the $\pm$ signs correspond to propagations in the directions $\pm x$ (atomic units are used, with $\hbar = e = m = 1$). We shall often use the symbol $\perp$ instead of x, e.g. $k_{\underline{g}\perp}^{\pm}$.

Those plane waves that emerge from the crystal produce spots observed on a screen; the spots are labeled by the corresponding values of $\underline{g}$, i.e. (0,0), (1,0), etc. An accurate LEED calculation requires, in addition to the emerging plane waves (and those that propagate through the crystal without emerging, because of the surface potential step), a number of additional plane waves that decay exponentially away from the center plane of each layer: these are called evanescent waves. Let us require that all those plane waves (evanescent or not) are included in the calculations that do not decay from one layer to the next to less than a fraction t of their original amplitude; i.e., let us include all plane waves for which

$$\exp\left(-\text{Im}\{k_{\underline{g}x}^{\pm}\}\, d_{min}\right) > t \quad , \tag{13}$$

where d_{min} is the smallest interlayer separation occuring in the crystal. This condition defines a "beam circle" in the two-dimensional reciprocal space, within which all points (k_{gy}, k_{gz}) should be included for an accurate description of the wavefield.

As input, the user must select a set of reciprocal lattice vectors $\underline{g}$ for use in the calculation. This set must include at least the beams actually needed. The dynamical LEED programs automatically choose at each energy only those beams of the input list that satisfy (13), based on a user-supplied value of t. This minimizes the size of the matrices used at each energy. We note that this set of vectors $\underline{g}$ depends on the value of (k_{0y}^V, k_{0z}^V) and therefore on the angle of incidence. This set also depends on the electron energy E, both through (k_{0y}^V, k_{0z}^V) and through $\text{Im}\{k_{gx}\}$. Therefore, the input list of vectors $\underline{g}$ should be chosen so as to accommodate all planned variations of energy and angle of incidence (note that the vector $\underline{g} = (0,0)$ is required to be the first element in the input list). The radius k_b of the "beam circle" and the number n_b of beams falling within it can be estimated by the following formulas:

$$k_b \simeq \sqrt{2(E - V_0) + \left(\frac{\log(t)}{d_{min}}\right)^2} \ , \tag{14}$$

$$n_b \simeq \frac{A}{4\pi} \left| 2(E - V_0) + \left(\frac{\log(t)}{d_{min}}\right)^2 \right| \ . \tag{15}$$

Here A is the area of the surface unit cell (atomic units are used). Examples of beam set lists used in different calculations are given in Chap. 10.

Note that for very small values of d_{min}, the number n_b becomes very large; this implies proportionally large matrices in the computation. It is to avoid this occurence that composite layers are envisaged: the wavefield between their closely-spaced subplanes will be expanded not in plane waves, but in spherical waves (cf. Chaps.6,7; see also Sect.10.1 for a discussion of the choice of type of layer).

In connection with layers it should be mentioned that the dynamical LEED programs allow a certain layer-dependence of the potentials V_0 and V_{0i} (muffin-tin constant and imaginary potential, respectively): the programs will accept different values of V_0 (and V_{0i}) in a substrate and a top layer. Further details will be found in Chap. 10.

3.3 Superlattices

Overlayers and surface reconstructions often lead to non-bulk periodicities parallel to the surface, i.e., superlattices. The main effect of super-lattices is to produce "extra beams" in addition to the beams due to the bulk periodicity. We only consider surface unit cells whose area is an integral multiple of the bulk-layer unit cell area. Fractionally related unit cells can be reduced to this case by considering the "coincidence lattice" of the bulk and the surface unit cell. Irrationally related unit cells require additional treatment [10], and are not considered here. It should be noted that the subroutines that generate diffraction matrices for a *composite* layer (cf. Chaps.6,7) assume that the subplanes of the composite layer have a lattice equal to the superlattice of the surface, or equal to the lattice of the bulk if there is no superlattice.

It is convenient (and therefore it is systematically done in our programs) to subdivide the original and extra beams into sets in such a way that each extra set is obtained from the original set by a different and rigid shift in (k_{gy}, k_{gz}): each extra set, within itself, has the periodicity corresponding to the bulk layer periodicity and the shift relating it to the original set is a reciprocal lattice vector of the superlattice. For example, a p(2×2) superlattice generates reciprocal lattice vectors (0,0), (±1/2,0), (0,±1/2), (±1/2,±1/2), (±1,0), (±3/2,0), etc. These can be grouped into 4 sets: the original set (0,0), (±1,0) (0,±1), etc.; a set obtained by adding to it $\underline{g}_{ov}$ = (1/2,0): (1/2,0), (3/2,0), (1/2,1), etc.; a set obtained by adding $\underline{g}_{ov}$ = (0,1/2): (0,1/2), (1,1/2), (0,3/2), etc.; and a set obtained by adding $\underline{g}_{ov}$ = (1/2,1/2): (1/2,1/2), (3/2,1/2), (1/2,3/2), etc. (see also Fig.4.3 in Sect.4.2). In general, if the ratio of surface to bulk unit cell areas is P (an integer), there will be altogether P beam sets.

These beam sets have the property that bulk layers (which do not have the surface superlattice), can diffract beams *within* each set, but cannot diffract a beam from one set to another set (since that would require a non-bulk reciprocal lattice vector). This implies that the bulk diffraction matrices have a block-diagonalized form when the beams are ordered in such sets: one obtains one block per beam set.

This leads to big savings in computation effort in generating, storing and using these bulk diffraction matrices. In the programs, the block-diagonalized diffraction matrices for bulk layers are generated by building

individual blocks separately. Thus the main program calls a subroutine
(MSMF) that generates a given block for each bulk diffraction matrix; this
is done as many times as there are blocks (beam sets) to be considered
(this subroutine is actually unable to produce the vanishing matrix elements outside of the blocks). Similarly, the subroutines for Layer Doubling
(cf. Chap.8) and Renormalized Forward Scattering (cf. Chap.9) treat individual blocks for bulk layers separately. For that reason in the input
of beams $\underline{g}$ (cf. Sect.3.2) the beams *must* be grouped in sets. In addition,
one must remember to satisfy the requirement that $\underline{g} = (0,0)$ is the first
vector in the input list, and so the "bulk set" must be first in the list
of sets.

3.4 Atomic Scattering and Calculation of Phase Shifts

As discussed in Sect. 2.5, the interactions between LEED electrons and single atoms of the diffracting crystal is represented by the muffin-tin model of the potential: inside a radius R (the muffin-tin radius) the potential is taken to be spherically symmetrical. The scattering is generally
strong in the sense that a Born approximation is inadequate. Furthermore
the scattering characteristics (energy and angle dependence) vary considerably from one chemical species to another and to some extent from one atomic environment to another. For LEED calculations it is therefore necessary
to use different atomic potentials for different atomic species and it is
customary to include some gross features of the atomic environment. No
procedure is known that will produce the best potential in all cases: e.g.,
one procedure may be best for one metal, but another may be better for a
different metal. For the purpose of surface crystallography, this uncertainty is, however, not serious, if some care is taken in the choice of atomic
potentials and, in particular, if a few differently constructed potentials
are tried.

Atomic potentials can be constructed as follows. The simplest way of
taking into account the fact that an atom is embedded in a lattice is
through the muffin-tin constant: this represents, roughly speaking, the interaction of the LEED electrons with the relatively delocalized conduction
electrons (in metals) and bonding electrons (in most materials). Then the
atomic potential represents the interaction with the atom *cores*. To obtain

the atomic charge density, one can, for example, start from free-atom wave-functions for the bound electrons, truncate them at the muffin-tin radius and spread some charge through the muffin-tin sphere to represent conduction electrons. One can also, starting from the same free-atom wavefunctions, overlap spherically-symmetrical charge densities from neighboring atoms and spherically symmetrize the resulting charge density on each atom (instead one may also at a later stage of the calculation overlap and spherically symmetrize atomic potentials). The atomic potential consists of an electro-static term and an exchange-correlation term. The electrostatic term is ob-tained by integrating the Poisson equation with the above ion-core charge density. The exchange-correlation term can be generated with, for example, the self-consistent Hartree-Fock scheme, with an $X\alpha$ approximation, or with the Hedin-Lundqvist approximation, etc.

In practice, it appears to be quite adequate to use bulk atomic poten-tials for surface atoms of clean surfaces and for top substrate atoms when an overlayer is present. In particular, band-structure potentials have been quite successful in LEED (a very useful tabulation of band structure poten-tials for many metals exists [11]). In the case of overlayer atoms, atomic potentials must usually be generated specifically for LEED calculations by any of the above procedures: such potentials exist now for a fairly exten-sive set of chemical species, as can be seen from the list of structures investigated (cf. Chap.12). The authors of this book also have a collection of potentials (and phase shifts) available on request.

The scattering of waves by a spherically symmetrical potential is con-veniently expressed in terms of the atomic scattering amplitude $t(\theta)$, where θ is the scattering angle. For example, the scattering of a plane wave of wavevector $\underline{k}_0$ by such a potential has the following asymptotic form:

$$e^{i\underline{k}_0\cdot\underline{r}} + t(\theta)\,\frac{e^{ik_0r}}{r} \tag{16}$$

which is assumed in the LEED theory [r is the distance from the atomic nu-cleus; $k_0 = |\underline{k}_0|$ is normally taken complex in the crystal:
$k_0 = \sqrt{2(E-V_0-iV_{0i})}$]. The scattering amplitude is expanded in Legendre poly-nomials P_ℓ, yielding

$$t(\theta) = 4\pi \sum_{\ell=0}^{\infty} (2\ell+1)\, t_\ell\, P_\ell(\cos\theta) \quad , \tag{17}$$

where t_ℓ is a t-matrix element with

$$t_\ell = \frac{\exp(2i\delta_\ell) - 1}{4ik_0} = \frac{\exp(i\delta_\ell)\sin\delta_\ell}{2k_0} \quad . \tag{18}$$

In (18) appear phase shifts δ_ℓ ($\ell = 0,1,2,\ldots$) that are characteristic of the particular atomic potential used and that also depend on the energy of the scattered electrons. The sum over angular momenta ℓ can be truncated typically after 5 to 8 terms, depending on the electron energy and the scattering strength of the particular atom under consideration.

The phase shifts δ_ℓ are obtained by integration of the radial Schrödinger equation involving the atomic potential discussed above. If $\phi_\ell(r)$ is the radial wavefunction (for angular momentum ℓ) and $L_\ell = \phi_\ell'(R)/\phi_\ell(R)$ is the logarithmic derivative of ϕ_ℓ at the muffin-tin radius R, then the phase shift δ_ℓ satisfies the following relation:

$$\exp(2i\delta_\ell) = \frac{L_\ell h_\ell^{(2)}(\kappa R) - h_\ell^{(2)'}(\kappa R)}{h_\ell^{(1)'}(\kappa R) - L_\ell h_\ell^{(1)}(\kappa R)} \quad , \qquad \ell = 0,1,2,\ldots \quad , \tag{19}$$

where $\kappa = \sqrt{2(E-V_0)}$, and $h_\ell^{(1)}$, $h_\ell^{(2)}$ are spherical Hankel functions of the first and second kinds, respectively. The step from atomic potential to phase shifts involves basically only a straightforward numerical integration, for which, if necessary, a program is easily written or can be obtained from the authors of this book.

The phase shifts (and the muffin-tin constant) are the only quantities that describe the atomic scattering properties in a LEED calculation: the radial wavefunctions ϕ_ℓ and the muffin-tin radius R do not appear explicitly there, only implicitly through the phase shifts. Atoms are therefore effectively replaced in LEED calculations by point scatterers in a constant potential with scattering properties given by the phase shifts.

Note that the t-matrix elements of (18) are π-periodic in the phase shifts δ_ℓ: as a result, δ_ℓ may be reduced to the interval $[-\pi/2, \pi/2]$. Many phase shift programs therefore give only the reduced value of δ_ℓ (for example, (19) can only give δ_ℓ mod π). As a consequence the phase shifts can undergo as a function of energy discontinuous jumps by π as they go through $+\pi/2$ or $-\pi/2$. This can lead to errors with our LEED programs. There the phase shifts are read in at a certain set of energy values. The phase shifts are then interpolated from the input values to each actual energy of the calculation: therefore no discontinuity by π is allowed to occur. The user should be particularly careful about this, as it may not be obvious from the calculated IV-curves when an error of this type arises. A further necessary warning concerns the unit of the energies associated with the input phase shifts: most phase shift programs produce energies in rydbergs; however our LEED program assumes that at input hartrees are used instead (1 hartree = 2 rydbergs).

As mentioned in Sect. 2.5, the effect of thermal vibrations is represented in LEED by a Debye-Waller factor e^{-M} (cf. Sect.3.5) multiplying each atom's scattering amplitude $t(\theta)$:

$$t^T(\theta) = e^{-M} t(\theta) \quad . \tag{20}$$

The effective atomic scattering amplitude $t^T(\theta)$ can be expanded in Legendre polynomials P_ℓ if the thermal vibrations are isotropic. Then new, effective temperature-dependent phase shifts $\delta_\ell(T)$ appear:

$$t^T(\theta) = 4\pi \sum_{\ell=0}^{\infty} (2\ell+1) t_\ell^T P_\ell(\cos\theta) \quad , \tag{21}$$

$$t_\ell^T = \frac{\exp[2i\delta_\ell(T)]-1}{4ik_0} = \frac{\exp[i\delta_\ell(T)]\sin\delta_\ell(T)}{2k_0} \quad , \quad \ell = 0,1,2,... \tag{22}$$

The relation between the two sets of phase shifts $\delta_\ell(T)$ and δ_ℓ is given by:

$$\exp[i\delta_\ell(T)]\sin\delta_\ell(T) = \sum_{\ell'\ell''} i^{\ell'} \exp[-4\alpha(E-V_0)]j_{\ell'}[-4i\alpha(E-V_0)]$$

$$\times \exp(i\delta_{\ell''})\sin\delta_{\ell''} \left[\frac{4\pi(2\ell'+1)(2\ell''+1)}{(2\ell+1)}\right]^{1/2} B^{\ell''}(\ell'0,\ell0) \quad , \tag{23}$$

where $j_{\ell'}$ is a spherical Bessel function of the first kind,

$$\alpha = \frac{1}{6} <(\Delta \underset{\sim}{r})^2>_T \quad , \tag{24}$$

[cf. (26)], and

$$B^{\ell''}(\ell'm',\ell m) = \int Y_{\ell''m''}(\Omega)Y_{\ell'm'}(\Omega)Y_{\ell-m}(\Omega)d\Omega \quad . \tag{25}$$

3.5 Thermal Vibrations

Ísotropic thermal vibrations of the crystal atoms are taken into account
by the LEED programs through the generation of "temperature-dependent
phase shifts" (these are complex numbers). Correlations in the vibrations
of different atoms are neglected. The temperature effect is included
through the multiplication of each atom's scattering amplitude by a Debye-
Waller factor e^{-M} (cf. Sect.3.4), where

$$M = \frac{1}{2} <\Delta \underline{k} \cdot \Delta \underline{r})^2>_T = \frac{1}{6} |\Delta \underline{k}|^2 <(\Delta \underline{r})^2>_T \quad . \tag{26}$$

Here $\Delta \underline{k}$ is the momentum transfer resulting from the diffraction from one
beam into another. $\Delta \underline{k}$ is defined inside the surface, where the muffin-tin
constant is added to the outside electron energy; the real part of the complex number $\Delta \underline{k}$ is taken, since only the real part gives rise to the wave
interference on which the Debye-Waller factor is based. In the high-temperature limit ($T \gg \theta_D$, where θ_D is the Debye temperature) we have, in atomic units,

$$<(\Delta \underline{k})^2>_{T \to \infty} \simeq \frac{9T}{m k_B \theta_D^2} \quad , \tag{27}$$

where the atomic mass m is expressed in units of the electron mass,
$k_B = 3.17 \times 10^{-6}$ hartrees /kelvin, and the actual and Debye temperatures, T
and θ_D, are in kelvins.

To account for the fact that vibration amplitudes for surface layers may
be different from those of bulk layers, the programs accept as input an enhancement factor (with components parallel and perpendicular to the sur-
face, see below) that is multiplied into $<(\Delta \underline{k})^2>_{T \to \infty}$. The programs allow
this enhancement factor to depend on the atomic species; thus for layers of
the same atomic species, a single enhancement factor is to be used.

For low temperatures ($T \lesssim \theta_D$) the expression (27) is inadequate. In the
very-low temperature limit the correct expression is

$$<(\Delta \underline{k})^2>_{T \to 0} \simeq \frac{9}{m k_B \theta_D} \left(\frac{1}{4} + 1.642 \frac{T^2}{\theta_D^2} \right) \quad . \tag{28}$$

In the LEED programs we use a single functional form to represent the temperature dependence of $<(\Delta \underline{r})^2>$ for all T: setting T = 0 in (28), we write

$$<(\Delta \underline{r})^2>_T = \sqrt{\left[<(\Delta \underline{r})^2>_{T=0}\right]^2 + \left[<(\Delta \underline{r})^2>_{T\to\infty}\right]^2} \quad . \tag{29}$$

This expression has the correct asymptotic limits. Here the quantity $\sqrt{<(\Delta \underline{r})^2>_{T=0}}$ is to be input by the user and $<(\Delta \underline{r})^2>_{T\to\infty}$ is computed according to (27) for a given T. For most materials at room temperature the zero-temperature correction is not desired and an input of zero for $\sqrt{<(\Delta \underline{r})^2>_{T=0}}$ should be used.

The user inputs to the LEED programs temperature-independent phase shifts, from which atomic scattering amplitudes are generated. The Debye-Waller factor described above, valid only for isotropic vibrations, is then combined with these atomic scattering factors by the programs to yield complex (temperature-dependent) phase shifts, which are used to obtain the layer diffraction matrices [cf.(23)].

If the user so desires, he can include partial anisotropy in the thermal vibrations (except in composite layers): the vibration amplitudes perpendicular and parallel to the surface may be different, as far as an *interlayer* multiple scattering is concerned. To achieve this, Bravais-lattice layers of any given atomic species can be assigned an enhancement factor that has different values for vibration perpendicular and parallel to the surface. (No enhancement factors and, therefore, also no anisotropy are included for the zero-temperature vibration). Because of anisotropic vibration, (26) is replaced by

$$M = \frac{1}{6}\left[\left|\Delta \underline{k}_{/\!/}\right|^2 \cdot <(\Delta \underline{r}_{/\!/})^2>_T + \left|\Delta k_\perp\right|^2 \cdot <(\Delta r_\perp)^2>_T\right] \quad , \tag{30}$$

where $<(\Delta \underline{r}_{/\!/})^2>_T$ and $<(\Delta r_\perp)^2>_T$ are computed as in (29) with appropriate enhancement factors.

The anisotropic Debye-Waller factor e^{-M}, using M obtained in (30), is then multiplied directly into the completed layer diffraction matrix elements (of a Bravais-lattice layer) given in (34): cf. (39). Although vibrational anisotropy can be included in this way, we note that the phase shifts used in the *intralayer* multiple scattering matrix $X_{\ell m, \ell'm'}$ in (34) must still be generated from spherically-averaged vibration amplitudes, because the programs cannot take anisotropic vibrations into account as far as intralayer multiple scattering is concerned. More details are given in Sect. 5.4.

In most structural determinations, the usual mode of calculation is to ignore entirely anisotropies in the thermal vibrations and use temperature-dependent phase shifts throughout.

3.6 Ordering of (ℓ,m) Pairs

In dealing with the spherical harmonics $Y_{\ell m}$, it is useful to order them sequentially in what we shall call their "natural ordering", defined by the following sequence of (ℓ,m) pairs:

$$(0,0),\ (1,-1),\ (1,0),\ (1,1),\ (2,-2),\ (2,-1),\ (2,0),\ (2,1),\ (2,2),\ (3,-3),\ldots \tag{31}$$

However, in considering intralayer multiple scattering within a *coplanar* array of atoms, one finds that spherical waves with $\ell + m$ even and $\ell + m$ odd do not couple. This leads to the following ordering, which we shall call "coplanar ordering", where (ℓ,m) pairs with even $\ell + m$ are taken first:

$$(0,0),\ (1,-1),\ (1,1),\ (2,-2),\ (2,0),\ (2,2),\ (3,-3),\ldots; \tag{32}$$

$$(1,0),\ (2,-1),\ (2,1),\ (3,-2),\ (3,0),\ldots$$

In this ordering, if ℓ_{max} is the maximum value of ℓ used, there are $\ell_{ev} = (\ell_{max} + 1)(\ell_{max} + 2)/2$ pairs (ℓ,m) with even $\ell + m$, and $\ell_{od} = \ell_{max} \times (\ell_{max} + 1)/2$ pairs (ℓ,m) with odd $\ell + m$, making a total of $(\ell_{max} + 1)^2$ pairs.

Furthermore, when in a coplanar array of atoms there is sufficient symmetry (as occurs sometimes at normal incidence), multiple scattering does not couple spherical waves of even m with those of odd m. This gives rise to the following "symmetrized ordering", where the coplanar ordering is further split up according to even and odd m:

$$(0,0),\ (2,-2),\ (2,0),\ (2,2),\ldots;\ (1,-1),\ (1,1),\ (3,-3),\ (3,-1),\ldots;$$

$$(1,0),\ (3,-2),\ (3,0),\ldots;\ (2,-1),\ (2,1),\ (4,-3),\ldots \tag{33}$$

The number of elements in the four groups of pairs (ℓ,m) is now
$\ell_{ee} = (\ell_{max}/2 + 1)^2$, $\ell_{eo} = ((\ell_{max} + 1)/2 + 1)((\ell_{max} + 1)/2)$,
$\ell_{oe} = ((\ell_{max} - 1)/2 + 1)^2$ and $\ell_{oo} = (\ell_{max}/2 + 1)\ell_{max}/2$, respectively, in
which expressions integer division is to be understood.

Our LEED programs are built around the symmetrized ordering, (33), while
the natural ordering (31) and the coplanar ordering (32) are both used in
various subroutines as well. (Note: even though the symmetrized ordering is
used, the conditions of symmetry giving rise to this ordering are not assumed in the version of the programs presented in this book, since the sim-
plifications resulting from those conditions are not exploited.)

The use of three orderings of (ℓ,m) pairs requires the knowledge of the
relationships between them: these relationships are generated (by subroutine
LXGENT) and used in the programs in the form of permutations, stored as vectors LXM, LX, LXI and LT. Permutation LXM gives the symmetrized from the
natural ordering, LX is essentially its inverse and LXI gives the symmetrized from the coplanar ordering. The permutation LT is similar to LX, but
it leads to element $(\ell,-m)$ rather than (ℓ,m). See the listing of subroutine LXGENT for more details.

4. Symmetry and Its Use

The basic ideas involved in the use of available symmetries are outlined in
App. A. Here, we discuss the preparation of a program in terms of the use
of the symmetries that may exist in any particular application. The user
needs to go through this reasoning in order to set up the program properly,
if symmetry is to be exploited.

No symmetry in *angular* momentum space is exploited in the LEED programs,
if it depends on the primary electron beam being directed in a particular
direction (such as at normal incidence). This choice is made to avoid the
more complicated, case-dependent implementation of symmetries in angular
momentum space; one case-independent implementation is dealt with in
Sect. 3.6.

4.1 Symmetry and Registries

As a first step in the use of symmetries, an appropriate symmetry axis
and/or set of symmetry planes (mirror planes) must be selected. Any combination of n-fold axes (n = 2, 3, 4 or 6) and mirror planes can be handled by
the programs, as long as they intersect in one common line (glide symmetry
is not considered). Note that we are referring to the symmetries of the
wavefield (in which the electron incidence direction is an important determining factor) rather than those of the crystalline surface geometry
alone. With normal electron incidence direction these symmetries coincide,
while at off-normal incidence the wavefield has at most a mirror plane
(containing the incident beam direction and the surface normal), namely if
the incidence azimuth coincides with a mirror plane of the surface. Thus
the orientations (and existence) of axes and planes of symmetry are easily
determined from the incidence direction and the surface structure under
examination.

There is some freedom in the choice of the exact location of the axis
and/or plane(s) of symmetry with respect to the surface lattice. For a given

calculation, a choice must be made and adhered to. This is important in connection with the "registries" (i.e., relative positions parallel to the surface) of successive layers with respect to the axis and/or plane(s) of symmetry. As described in Appendix A, the symmetrized layer diffraction matrices are registry-dependent: the symmetrized matrices already incorporate part or all of the registry of each layer.

We introduce a "propagation track" to help ensure that all layer registries are properly taken into account. If we first disregard symmetry, we may imagine a set of sequential vectors r_{10}, r_{21}, r_{32}, $\cdots$ linking atoms in successive layers of the surface, as illustrated in Fig. 4.1a (composite layers are treated somewhat specially; see below). These vectors define the

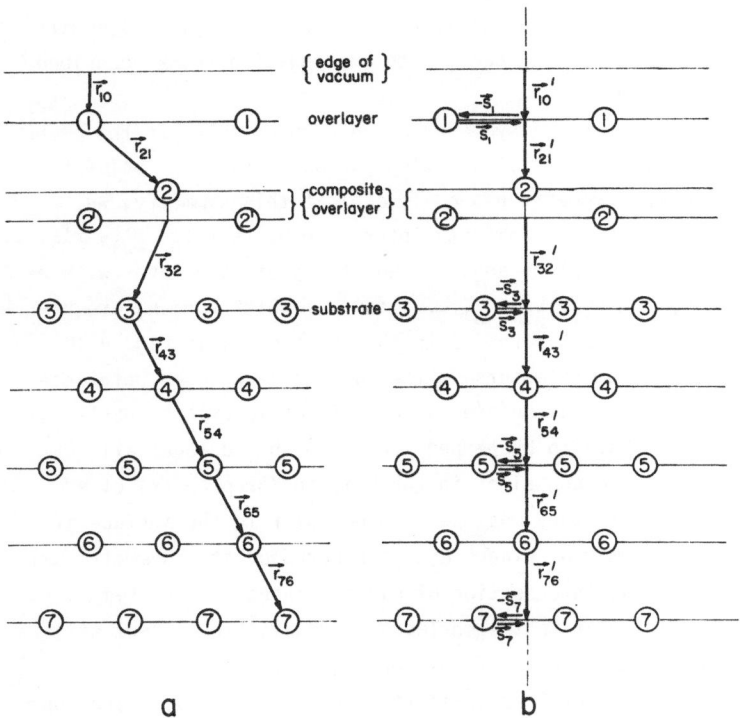

Fig. 4.1 Propagation track through an imaginary surface in 2 dimensions, seen in side view; the surface is composed of a substrate with a composite overlayer and a simple overlayer, both overlayers having double periodicity parallel to the surface. Case (a) (left-hand panel) does not consider symmetry, case (b) (right-hand panel) takes a mirror plane (dashed, perpendicular to the plane of the figure) into account. Atomic locations are labeled by digits that count the layers, starting from the surface

propagation track, along which we shall imagine that we propagate the wavefield (in particular the plane waves) into the crystal and back out again. The propagation track starts at a point on the crystal surface (where we consider inner potential and damping to set in abruptly), and penetrates as far into the crystal as the calculation is expected to have to go. The choice of this propagation track is, at this stage, entirely arbitrary, as the final result of the calculation does not depend on it. We choose it to link atomic centers because the diffraction matrices that our programs calculate normally assume atomic centers as points of reference, when no symmetry is taken into account (an exception occurs for composite layers, where two reference points are used, one in the topmost subplane and the other vertically under the first one in the deepest subplane of the composite layer (cf. Chaps.6,7); Fig. 4.1 includes an example of such a composite layer). If we make a calculation that does not exploit symmetry, e.g., if there is no symmetry, we can use the propagation track described above and shown in Fig. 4.1a.

The example of Fig. 4.1 has a symmetry plane perpendicular to the plane of the figure and we assume for illustration purposes that the incident beam is parallel to that symmetry plane. To exploit this symmetry, we change the propagation track so that its interlayer parts ($\underline{r}'_{10}$, $\underline{r}'_{21}$, $\underline{r}'_{32}$, ..) are contained in the mirror plane and the registries of each layer with respect to the mirror plane ($\underline{s}_1$, $\underline{s}_2$, $\underline{s}_3$, ..., some of which vanish) are represented by segments of track *parallel* to the surface, as in Fig. 4.1b. This track still links atomic centers in the various layers (or reference points on composite layers), as before. Note that the interlayer parts $\underline{r}'_{ji}$ of the new track do not have to be perpendicular to the surface: all that is required is that they be contained in the axis and/or plane(s) of symmetry. This does imply that they must be perpendicular to the surface if two or more intersecting mirror planes are considered or if a symmetry axis is present (since both the intersection of mirror planes and any symmetry axis are always perpendicular to the surface). Note also that there still remains some freedom in the choice of this propagation track.

In our programs, layers (of Bravais-lattice or composite type) are connected by plane wave propagators, $\exp(\pm i\underline{k}_{\underline{g}}^{\pm} \cdot \underline{r}_{ji})$, where i and j label points on two consecutive layers. If no symmetry is exploited, i and j are atomic centers (except with composite layers, as explained above) located on the propagation track and the propagators include the full interlayer registry. In the case of a composite layer, propagations within the composite layer (e.g., subplanes 2 and 2' in Fig. 4.1) are always taken care of internally by the composite layer diffraction matrices (cf. Chaps. 6,7).

When symmetry is exploited, the registries with respect to the symmetry axis and/or plane(s), i.e., the parts of the propagation track parallel to the surface ($\underline{s}_1$, $\underline{s}_2$, $\underline{s}_3$, ... in Fig.4.1), are included in the symmetrized layer diffraction matrices (cf. App.A), and only the remaining part of the interlayer vectors ($\underline{r}'_{10}$, $\underline{r}'_{21}$, $\underline{r}'_{32}$, ... in Fig.4.1) need be accounted for in the interlayer plane-wave propagators; these vectors $\underline{r}'_{10}$, $\underline{r}'_{21}$, ... are known as ASA, ASB or ASC in the programs (cf. Chap.10).

Note that this procedure for decomposing the interlayer vectors $\underline{r}_{10}$, $\underline{r}_{21}$, $\underline{r}_{32}$, ... applies whether there is a surface superlattice or not: a superlattice is actually included in the example of Fig. 4.1.

4.2 Symmetry Among Beams

In addition to the aspects of symmetry discussed above (the propagation track), we must supply the dynamical LEED programs with information regarding the symmetry relationships among the plane waves (beams). As explained in Appendix A, group theory allows us to consider a reduced set of plane waves, namely, the set of symmetrical linear combinations of symmetry-related plane waves. To that effect equation (A.4) of App. A is implemented in our programs; included also are the registry-dependent factors (involving $\underline{s}_j$ of Fig.4.1) of (A.5), and anisotropic Debye-Waller factors, when those are desired (see Sects.3.5,5.4). The vectors $\underline{g}$ for all beams are needed in (A.4). But rather than having the user input all those vectors (as he must do when not exploiting symmetry), he is required to input only one vector $\underline{g}$ (any one) for each set of symmetry-related vectors $\underline{g}$, together with a code number. The code number automatically informs the programs about those symmetry-related vectors $\underline{g}$ that were omitted in the input. This code number also instructs the programs to use the appropriate symmetrical wavefunctions (rather than the simple plane waves labelled $\underline{g}$) and to assign only one row and one column in diffraction matrices for each symmetrical wavefunction (rather than one each for the various symmetry-related plane waves).

The available symmetry code numbers used in the programs are defined in the caption of Fig. 4.2. Each code number corresponds to a different combination of symmetry-related plane waves (beams) using various symmetry

38

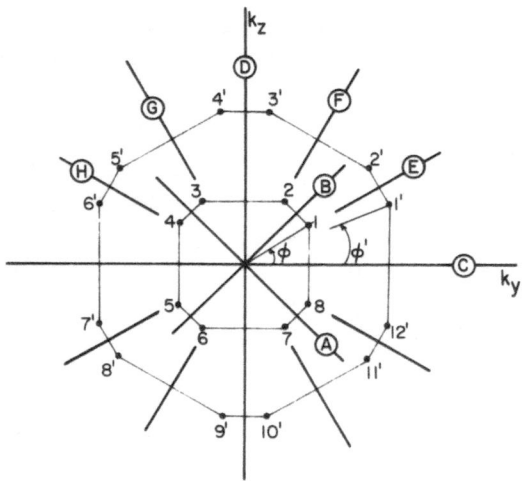

Fig. 4.2 Symmetry relationships of beams (numbered in this diagram 1,2,...,8,1',2', ...,12'), relative to an axis O and mirror planes at azimuths -45°, 0°, 45° and 90° (denoted A, C, B and D, resp.) and at azimuths 30°, 60°, 120° and 150° (denoted E, F, G and H, resp.); ϕ and ϕ' are arbitrary angles, not confined to $0 < \phi \leq 45°$, $0 < \phi' \leq 30°$, defining the azimuth of a beam. Groups of symmetry-related beams are given code numbers as defined in the following list:

Symmetry code number	Description	Examples of groups of beams
1	single or unsymmetrized beam	(1),(5')
2	2 beams, 2-fold axis	(1,5),(2',8')
3	2 beams, mirror plane A	(2,5),(3,4)
4	2 beams, mirror plane B	(1,2),(4,7)
5	2 beams, mirror plane C	(1,8),(3',10')
6	2 beams, mirror plane D	(6,7)(7',12')
7	4 beams, 4-fold axis	(1,3,5,7),(2,4,6,8)
8	4 beams, 2 mirror planes A, B	(1,2,5,6),(3,4,7,8)
9	4 beams, 2 mirror planes C, D	(1,4,5,8),(2',5',8',11')
10	8 beams, 4 mirror planes A, B, C, D	(1,2,3,4,5,6,7,8)
11	3 beams, 3-fold axis	(1',5',9'),(2',6',10')
12	6 beams, 3-fold axis, mirror plane C	(1',4',5',8',9',12')
13	6 beams, 3-fold axis, mirror plane D	(3',4',7',8',11',12')
14	6 beams, 6-fold axis	(1',3',5',7',9',11')
15	12 beams, 6-fold axis, 2 mirror planes C, D	(1',2',3',4',...,11',12')

axes and/or plane(s). For increased computation speed the orientation of the mirror planes is assumed in our programs to be as indicated in Fig. 4.2: at least one of the mirror planes must be at azimuths 0°, ±45° or 90°, while additional mirror planes should occur at any of the azimuths 0°, ±30°, ±45°, ±60° or 90°. It may happen that a given group of symmetry-related beams can be described by more than one symmetry code (for example, at beam azimuth $\phi = 45°$ in Fig.4.2, the code numbers 7 and 9 are equivalent in describing beams 1, 3, 5 and 7). Then either one of the equivalent symmetry codes may be used.

It is permissible to exploit less than the complete available symmetry, such as, for example, only one of two mirror planes.

As an illustration, consider a surface that has a structure with a 4-fold rotational axis of symmetry. At normal incidence the wavefield will also have that symmetry axis. To exploit that symmetry, the user could (in addition to taking care of the layer registries with respect to that axis and of the interlayer vectors) input only those vectors $\underline{g}$ that appear in a quadrant, each accompanied by the code number 7 appropriate for 4-fold symmetry; an exception is the $\underline{g} = (0,0)$ vector (representing the incident and specularly-reflected beams), which is not to be symmetrized (it is already symmetrical) and therefore requires the code number 1. (This number 1 is also the "default" code number to be input when no symmetry is used.) Which quadrant is chosen for this purpose is arbitrary, as is even the choice of quadrant as opposed to another shape: it is sufficient to input, for each foursome of symmetrical vectors $\underline{g}$, any one of the four. Further illustrations will be found in Chap. 10.

A slight complication can arise in the foregoing discussion when we consider superlattices. It can occur that different subsets of plane waves are mutually symmetry-related: we should then distinguish between the symmetry relationships among plane waves in the bulk and in the superlattice region of the crystal. Consider the example of Fig. 4.3. Here we have a square bulk lattice [as for an fcc (100) surface] that produces the beams labelled 1 and a p(2×2) overlayer that produces three additional sets of beams, labelled 2, 3 and 4. If we assume the planes (xy) and (xz) to be mirror planes and the x-axis to be a 4-fold symmetry axis, then the beam sets labelled 3 and 4 are (at normal incidence) mutually symmetrical under rotation by $90°$ about the x-axis. In the overlayer, the 8 circled beams in Fig. 4.3a can be combined into a symmetrical wavefunction (symmetry code 10). The symmetrized diffraction matrix elements can then be computed according to (A.2), where the index i ranges over the 8 circled beams of Fig. 4.3a. But in the substrate the beams of set 3 do not couple with the beams of set 4, so that in (A.2), for the substrate, some of the matrix elements $M_{\underline{g}'1,\underline{g}i}$ ($i = 1,2,\ldots,8$) vanish (namely, those for which $\underline{g}'1$ and $\underline{g}i$ belong to different beam sets). Since our programs are designed to avoid generating these vanishing matrix elements, it is necessary to limit the summation over i in

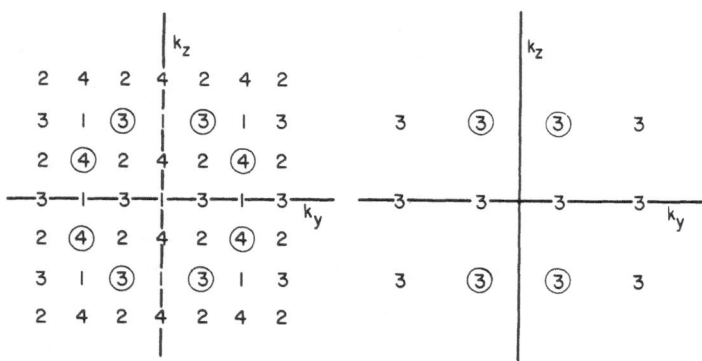

Fig. 4.3. Reciprocal lattice for a p (2×2) superlattice on a square sub-strate net. Beam sets are numbered 1, 2, 3 and 4 (1 is the bulk layer re-ciprocal lattice). Panel (a) illustrates symmetry in the superlattice re-gion of the crystal (2 mirror planes and a 4-fold rotation axis). Panel (b) illustrates symmetry within beam set 3 in the substrate (2 mirror planes only)

(A.2) to beams within the set to which $\underline{g}'1$ belongs, say set 3. This is sim-ply done by specifying for such beams a different symmetry code in the sub-strate and in the overlayer [the factor $\sqrt{J/I}$ of (A.2) also comes out right automatically in this way]. Thus, in our example, the symmetrization of the 8 circled beams of Fig. 4.3a should, in the substrate, limit itself to, e.g., the 4 beams belonging to set 3: these 4 beams have the symmetry code 9 rather than 10; cf. Fig. 4.3b. It is therefore sufficient in the input list of beams to omit beam set 4, and to specify for each beam of set 3 ap-pearing in the input list two different symmetry codes: one for the over-layer (e.g. 10), and one for the substrate (e.g. 9).

This is why *always* two symmetry codes are input with each beam appearing in the input list: the two codes are, however, different only in the case of mutually symmetrical beam sets. Further illustrations will be found in Chap. 10.

A different way of describing this situation is as follows. If we first fold together the two mutually symmetrical beam sets (by forming symmetrical and antisymmetrical waves with each pair of symmetrical beams belonging to the third and fourth beam sets, respectively), (A.1) shows that, in the sub-strate, the block-diagonalized diffraction matrix M obeys the relation

$$M' = TMT^{-1} = M \quad ,$$

i.e., M is unchanged under this symmetrization. The third and fourth beam sets are now, respectively, the sets of symmetrical and antisymmetrical waves formed by combining the original third and fourth beam sets: the fourth set (and block) can now be dropped (cf. App.A), leaving the third set (and block). But, since M' = M, we can effectively skip this symmetrization step altogether, and, in the substrate, drop the fourth beam set from the start; the correct symmetrized diffraction matrices will still be produced. So, as far as the overlayer is concerned, sets 3 and 4 are both present (but mixed by symmetry), while, as far as substrate layers are concerned, set 4 can be ignored: this fact is communicated to the programs through a difference in the beam symmetry codes of set 3 for the overlayer and the substrate.

A further note: a reason for not letting the program automatically generate the list of vectors $\underline{g}$ that the user is asked to input, is to allow the user greater latitude as to which beams he wants to include. This is useful when the user has to fine-tune the core-size of a program by adding or subtracting a few beams, or when special test runs are made (for example, to test symmetries on just a few beams individually instead of on all beams simultaneously), or when the number of beams is consciously desired to be less than strictly necessary.

4.3 Some Formulas

We consider here the expressions used in the programs to work out the sum over symmetry-related beams of (A.4). Rather than performing the sum over exponentials indicated in (A.4), we actually sum over the quantities $Y_{\ell-m}[\Omega(\underline{k}_{gi})]$, each one corrected for a registry shift and a Debye-Waller factor if anisotropic thermal effects are desired. We list in Table 4.1 the symmetry relations for spherical harmonics used to compute $Y_{\ell-m}[\Omega(\underline{k}_{gi})]$ efficiently, given the value for i = 1 or 1' (the index i runs over the beams numbered 1,2,...,8 and 1', 2', ...,12' of Fig.4.2). These symmetry relations assume that any mirror planes are oriented as shown in Fig. 4.2. Note that in the case of composite layers we need the relations of Table 4.1 for $Y_{\ell m}(i)$ instead of $Y_{\ell-m}(i)$, cf. Sect. 6.1: for that purpose it is sufficient to change the sign of m in the given relations.

Table 4.1. Symmetry relations of spherical harmonics (i labels the points given in Fig. 4.2). A short notation is used: $Y_{\ell-m}(i) = Y_{\ell-m}[\Omega(\underline{k}_{\underline{g}i})]$. $Y_{\ell-m}(i)$ is expressed in terms of $Y_{\ell-m}(i=1)$ or $Y_{\ell m}(i=1)$

i	$Y_{\ell-m}(i)$	i	$Y_{\ell-m}(i)$
1	$Y_{\ell-m}(1)$	3'	$\exp(-im\pi/3)Y_{\ell-m}(1')$
2	$i^m Y_{\ell m}(1)$	4'	$(-1)^m \exp(-im2\pi/3)Y_{\ell m}(1')$
3	$(-i)^m Y_{\ell-m}(1)$	5'	$\exp(-im2\pi/3)Y_{\ell-m}(1')$
4	$Y_{\ell m}(1)$	6'	$(-1)^m \exp(-im3\pi/3)Y_{\ell m}(1')$
5	$(-1)^m Y_{\ell-m}(1)$	7'	$\exp(-im3\pi/3)Y_{\ell-m}(1')$
6	$(-i)^m Y_{\ell m}(1)$	8'	$(-1)^m \exp(-im4\pi/3)Y_{\ell m}(1')$
7	$i^m Y_{\ell-m}(1)$	9'	$\exp(-im4\pi/3)Y_{\ell-m}(1')$
8	$(-1)^m Y_{\ell m}(1)$	10'	$(-1)^m \exp(-im5\pi/3)Y_{\ell m}(1')$
1'	$Y_{\ell-m}(1')$	11'	$\exp(-im5\pi/3)Y_{\ell-m}(1')$
2'	$(-1)^m \exp(-im\pi/3)Y_{\ell m}(1')$	12'	$(-1)^m Y_{\ell m}(1')$

4.4 Summary

To use symmetry one first chooses a location for the symmetry axis and/or plane(s). One then draws a propagation track from layer to layer, linking atom centers (and reference points of composite layers), in such a way that layer registries separate out, leaving only segments of track that lie in the symmetry axis and/or plane(s). The registries are incorporated in the symmetrized layer diffraction matrices (requiring different matrices for different registries), while the interlayer segments of track will go into the interlayer propagators.

A list of vectors $\underline{g}$ is input (broken up into beam sets, if a superlattice is present), one vector for each symmetry-related group of beams. Each vector is accompanied by two symmetry code numbers, the first relating to a superlattice layer, the second to a bulk layer; these two numbers are equal, unless one is dealing with symmetry-related beam sets. A beam set symmetrical to another one can be effectively dropped from the input list.

5. Calculation of Diffraction Matrices for Single Bravais-Lattice Layers

Here we consider the diffraction of plane waves by an ordered layer of atoms that has one atom per unit cell (i.e., has a Bravais lattice). When more than one atom per unit cell are present, the layer should either be considered as a composite layer (cf. Chaps.6,7) or it should be taken to be several Bravais-lattice layers if their perpendicular separation distances are large (see Sects.3.2,11.1 for the appropriate criteria).

5.1 Layer Diffraction Matrices

The diffraction matrices are calculated in our programs according to the following formula [2]. The single-layer diffraction amplitude between the incident wave $\exp(i\underline{k}_{\underline{g}}^{\pm} \cdot \underline{r})$ and the scattered wave $\exp(i\underline{k}_{\underline{g}'}^{\pm} \cdot \underline{r})$ is given by

$$
M_{\underline{g}'\underline{g}}^{\pm\pm} = \frac{8\pi^2 i}{A|\underline{k}_0|k_{\underline{g}'\perp}^{+}} \sum_{\substack{\ell'm' \\ \ell m}} \left\{ i^{\ell}(-1)^{m} Y_{\ell-m}[\Omega(\underline{k}_{\underline{g}}^{\pm})] \right\} (1-X)_{\ell m, \ell'm'}^{-1}
$$

$$
\times \left\{ i^{-\ell'} Y_{\ell'm'} [\Omega(\underline{k}_{\underline{g}'}^{\pm})] \right\} \exp(i\delta_{\ell'})\sin\delta_{\ell'} \quad .
$$

(34)

Equation (34) is valid only if the origin of coordinates within the layer is an atom center (see below for more generality). A is the area of the layer unit cell, $|\underline{k}_0| = \sqrt{2(E-V_0)}$ is real and δ_{ℓ} ($\ell = 0,1,2,\ldots$) is the set of phase shifts that describes the electron-atom scattering. The matrix X describes multiple scattering within the atomic layer in angular momentum space and its elements are (see [2])

$$
X_{\ell m, \ell''m''} = \sum_{\ell'+m'=\text{even}} C^{\ell}(\ell'm', \ell''m'')F_{\ell'm'} \exp(i\delta_{\ell'})\sin\delta_{\ell'} \quad , \quad (35)
$$

where

$$C^{\ell}(\ell'm',\ell''m'') = 4\pi(-1)^{(\ell-\ell'-\ell'')/2}(-1)^{m'+m''}Y_{\ell'-m'}(\tfrac{\pi}{2},0)$$

$$\times \int Y_{\ell m}(\Omega)Y_{\ell'm'}(\Omega)Y_{\ell''-m''}(\Omega)d\Omega$$

(36)

are crystal- and energy-independent quantities, and

$$F_{\ell'm'} = \sum_{j}{}' \exp(i\underline{k}_{0//} \cdot \underline{R}_j)h^{(1)}_{\ell'}(k_0|\underline{R}_j|)(-1)^{m'}\exp[-im'\phi(\underline{R}_j)]$$

(37)

is a sum over the layer's lattice vectors $\underline{R}_j$ ($\sum'$ excludes the point $\underline{R}_j = \underline{0}$); $h^{(1)}_{\ell}$ is a Hankel function of the first kind; $\phi(\underline{R}_j)$ is the azimuthal angle of the two-dimensional vector $\underline{R}_j$, and $k_0 = \sqrt{2(E-V_0-iV_{0i})}$. We shall distinguish between reflected and transmitted plane waves and use the following notation for layer reflection and transmission matrices, respectively:

$$r^{+-} = M^{+-} \quad , \quad r^{-+} = M^{-+} \quad , \quad t^{++} = M^{++} + I \quad , \quad t^{--} = M^{--} + I \, , \, (38)$$

where I is the unit matrix (representing the unscattered plane wave).

The basic formula (34), in the form of (38), is further modified in our programs to include the beam symmetries discussed in Chap. 4 and App. A, as well as temperature effects as discussed in Sect. 3.5 and below in Sect. 5.4. The resulting final formula then becomes:

$$M^{\pm\pm}_{\underline{g1}',\underline{g1}} = \delta_{\underline{g1}',\underline{g1}}\,\delta_{\pm\pm} + \frac{8\pi^2 i}{A|k_0|k^+_{g1'\perp}}\sum_{\substack{\ell'm'\\\ell m}}\left\{i^{\ell}(-1)^m\sqrt{\frac{J}{I}}\sum_{i=1}^{I}\exp[i(\underline{g1}'-\underline{gi})\underline{s}],\right.$$

$$\times \left. \exp[-im\phi(\underline{gi})+im\phi(\underline{g1})]Y_{\ell-m}[\Omega(\underline{k}^{\pm}_{g1})]\right\}(1-X)^{-1}_{\ell m,\ell'm'}$$

(39)

$$\times \left\{i^{-\ell'}Y_{\ell'm'}[\Omega(\underline{k}^{\pm}_{g1'})]\right\}\exp[i\delta_{\ell'}(T)]\sin\delta_{\ell'}(T)\,e^{-M} \quad ,$$

using the notations of the appropriate sections (the Debye-Waller factor e^{-M} is optional: cf. Sect.5.4).

5.2 Subroutine MSMF

The matrices of (39) are computed by subroutine MSMF (and other subroutines called by MSMF; for programming details, see the subroutine listings given in Chap.11) Separate matrices are produced for reflection by and transmission through the layer. For Bravais-lattice layers, there is symmetry relative to the plane of nuclei, so that $M^{+-} = M^{-+}$ and $M^{++} = M^{--}$. Symmetry as discussed in Chap. 4 is built in, if specified by the user through the input to the programs and the calling parameters of the subroutine. Since the diffraction matrices are registry-dependent when symmetry is used, subroutine MSMF has been made capable of simultaneously producing several (up to 4) sets of different diffraction matrices, each set corresponding to one particular registry. The registry shift is included by subroutine TRANS as indicated in (A.5).

The same subroutine MSMF is used whether the layer under consideration is a substrate layer or a superlattice overlayer. The only condition is that the layer have a Bravais lattice. To that effect the lattice sums $F_{\ell m}$ of (37) are computed by subroutine FMAT in the form of "sublattice sums" $F_{\ell ms}$ (cf. App.B); depending on whether a substrate layer or an overlayer is considered, the appropriate lattice sums are selected for use in MSMF by setting NLL = NL or = 1, respectively (NLL is the number of required sublattices). Also, to inform the subroutine of which registries should be used, one sets LAY = 1 for an overlayer and LAY = 2 for a substrate layer.

With a surface superlattice the substrate layer diffraction matrices block-diagonalize (cf. Sect.3.3). Subroutine MSMF is called separately to generate each separate block (corresponding to each beam set; see the examples in Chap.10). For each block the appropriate combination of sublattice sums $F_{\ell ms}$ is formed, so that the partial sums $F_{\ell ms}$ need not be recomputed from one beam set to the next. In the case where there is symmetry between beam sets (cf. Sect.4.2 and, in particular, Fig.4.3), the symmetry code numbers are different for an overlayer and for a substrate layer. Subroutine MSMF uses the correct code numbers by checking the value of LAY: if LAY = 1 an overlayer is implied, if LAY = 2 a substrate layer is implied.

A few words are necessary here about the dimensioning and organization of the diffraction matrices output from MSMF. First, because the diffraction matrix dimensions are made variable in the subroutines (this is due to the fact that the number of beams used is energy-dependent) and because in Fortran matrices are stored as vectors, individual matrix elements cannot easily be correctly accessed in the main (calling) program. The index values of any matrix element are in general different in the main program from the index values used in the called subroutine with variable dimensions. One must in fact beware of manipulating such matrices in any way in the main program; any manipulations should be done inside a subroutine called from the main program, using variable dimensions. This applies in particular to the printing of matrices.

Secondly, the block-diagonalization of the substrate matrices under a superlattice surface implies a waste of storage space (cf. Fig.5.1a), unless savings are introduced. Our programs use two types of savings, depending on whether Layer Doubling (cf. Chap.8) or RFS (cf. Chap.9) is used for

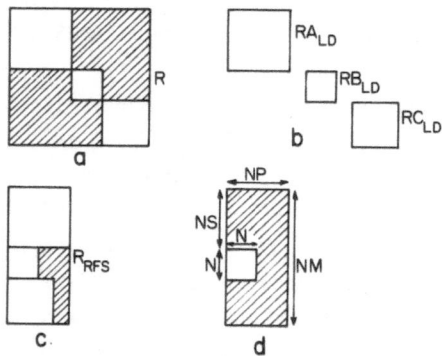

Fig. 5.1 Organization of diffraction matrices output by subroutine MSMF. Panel (a) shows a block-diagonalized matrix R, which is separated into its blocks RA_{LD}, RB_{LD}, and RC_{LD} in panel (b) as used with Layer Doubling, or compacted into one matrix R_{RFS} in panel (c) as used with RFS. Hatched areas contain only vanishing matrix elements. Panel (d) indicates how a block is positioned by subroutine MSMF in an output matrix to produce either case (b) or case (c); the numbers NM,NP (matrix dimensions), N (block dimension) and NS (shift in columns) are named as in the subroutine argument list (cf. Chapter 11). The particular call illustrated in panel (d) evaluates only the N×N elements in the unhatched area. In panel (c), the largest block (whose dimension is equal to NP) does not always have to be the first beam set, as is drawn here

stacking together layers. With Layer Doubling in the substrate, each block in the diagonal is a separate matrix in the main program and in the subroutines, so that each initial big matrix is represented by several smaller matrices (cf. Fig.5.1b). With RFS the blocks are left-justified into the first column of a single matrix, whose right-hand columns are trimmed off, so that the original matrices have a second dimension reduced to that of the largest block (cf. Fig.5.1c; some storage space is still wasted whenever the blocks have unequal sizes). As each call to MSMF produces only one block, MSMF will have to be called once for each block and it will position its output in the appropriate section of the output matrices. Fig. 5.1d illustrates how this is accomplished by a suitable choice of the numbers NM, NP, N and NS.

5.3 The Intralayer Multiple-Scattering Matrix X

The intralayer multiple-scattering matrix X given by (35), computed by subroutine XM, block-diagonalizes into two blocks under a suitable permutation of the (ℓ,m)-pair sequence. This is a consequence of the planar arrangement of atoms in a Bravais lattice. The two blocks are stored in separate matrices and the inversion step of (39) is performed on each block separately. The (ℓ,m) sequence used for the matrix X is the "symmetrical ordering" (cf. Sect.3.6; at normal incidence X may be broken further into four blocks along the diagonal, but, due to its limited validity, this feature is not taken advantage of in the programs presented here).

To compute (35), the quantities $C^{\ell}(\ell'm',\ell''m'')$ of (36) are needed: these are Clebsch-Gordon coefficients with prefactors. They are energy- and crystal-structure-independent and are therefore generated only once in a calculation by subroutine CELMG (which produces them in the sequence in which they are used in subroutine XM called by MSMF). Also needed in the generation of X are the lattice sums $F_{\ell m}$ of (37). These are produced by subroutine FMAT, which, when a superlattice is present, performs the lattice summation over different sublattices separately (cf. App.B). Note that $F_{\ell m}$ is energy-dependent, and, therefore, if the muffin-tin constant V_0 or inelastic damping V_{0i} changes from layer to layer, $F_{\ell m}$ must be recalculated. In normal cases, we assume V_0 and V_{0i} to be identical for all layers, so that $F_{\ell m}$ need be calculated only once (at each energy) for all layers.

If, for any reasons, one desires to remove all intralayer multiple-scattering, it is sufficient to set the matrix X to zero in (39).

5.4 Scattering Amplitudes and Temperature Effects

The atomic scattering amplitudes are input into subroutine MSMF as AF(L) = CAF(L) = $\exp(i\delta_{\ell})\sin\delta_{\ell}$ where δ_{ℓ} ($\ell = 0,1,2,\ldots,\ell_{max}$; L = ℓ + 1) are the phase shifts. With isotropic thermal vibration (the normally recommended choice) these phase shifts must include the correction due to the Debye-Waller factor: the "temperature-dependent phase shifts" (cf. Sect.3.5) are computed by subroutine PSTEMP called by subroutine TSCATF in the main program. Once the temperature-dependent phase shifts are generated and input to MSMF, no further temperature correction is needed. This isotropic thermal vibration correction is signified by the input parameter choice IT = 0 in the calling of MSMF from the main program (AF should then be equal to CAF, as is shown in all the examples of Chap.10).

In the case of anisotropic thermal vibrations, an explicit Debye-Waller factor e^{-M} [computed by subroutine DEBWAL according to (30)] is multiplied into the layer diffraction amplitudes as shown in (39). The phase shifts δ_ℓ, explicitly appearing in (39) are now *not* corrected for thermal vibrations. This is done by setting IT = 1, which choice uses the non-temperature-corrected scattering amplitudes AF(L). However, the matrix X also contains phase shifts [cf. (35)]. Equation (35) is not general enough to accept anisotropic vibration corrections and so we use temperature-dependent phase shifts here, obtained (by subroutine TSCATF) from an angle-averaged vibration amplitude [these temperature -corrected values must enter MSMF through CAF(L) $\neq$ AF(L)]. The choice IT = 1 also achieves this operation of MSMF.

Zero-temperature vibration is included as described in Sect. 3.5 and taken into account in subroutine PSTEMP (and DEBWAL, if IT = 1). Setting this vibration amplitude to zero in the input is sufficient to remove its influence.

6. The Combined Space Method for Composite Layers: by Matrix Inversion

6.1 The Formalism

To treat an atomic layer with more than one atom per unit cell (a composite layer) one may use a matrix inversion formalism, expressed in angular momentum space, applied to the subplanes constituting that composite layer [12], generalized [13] to include incidence of the plane waves $\underline{k}_{\underline{g}}^{\pm}$ for all beams $\underline{g}$.

Let us consider a composite layer with N subplanes (the subplanes all have identical Bravais lattices, but their atoms may be of different species). We define the following quantities in spherical wave [L = (ℓ,m)] representation (atomic units):

$$t_\ell^i = \frac{1}{2k_0} e^{i\delta_\ell} \sin\delta_\ell : \text{ scattering t-matrix of a single atom in} \tag{40}$$
$$\text{subplane i; here } k_0 = \sqrt{2(E-V_0-iV_{0i})};$$

$\tau_{LL'}^i$: scattering matrix containing all scattering paths within subplane i;

$T_{LL'}^i$: scattering matrix including all those scattering paths within the composite layer that terminate at subplane i;

$G_{LL'}^{ji}$: structural propagator describing all unscattered propagations from atoms in subplane i to atoms in subplane j (superscripts to be read from right to left).

The propagator $G_{LL'}^{ji}$ can be expressed either as a sum over lattice points in real space or as a sum over reciprocal-lattice points [9]:

$$G_{LL'}^{ji} = e^{-i\underline{k}_{\underline{g}}^{\pm}\cdot(\underline{r}_j-\underline{r}_i)} \hat{G}_{LL'}^{ji}, \tag{41}$$

with

$$\hat{G}^{ji}_{LL'} = -8\pi i \; k_0 \sum_{L_1} {\sum_{\underline{P}}}' \; i^{\ell_1} a(L,L',L_1) h^{(1)}_{\ell_1}(k_0|\underline{P} + \underline{r}_j - \underline{r}_i|)$$

$$\times Y_{L_1}(\underline{P} + \underline{r}_j - \underline{r}_i) \; e^{-i\underline{k}_0 \cdot \underline{P}} \tag{42}$$

or

$$\hat{G}^{ji}_{LL'} = -\frac{16\pi^2 i}{A} \sum_{\underline{g}_1} \frac{e^{i\underline{k}^{\pm}_{\underline{g}_1}(\underline{r}_j - \underline{r}_i)}}{k^+_{\underline{g}_1 \perp}} Y^*_L(\underline{k}^{\pm}_{\underline{g}_1}) Y_{L'}(\underline{k}^{\pm}_{\underline{g}_1}) \quad . \tag{43}$$

Here, $\underline{P}$ runs over all lattice points of one subplane, excluding the term with $\underline{P} + \underline{r}_j - \underline{r}_i = \underline{0}$, $\underline{r}_i$ and $\underline{r}_j$ are positions of arbitrary reference atoms in subplanes i and j, respectively, $h^{(1)}_{\ell}$ is a Hankel function of the first kind, $Y_{\ell m}$ is a spherical harmonic and A is the unit cell area. The Clebsch-Gordon coefficients are defined as follows [this is the convention used in [9], somewhat different from (36)]:

$$a(L_1,L_2,L_3) = \int Y^*_{L_1}(\Omega) \; Y_{L_2}(\Omega) \; Y^*_{L_2}(\Omega) \; d\Omega \quad . \tag{44}$$

The subplane τ-matrix has the following expression [9] (I is the unit matrix):

$$\tau^i_{LL'} = [(I - t^i G^{ii})^{-1}]_{LL'} t^i_{\ell}. \tag{45}$$

Since $G^{ii} = \hat{G}^{ii}$, there is no g-dependence in $\tau^i_{LL'}$.

According to Beeby [12], the matrices $T^1, \ldots, T^N$ are obtained exactly from τ^i and G^{ji} through the equation

$$
\begin{pmatrix} T^1 \\ T^2 \\ \cdot \\ \cdot \\ \cdot \\ T^N \end{pmatrix}
=
\begin{pmatrix}
I & -\tau^1 G^{12} & \cdots & -\tau^1 G^{1N} \\
-\tau^2 G^{21} & I & \cdots & -\tau^2 G^{2N} \\
\cdot & \cdot & & \cdot \\
\cdot & \cdot & & \cdot \\
\cdot & \cdot & & \cdot \\
-\tau^N G^{N1} & -\tau^N G^{N2} & \cdots & I
\end{pmatrix}^{-1}
\begin{pmatrix} \tau^1 \\ \tau^2 \\ \cdot \\ \cdot \\ \cdot \\ \tau^N \end{pmatrix} \quad . \tag{46}
$$

At first glance, since G^{ji} depends on the beam $\underline{g}$, it seems that one needs to do the large-matrix inversion in (46) once for each incidence vector $\underline{k}\underline{g}^{\pm}$. However, using $\hat{G}^{ji}$ of (41) rather than G^{ji} and factoring out the quantities

$$R_{\underline{g}}^{i\pm} = e^{\pm i\underline{k}_{\underline{g}}^{\pm}\cdot\underline{r}_i} \tag{47}$$

allows (46) to be rewritten as follows:

$$
\begin{pmatrix} T^1 \\ T^2 \\ \cdot \\ \cdot \\ \cdot \\ T^N \end{pmatrix}
=
\begin{pmatrix}
R_{\underline{g}}^{1\pm}I & 0 & \cdots & 0 \\
0 & R_{\underline{g}}^{2\pm}I & \cdots & 0 \\
\cdot & \cdot & & \cdot \\
\cdot & \cdot & & \cdot \\
\cdot & \cdot & & \cdot \\
0 & 0 & \cdots & R_{\underline{g}}^{N\pm}I
\end{pmatrix}^{-1}
\begin{pmatrix}
I & -_\tau{}^1\hat{G}12 & \cdots & -_\tau{}^1\hat{G}1N \\
-_\tau{}^2\hat{G}21 & I & \cdots & -_\tau{}^2\hat{G}2N \\
\cdot & & & \cdot \\
\cdot & & & \cdot \\
\cdot & & & \cdot \\
-_\tau{}^N\hat{G}N1 & -_\tau{}^N\hat{G}N2 & \cdots & I
\end{pmatrix}^{-1}
$$

$$
\times
\begin{pmatrix}
R_{\underline{g}}^{1\pm}I & 0 & \cdots & 0 \\
0 & R_{\underline{g}}^{2\pm}I & \cdots & 0 \\
\cdot & \cdot & & \cdot \\
\cdot & \cdot & & \cdot \\
\cdot & \cdot & & \cdot \\
0 & 0 & \cdots & R_{\underline{g}}^{N\pm}I
\end{pmatrix}
\begin{pmatrix} \tau^1 \\ \tau^2 \\ \cdot \\ \cdot \\ \cdot \\ \tau^N \end{pmatrix}
\tag{48}
$$

Since $\hat{G}^{ji}$ is independent of $\underline{g}$, the time-consuming matrix inversion step in (48) needs to be done only once (the first matrix on the R.H.S. is diagonal and its inversion trivial).

The diffraction matrix elements $M_{\underline{g}'\underline{g}}^{\pm\pm}$ for the entire composite layer are obtained from the quantities T^i:

$$M_{\underline{g}'\underline{g}}^{\pm\pm} = -\frac{16\pi^2 i}{A} \sum_{LL'} \frac{Y_L(\underline{k}_{\underline{g}'}^{\pm})Y_{L'}^*(\underline{k}_{\underline{g}}^{\pm})}{k_{\underline{g}'\perp}^+} \sum_{i=1}^{N}\left[R_{\underline{g}}^{i\pm}(R_{\underline{g}'}^{i\pm})^{-1}T_{LL'}^i\right] \tag{49}$$

As in (39), a unit matrix is subsequently added to M^{++} and M^{--} to form transmission matrices t^{++} and t^{--}. Equation (49) assumes a single arbitrary reference point: the origin of the vectors $\underline{r}_i$. This origin may be a point

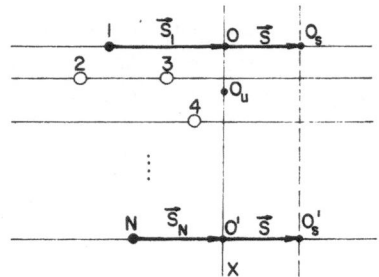

Fig. 6.1 Geometry in a composite layer. O_u is an arbitrary local origin of coordinates, with respect to which the positions of the N atoms of the unit cell are to be given (vectors POS in the programs). The first reference point O will be coplanar with the topmost subplane 1, while the second reference point O' will be coplanar with subplane N, the subplane farthest from the surface. O_u, O and O' have identical y- and z-coordinates (parallel to the surface). They have a registry $\underline{s}$ with respect to an axis or plane of symmetry (broken line) perpendicular to the surface. O_s and O_s' are reference points used when the symmetry is used: in that case the interlayer propagation track (cf. Fig.4.1) arrives at O_s and leaves at O_s', rather than at O and O'

inside or outside the composite layer: in Fig. 6.1, we represent this origin as the point O_u. The program actually uses two reference points instead of one. If we assume these reference points to be, at first, atoms 1 and N of Fig. 6.1, then (49) is replaced by

$$M_{\underline{g}'\underline{g}}^{\pm\pm} = -\frac{16\pi^2 i}{A} \sum_{LL'} \frac{Y_L(k_{\underline{g}'}^{\pm})Y_{L'}^*(k_{\underline{g}}^{\pm})}{k_{\underline{g}'\perp}^+} \sum_{i=1}^{N} \left[\begin{pmatrix} 1+ \\ R_{\underline{g}}^{N-} \end{pmatrix}^{-1} R_{\underline{g}}^{i\pm} T_{LL'}^i (R_{\underline{g}'}^{i\pm})^{-1} R_{\underline{g}'}^{\begin{smallmatrix}N+\\1-\end{smallmatrix}} \right]. \quad (50)$$

(The quantities R are those of (47); the notation, e.g., $R_{\underline{g}}^{\begin{smallmatrix}1+\\N-\end{smallmatrix}}$ means that 1 is chosen if + applies, and N is chosen if - applies.) The unit matrix added to M^{++} and M^{--} of (49) for unscattered transmission is now replaced by diagonal matrices whose elements are

$$\delta_{\underline{g}'\underline{g}} \, \delta_{\pm\pm} \begin{pmatrix} 1+ \\ R_{\underline{g}}^{N-} \end{pmatrix}^{-1} R_{\underline{g}}^{\begin{smallmatrix}N+\\1-\end{smallmatrix}}.$$

The two reference points actually used by the programs are O and O' if no symmetry is used, or O_s and O_s', if symmetry is used: these points are defined in Fig. 6.1. The user only needs to choose a convenient (and arbitrary) point of reference O_u, from which the program automatically generates the reference points O and O' when no symmetry is used. If symmetry is used, the user inputs $\underline{s}$ and O_u, from which the program generates O_s and O'_s. Using the notation of Fig. 6.1, the diffraction matrices $M^{\pm\pm}_{\underline{g}'\underline{g}}$ of (50) then change as follows [in direct analogy with (A.5)]:

$$\tilde{M}^{++}_{\underline{g}'\underline{g}} = e^{i\underline{k}_{0//} \cdot (\underline{s}_N - \underline{s}_1)} \; e^{i(\underline{g}' \cdot \underline{s}_N - \underline{g} \cdot \underline{s}_1)} \; e^{i(\underline{g}' - \underline{g}) \cdot \underline{s}} \; M^{++}_{\underline{g}'\underline{g}} \quad,$$

$$\tilde{M}^{--}_{\underline{g}'\underline{g}} = e^{i\underline{k}_{0//} \cdot (\underline{s}_1 - \underline{s}_N)} \; e^{i(\underline{g}' \cdot \underline{s}_1 - \underline{g} \cdot \underline{s}_N)} \; e^{i(\underline{g}' - \underline{g}) \cdot \underline{s}} \; M^{--}_{\underline{g}'\underline{g}} \quad,$$

$$\tilde{M}^{-+}_{\underline{g}'\underline{g}} = e^{i(\underline{g}' - \underline{g}) \cdot (\underline{s}_1 + \underline{s})} \; M^{-+}_{\underline{g}'\underline{g}} \quad,$$

$$\tilde{M}^{+-}_{\underline{g}'\underline{g}} = e^{i(\underline{g}' - \underline{g})(\underline{s}_N + \underline{s})} \; M^{+-}_{\underline{g}'\underline{g}} \quad. \tag{51}$$

The exponential factors of this transformation are computed in subroutine TRANSP. With the use of symmetry, the matrices $\tilde{M}^{\pm\pm}_{\underline{g}'\underline{g}}$ depend on $\underline{s}$, which may vary from layer to layer, depending on the registries of the layers. When no symmetry is used, the vector $\underline{s}$ is set equal to $\underline{0}$ in (51). Symmetry is further exploited as described in Appendix A, quite analogously to the case of Bravais-lattice layers [cf. (39)]. The only difference is that, for computational convenience, (A.2) is changed to the equivalent form

$$M'_{\underline{g}'\underline{g}} = \sqrt{I/J} \sum_{i=1}^{J} M_{\underline{g}'j,g1} \quad,$$

i.e., the sum runs through columns rather than rows of M.

6.2 Subroutine MTINV

The method of Sect. 6.1 is implemented in subroutine MTINV (and subroutines called by MTINV). Many comments made in Sect. 5.1 concerning subroutine MSMF apply here as well, in particular regarding the use of symmetry and the registry-dependence of the diffraction matrices. However, subroutine MTINV works only for layers that have the surface superlattice, or the bulk lattice if there is no surface superlattice (this is because no summation over sublattices is programmed here). Therefore also no block-diagonalization has to be considered.

To generate $\tilde{M}^{\pm\pm}_{g'g}$ of (51), subroutine MTINV first obtains τ^i of (45) by calling subroutine TAUMAT. For this purpose the quantity $t^i G^{ii}$ of (45), [where t^i, given in (40), is considered to be a diagonal matrix] is related to the quantity $X_{LL'}$ of (35) through (cf. Sect.2.2)

$$[t^i \ G^{ii}]_{LL'} = i^{\ell'-\ell}(-1)^{\ell'+\ell} \frac{t_{\ell'}}{t_\ell} X_{LL'} \ . \tag{52}$$

This relation is exploited in subroutine XMT, whose mechanism closely follows that of subroutine XM mentioned in Sect. 5.3 (including the use of lattice sums produced by FMAT and of Clebsch-Gordon coefficients produced by CELMG). Instead of producing one τ^i for each subplane i, TAUMAT produces only as many τ's as there are different chemical elements in the composite layer. These different τ's are stored (stacked under each other) in a single matrix called TAU. Because a subplane is always planar, τ block-diagonalizes ($\ell+m$ even decouples from $\ell+m$ odd) in the "symmetrized ordering" of the (ℓ,m) pairs used in subroutine TAUMAT (cf. Sect.3.6). To reduce storage space, the individual blocks are left-justified to the left column of the matrix and the unused right-hand columns are trimmed off.

The next operation in MTINV is performed by subroutine SRTLAY, which examines the geometry of the layer. SRTLAY orders the subplanes according to increasing distance from the surface: this is needed to correctly position the two reference points of the composite layer and has an important function in the Reverse Scattering Perturbation program (cf. Chap.7), which also uses this subroutine. SRTLAY also generates the interplanar vectors pointing from one subplane to another. These are mutually compared for any duplications. In addition, if the program scans through a series of different geometries within the composite layer, the set of vectors is compared with the equivalent set used in the previous geometry at the same energy. This is important in the calculation of the quantities $\hat{G}^{ji}$ [see (42,43)], since their generation is time-consuming and they differ only through the interplanar vectors: the program avoids recomputing matrices $\hat{G}^{ji}$ that have already been computed. To this end SRTLAY produces a vector NUGH containing a set of numbers that indicate which matrices $\hat{G}^{ji}$ need to be computed, a vector NGEQ that indicates which matrices $\hat{G}^{ji}$ can be copied from which other ones, and a vector NGOL that indicates which matrices $\hat{G}^{ji}$ are unchanged from the previous call to MTINV (for more details see the listing of subroutine SRTLAY in Chap.11). By setting NEW = -1 in the main program, the user signals that the energy was not changed since the last call to MTINV. When NEW = +1 is set, all $\hat{G}^{ji}$, spherical harmonics, τ matrices, etc., are to be recomputed: This usually signals a new energy point.

The matrices $\hat{G}^{ji}$ are calculated in subroutine GHMAT, called by MTINV. GHMAT selects either the real-space summation (42) or the reciprocal-space summation (43), depending on their relative speed of computation. The number of terms required by the reciprocal-space summation depends on the separation distance between subplanes. This mode of summation is often more favorable than the real-space summation. Note that the real-space summation for $\hat{G}^{ji}$ requires a set of Clebsch-Gordon coefficients, defined in (44). This set is generated by subroutine CAAA in the main program in parallel with the other set [defined by (36) and used in the matrix X] generated by subroutine CELMG.

The matrices $\hat{G}^{ji}$ are stored in a single matrix GH stacked one under the other according to the following scheme [$\hat{G}^{ji}$ for j = i is not needed, since it can be obtained from the matrix X: cf. (52)]. Choosing 4 subplanes for illustrative purposes, each matrix $\hat{G}^{ji}$ (i,j $\le$ 4; i $\ne$ j) is assigned a serial number by assuming a one-to-one relationship between the elements of the following two matrices, whose structures are self-explanatory:

$$\begin{pmatrix} - & \hat{G}^{12} & \hat{G}^{13} & \hat{G}^{14} \\ \hat{G}^{21} & - & \hat{G}^{23} & \hat{G}^{24} \\ \hat{G}^{31} & \hat{G}^{32} & - & \hat{G}^{34} \\ \hat{G}^{41} & \hat{G}^{42} & \hat{G}^{43} & - \end{pmatrix} \leftrightarrow \begin{pmatrix} - & 1 & 2 & 3 \\ 7 & - & 4 & 5 \\ 8 & 10 & - & 6 \\ 9 & 11 & 12 & - \end{pmatrix} = MGH \quad . \tag{53}$$

The serial numbers contained in the matrix MGH on the right in (53) simply indicate the location of each matrix $\hat{G}^{ji}$ in the stack contained in the following big matrix GH

$$GH = \begin{pmatrix} \hat{G}^{12} \\ \hat{G}^{13} \\ \hat{G}^{14} \\ \hat{G}^{23} \\ \hat{G}^{24} \\ \hat{G}^{34} \\ \hat{G}^{21} \\ \hat{G}^{31} \\ \hat{G}^{41} \\ \hat{G}^{32} \\ \hat{G}^{42} \\ \hat{G}^{43} \end{pmatrix} . \tag{54}$$

With the knowledge of τ^i and $\hat{G}^{ji}$, MTINV produces, through subroutine THMAT, the large matrix that is to be inverted in (48). Rather than multiply the inverted matrix into the matrices $\tau^i_{LL'}$, we first perform on $\tau^i_{LL'}$ the summation over L' required by (50). For that purpose, we define the quantities

$$\tau^i_L(\underline{g}^\pm) = \sum_{L'} \tau^i_{LL'} \, Y^*_{L'}(\underline{k}^\pm_{\underline{g}}), \tag{55}$$

called TAUG and TAUGM (for + and -) in the programs (they are produced by subroutine TAUY).

The reason for the above step is that now the inverted matrix of (48) is multiplied into vectors $\tau^i_L(\underline{g}^\pm)$ (for each given $\underline{g}$) rather than into matrices $\tau^i_{LL'}$ (of course, the factors $R^{i\pm}_{\underline{g}}$ are also included), which on balance saves computation time. This is especially true in the frequent case where the number of beams is less than $(\ell_{max}+1)^2$.

Finally the summation over L in (50) is performed by subroutine MFOLT, which also introduces any required symmetrization and registry shifts. The resulting diffraction matrices are output under the names RA1 = $\tilde{M}^{-+}$, TA1 = $\tilde{M}^{++} + (R^{N+})^{-1}R^{1+}$, RA2 = $\tilde{M}^{+-}$, TA2 = $\tilde{M}^{--} + (R^{1-})^{-1}R^{N-}$, where the added diagonal matrices represent the unscattered plane waves. Since these matrices are registry-dependent [cf. (51)], they are distinguished by the second letters, e.g., RA1, RB1 and RC1 for three different registries.

It will be realized that in special cases a composite layer may be symmetrical with respect to a mirror plane *parallel* to the surface, in which case $\tilde{M}^{++} = \tilde{M}^{--}$ and $\tilde{M}^{-+} = \tilde{M}^{+-}$. To reduce the computational effort in such

cases, subroutine MTINV does not compute $\tilde{M}^{--}$ and $\tilde{M}^{+-}$ if the user sets
NOPT = 2 in the main program. The more general (asymmetrical) case is sig-
nalled by setting NOPT = 1. A further option (NOPT = 3) allows the user to
have MTINV compute only the diffraction matrix elements for the primary in-
cident beam $\underline{k}_{\underline{0}}^{\pm}$, neglecting the other incident beams $\underline{k}_{\underline{g}}^{\pm}$ for $\underline{g} \neq \underline{0}$, i.e. pro-
producing only $\tilde{M}_{\underline{g}'\underline{0}}^{\pm\pm}$. This special case includes the method of BEEBY [12]
where only $M_{\underline{g}'\underline{0}}^{-+}$ is used.

7. The Combined Space Method for Composite Layers: by Reverse Scattering Perturbation

7.1 The Formalism of Reverse Scattering Perturbation (RSP) Theory

The purpose of the Reverse Scattering Perturbation (RSP) method [14] is to replace the inversion of a large matrix in (48) by a perturbation expansion over orders of scattering. In our version of RSP [13] we take into account that forward scatterings may not be treated as weak events, but must always be included exactly: only back-scatterings are assumed to be weak. This feature requires a logical ordering of the subplanes in terms of increasing distance from the surface.

Every perturbation method has its limits of validity. In this case it may happen that the atoms of a composite layer are too strong scatterers for proper convergence of the RSP method: then matrix inversion by the method of Chap. 6 may be necessary. If only atoms in *some* subplanes of the composite layer are too strong scatterers, the multiple scattering between these subplanes must first be computed by matrix inversion. Then the Reverse Scattering Perturbation can be applied to add the remaining subplanes (made up of weaker scatterers) to a composite slab which is the matrix inversion result of the set of strongly scattering subplanes. This implies a combined matrix-inversion/RSP treatment, which will be dealt with in Sect. 7.2. In this section, we outline the procedure of Reverse Scattering Perturbation.

As in (55), we define

$$T_L^i(\underline{g}^{\pm}) = \sum_{L'} T_{LL'}^i \cdot Y_{L'}^*(\underline{k}_{\underline{g}}^{\pm}) \quad , \qquad i = 1,2,\ldots,N \quad , \tag{56}$$

and, from (55) we have

$$\tau_L^i(\underline{g}^{\pm}) = \sum_{L'} \tau_{LL'}^i \cdot Y_{L'}^*(\underline{k}_{\underline{g}}^{\pm}) \quad , \qquad i = 1,2,\ldots,N \quad .$$

The objective is to obtain $T_L^i(\underline{g}^\pm)$ perturbationally and to then define diffraction matrices using (50). With (56), (50) becomes (N is the number of subplanes in the composite layer):

$$M_{\underline{g}'\underline{g}}^{\pm\pm} = - \frac{16\pi^2 i}{A} \sum_L \frac{Y_L(\underline{k}_{\underline{g}'}^\pm)}{k_{\underline{g}'\perp}^+} \sum_{i=1}^N \left[\begin{pmatrix} 1+ \\ R_{\underline{g}}^{N-} \end{pmatrix}^{-1} R_{\underline{g}}^{i\pm} \, T_L^i(\underline{g}^\pm)(R_{\underline{g}'}^{i\pm})^{-1} \, R_{\underline{g}'}^{\begin{subarray}{l}N+ \\ 1-\end{subarray}} \right] \quad . \tag{57}$$

Further treatment of this expression (for symmetry and registry shifts) is identical to that in Chap. 6.

The perturbation expansion for $T_L^i(\underline{g}^\pm)$ is taken as follows:

$$T_L^i(\underline{g}^\pm) = \tau_L^i(\underline{g}^\pm) + \sum_{n=0}^\infty [T_L^{i+(n)}(\underline{g}^\pm) + T_L^{i-(n)}(\underline{g}^\pm)] \quad . \tag{58}$$

The meaning of the quantities in (58) is the following: $T_L^i(\underline{g}^\pm)$ contains all scattering paths terminating at subplane i (from an incident beam $\underline{k}_{\underline{g}}^\pm$); $\tau_L^i(\underline{g}^\pm)$ contains all paths with subplane i; $T^{i+(n)}$ contains those paths terminating at subplane i that arrive from the substrate side of this subplane (i.e., from below in Fig. 4.1) and that have experienced n reversals of direction (from +x to -x or vice versa), not counting the possible change of direction at the first scattering by a subplane; $T^{i-(n)}$ is identical to $T^{i+(n)}$ except that it includes only paths arriving from the vacuum side of subplane i (i.e., from above Fig. 4.1).

The following iteration relations hold [13]:

$$T_L^{i+(n)}(\underline{g}^\pm) = \sum_{L_1 L_2} \tau_{LL_1}^i \sum_{j>i} G_{L_1 L_2}^{ij} [T_{L_2}^{j+(n)}(\underline{g}^\pm) + T_{L_2}^{j-(n-1)}(\underline{g}^\pm)] \quad , \tag{59}$$

$$T_L^{i-(n)}(\underline{g}^\pm) = \sum_{L_1 L_2} \tau_{LL_1}^i \sum_{j<i} G_{L_1 L_2}^{ij} [T_{L_2}^{j-(n)}(\underline{g}^\pm) + T_{L_2}^{j+(n-1)}(\underline{g}^\pm)] \quad . \tag{60}$$

The initial conditions are:

$$T_L^{1-(n)}(\underline{g}^\pm) = 0 \quad , \qquad n = -1,0,1,2,\dots \tag{61}$$

$$T_L^{N+(n)}(\underline{g}^\pm) = 0 \quad , \qquad n = -1,0,1,2,\dots \tag{62}$$

$$T_L^{i-(-1)}(\underline{g}^\pm) = T_L^{i+(-1)}(\underline{g}^\pm) = \tau_L^i(\underline{g}^\pm) \quad , \qquad i = 1,2,...,N \quad . \tag{63}$$

The iterations(59,60) compute in succession $T^{1-(0)}$, $T^{2-(0)}$,..., $T^{N-(0)}$, $T^{N+(0)}$,..., $T^{2+(0)}$, $T^{1+(0)}$ and repeat this sequence with n = 1,2,... until convergence of (58).

7.2 Combining RSP and Matrix Inversion

If in Reverse Scattering Perturbation, the multiple scattering between some subplanes is too strong for a perturbation treatment, then the full multiple scattering between these subplanes can be separately included to infinite order by matrix inversion (cf. Chap. 6). The result of this matrix inversion is subsequently mixed in with the more weakly scattering subplanes, using RSP [13].

To this effect we need quantities $T_{LL'}^{\mu\nu}$ that describe all multiple scattering paths (to infinite order) that start at subplane ν, terminate at subplane μ and never scatter off the weakly scattering subplanes (ν and μ are themselves strongly scattering subplanes, as are all subplanes that we shall label with Greek letters). These are obtained by a small modification of (46)(the user should take note of the restriction on the ordering of strongly and weakly scattering subplanes: Sect. 7.3):

$$
\begin{pmatrix}
T^{\alpha\alpha} & T^{\alpha\beta} & \cdots & T^{\alpha\omega} \\
T^{\beta\alpha} & T^{\beta\beta} & & T^{\beta\omega} \\
\cdot & \cdot & & \cdot \\
\cdot & \cdot & & \cdot \\
\cdot & \cdot & & \cdot \\
T^{\omega\alpha} & T^{\omega\beta} & \cdots & T^{\omega\omega}
\end{pmatrix}
=
\begin{pmatrix}
I & -\tau^\alpha G^{\alpha\beta} & \cdots & -\tau^\alpha G^{\alpha\omega} \\
-\tau^\beta G^{\beta\alpha} & I & \cdots & -\tau^\beta G^{\beta\omega} \\
\cdot & \cdot & & \cdot \\
\cdot & \cdot & & \cdot \\
\cdot & \cdot & & \cdot \\
-\tau^\omega G^{\omega\alpha} & -\tau^\omega G^{\omega\beta} & \cdots & I
\end{pmatrix}^{-1}
\begin{pmatrix}
\tau^\alpha & 0 & \cdots & 0 \\
0 & \tau^\beta & \cdots & 0 \\
\cdot & \cdot & & \cdot \\
\cdot & \cdot & & \cdot \\
\cdot & \cdot & & \cdot \\
0 & 0 & \cdots & \tau^\omega
\end{pmatrix} . \tag{64}
$$

Here $\alpha,\beta,...,\omega$ label the strongly scattering subplanes. The matrix inversion is made independent of the incident beam $\underline{k}_g^\pm$ in the same way as in (48), namely by defining quantities $\hat{T}^{\mu\nu}$ as follows:

$$T_{LL'}^{\mu\nu} = e^{-i\underline{k}_g^\pm \cdot (\underline{r}_\mu - \underline{r}_\nu)} \hat{T}_{LL'}^{\mu\nu} \quad . \tag{65}$$

Then (64) holds with $\hat{T}^{\mu\nu}$ and $\hat{G}^{\mu\nu}$ replacing $T^{\mu\nu}$ and $G^{\mu\nu}$, respectively. Now we must modify RSP to accommodate the fact that all multiple scattering within a set of strongly scattering subplanes has already been accounted for by matrix inversion. Thus we define a quantity $\bar{T}_L^{\nu}(\underline{g}^{\pm})$ for the strongly scattering subplanes involved in the matrix inversion:

$$\bar{T}_L^{\nu}(\underline{g}^{\pm}) = \sum_{L'} \left[\sum_{\mu=\alpha}^{\omega} T_{LL'}^{\nu\mu} \right] Y_{L'}^*(\underline{k}_{\underline{g}}^{\pm}) \quad , \qquad \nu = \alpha,\beta,\ldots,\omega \quad . \tag{66}$$

This represents all those multiple scattering paths that are confined to the group of strongly scattering subplanes and that terminate at subplane ν. Now RSP for weakly scattering subplanes is still given by (58):

$$T_L^i(\underline{g}^{\pm}) = \tau_L^i(\underline{g}^{\pm}) + \sum_{n=0}^{\infty} \left[T_L^{i+(n)}(\underline{g}^{\pm}) + T_L^{i-(n)}(\underline{g}^{\pm}) \right] \quad , \qquad i \neq \alpha,\beta,\ldots,\omega \quad .$$

For strongly scattering subplanes, this relation is modified to:

$$T_L^{\nu}(\underline{g}^{\pm}) = \bar{T}_L^{\nu}(\underline{g}^{\pm}) + \sum_{n=0}^{\infty} \left[T_L^{\nu+(n)}(\underline{g}^{\pm}) + T_L^{\nu-(n)}(\underline{g}^{\pm}) \right] \quad , \qquad \nu = \alpha,\beta,\ldots,\omega \quad . \tag{67}$$

Here $T^{i\pm(n)}$ and $T^{\nu\pm(n)}$ include more paths than in Sect. 7.1, namely those paths (up to infinite order) that the matrix inversion of (64) has added.

The iteration relations of (59,60) remain valid for weakly scattering subplanes (i.e., $i \neq \alpha,\beta,\ldots,\omega$). For the strongly scattering subplanes (i.e., $\nu = \alpha,\beta,\ldots,\omega$), the new relations are

$$T_L^{\nu+(n)} = \sum_{L'} \sum_{j>\nu}' \left[\sum_{\lambda=\alpha}^{\omega} T^{\nu\lambda} G^{\lambda j} \right]_{LL'} \left[T_{L'}^{j+(n)}(\underline{g}^{\pm}) + T_{L'}^{j-(n-1)}(\underline{g}^{\pm}) \right] \quad , \tag{68}$$

$$T_L^{\nu-(n)} = \sum_{L'} \sum_{j<\nu}' \left[\sum_{\lambda=\alpha}^{\omega} T^{\nu\lambda} G^{\lambda j} \right]_{LL'} \left[T_{L'}^{j-(n)}(\underline{g}^{\pm}) + T_{L'}^{j+(n-1)}(\underline{g}^{\pm}) \right] \quad . \tag{69}$$

Here $\sum'$ excludes summation over the strongly scattering subplanes (i.e., $j = \alpha,\beta,\ldots,\omega$ are excluded). This is to avoid duplicating paths already included by the matrix inversion.

The initial conditions (61-63) are unchanged, with one exception: for the strongly scattering subplanes, (63) becomes

$$T_L^{\nu-(-1)}(\underline{g}^{\pm}) = T_L^{\nu+(-1)}(\underline{g}^{\pm}) = \bar{T}_L^{\nu}(\underline{g}^{\pm}) \quad , \qquad \nu = \alpha, \beta, \ldots, \omega \quad . \tag{70}$$

7.3 Subroutine MPERTI

Both RSP and the combination of RSP and matrix inversion are implemented in subroutine MPERTI. A few restrictions apply:

- the case where no RSP is desired is treated not by MPERTI but by MTINV (cf. Sect. 6.2); all other cases are handled by MPERTI;
- it is important to note: the strongly scattering subplanes are assumed to be arranged in such a way that no weakly scattering subplane is intercalated between them in the order that subroutine SRTLAY gives these subplanes (SRTLAY orders the subplanes by increasing distance from the surface; it leaves coplanar subplanes in the order defined by the user).

Many features of subroutine MPERTI are identical to those of subroutine MTINV (cf. Sect. 6.2), including: treatment of subplane ordering, of τ^i, of G^{ji} and of symmetries. We discuss now some special features of MPERTI.

The user must inform subroutine MPERTI about which subplanes are to be treated by matrix inversion and which not: this is done by setting the vector INV(i) = 1 or 0, respectively (i = 1,2,...,N; N, called NLAY in the programs, is the number of subplanes in the composite layer). The inversion step is performed in subroutine THINV, which produces the quantities $\hat{T}^{\mu\nu}$ of (65). Then both $\tau_L^i(\underline{g}^{\pm})$, [see (55)] and $\bar{T}_L^{\nu}(\underline{g}^{\pm})$, [see (66)], are generated in subroutine TAUY and stored in TAUG (for +) and TAUGM (for -).

The perturbation interation is controlled by subroutine TPERTI; each individual step [propagation to the next subplane, i.e., evaluation of either (59,60) or (68,69)] is done by subroutine TPSTPI. The number of iterations is limited to NPERT, but a lack of convergence does not terminate execution (a message is, however, output).

The quantities $T_L^{i\pm(n)}(\underline{g}^{\pm})$ (here, i runs over all subplanes) are stored for two successive values of n in a double-column matrix called TG, while the cumulative results $T_L^i(\underline{g}^{\pm})$ [see (58), (67)] are stored in a single-column vector called TS, as follows ($\underline{g}^+$ and $\underline{g}^-$ are handled separately in successive stages of the computation):

$$
TG = \begin{vmatrix} T^{1-(n)}(\underline{g}^{\pm}) & T^{1-(m)}(\underline{g}^{\pm}) \\ T^{2-(n)}(\underline{g}^{\pm}) & T^{2-(m)}(\underline{g}^{\pm}) \\ \vdots & \vdots \\ T^{N-(n)}(\underline{g}^{\pm}) & T^{N-(m)}(\underline{g}^{\pm}) \\ \hline T^{N+(n)}(\underline{g}^{\pm}) & T^{N+(m)}(\underline{g}^{\pm}) \\ \vdots & \vdots \\ T^{2+(n)}(\underline{g}^{\pm}) & T^{2+(m)}(\underline{g}^{\pm}) \\ T^{1+(n)}(\underline{g}^{\pm}) & T^{1+(m)}(\underline{g}^{\pm}) \end{vmatrix} , \qquad TS = \begin{vmatrix} T^{1}(\underline{g}^{\pm}) \\ T^{2}(\underline{g}^{\pm}) \\ \vdots \\ T^{N}(\underline{g}^{\pm}) \end{vmatrix} \qquad (71)
$$

Note that each $T^{i\pm(n)}(\underline{g}^{\pm})$ and each $T^{i}(\underline{g}^{\pm})$ is a vector of length $(\ell_{max} + 1)^{2}$. The left-hand column of TG contains at first the initial values; the perturbation steps move down the columns, putting the first iteration in the right-hand column (with m = n + 1), the second iteration in the left-hand column again (with n = m + 1), etc., alternating between the columns. As each column is completed, the current iteration's contribution is added to TS.

The initialization of TG, when no matrix inversion is present, gives, according to (61-63):

$$
TG_{initial} = \begin{vmatrix} 0 & 0 \leftarrow \\ \tau^{2}(\underline{g}^{\pm}) & 0 \\ \vdots & \vdots \\ \tau^{N}(\underline{g}^{\pm}) & 0 \\ \hline 0 & 0 \leftarrow \\ \tau^{N-1}(\underline{g}^{\pm}) & 0 \\ \vdots & \vdots \\ \tau^{1}(\underline{g}^{\pm}) & 0 \end{vmatrix} \qquad (72)
$$

The iteration (59,60) starts, for a beam incident from the vacuum side ($k_{\underline{g}}^{+}$), at the top right-hand corner of TG, while for a beam incident from the substrate side ($k_{\underline{g}}^{-}$), it starts just below the midpoint of the right-hand column of TG [the two starting positions are indicated by arrows in (72)].

In the presence of matrix inversion, the quantities $\bar{T}_L^\nu(\underline{g}^\pm)$ of (70) replace in (72) the quantities $\tau_L^\nu(\underline{g}^\pm)$ for $\nu=\alpha,\beta,\ldots,\omega$. A simplification occurs and is exploited in the program when the strongly scattering subplanes are all on one side of the composite layer, the weakly scattering subplanes being all on the other side. Suppose that the strongly scattering subplanes ($\nu=\alpha,\beta,\ldots,\omega$) are either on the vacuum side (v.s.) or on the substrate side (s.s.) of the composite layer, then the initial TG is, depending on the case:

$$
TG_{initial}^{v.s.} =
\begin{pmatrix}
0 & 0 \\
\vdots & \vdots \\
0 & 0 \\
\tau^{\omega+1}(\underline{g}^\pm) & 0 \\
\vdots & \vdots \\
\tau^N(\underline{g}^\pm) & 0 \\
\hline
0 & 0 \\
\tau^{N-1}(\underline{g}^\pm) & 0 \\
\vdots & \vdots \\
\bar{T}^\omega(\underline{g}^\pm) & 0 \\
\vdots & \vdots \\
\bar{T}^\alpha(\underline{g}^\pm) & 0
\end{pmatrix}
\quad ; \quad
TG_{initial}^{s.s.} =
\begin{pmatrix}
0 & 0 \\
\tau^2(\underline{g}^\pm) & 0 \\
\vdots & \vdots \\
\bar{T}^\alpha(\underline{g}^\pm) & 0 \\
\vdots & \vdots \\
\bar{T}^\omega(\underline{g}^\pm) & 0 \\
\hline
0 & 0 \\
\vdots & \vdots \\
0 & 0 \\
\tau^{N-\omega}(\underline{g}^\pm) & 0 \\
\vdots & \vdots \\
\tau^1(\underline{g}^\pm) & 0
\end{pmatrix}
. \quad (73)
$$

Those rows in (71-73) which initially have only zeroes, will in the perturbation iteration always retain zeroes. The program is therefore made never to compute those rows.

8. Stacking Layers by Layer Doubling

Once the individual layer diffraction matrices are known (as obtained by
any of the methods described in Chaps. 5-7), the total diffraction by a
stack of such layers can be calculated by either the Layer Doubling method
described below or the Renormalized Forward Scattering method (Chap. 9).

8.1 The Formalism

It is possible, for the purpose of calculating LEED reflectivities, to con-
struct the crystal surface layer by layer, starting from the layers whose
diffraction amplitudes we know: this is done by calculating exactly the
diffraction matrices for a pair of (in general unequal) layers and iterating
the process of stacking layers onto a growing slab until convergence of the
reflected intensities. Convergence is ensured by electron absorption which
limits the required thickness of the final slab to, typically, eight mono-
atomic layers to obtain a result within 1% from the exact result. The iteration can be considerably speeded up when identical layers are being stacked,
as happens with clean surfaces and substrates as a result of bulk periodicity
perpendicular to the surface. Then the iteration can at each step double the
thickness of the slab of layers by repeatedly combining two identical slabs:
thus the n-th iteration step combines two identical slabs consisting of
2^{n-1} layers each into one slab consisting of 2^n layers. This is the "Layer
Doubling" method. For a stack of 8 layers, three iterations are sufficient:
a fourth iteration would produce a stack of 16 layers in only 33% more time,
which illustrates the effectiveness of this method for dealing with relatively deep electron penetration.

   The formulae required for the Layer Doubling calculation are those for reflection and transmission from a pair of diffracting layers A and B, each hav-
ing diffraction matrices r_A^{+-}, r_A^{-+}, t_A^{++}, t_A^{--} and r_B^{+-}, r_B^{-+}, t_B^{++}, t_B^{--}, respectively

(cf. Chaps. 5-7). The combined pair has diffraction matrices given by [2]:

$$R^{-+} = r_A^{-+} + t_A^{--} P^- r_B^{-+} P^+ (I - r_A^{+-} P^- r_B^{-+} P^+)^{-1} t_A^{++} \quad ,$$

$$T^{++} = t_B^{++} P^+ (I - r_A^{+-} P^- r_B^{-+} P^+)^{-1} t_A^{++} \quad ,$$

$$R^{+-} = r_B^{+-} + t_B^{++} P^+ r_A^{+-} P^- (I - r_B^{-+} P^+ r_A^{+-} P^-)^{-1} t_B^{--} \quad , \tag{74}$$

$$T^{--} = t_A^{--} P^- (I - r_B^{-+} P^+ r_A^{+-} P^-)^{-1} t_B^{--} \quad .$$

Here P^+ and P^- are diagonal matrices propagating the plane waves from a reference point in layer A to a reference point in layer B and vice versa: if r_{BA} relates the two reference points in layers A and B we have the diagonal elements:

$$P_g^{\pm} = \exp(\pm i \, \underline{k}_g^{\pm} \cdot \underline{r}_{BA}) \quad . \tag{75}$$

(The vector $\underline{r}_{BA}$ is shown in Fig. 4.1a as being the interlayer vector. If symmetry is used, $\underline{r}_{BA}$ is to be replaced by $\underline{r}_{BA}'$ of Fig. 4.1b, i.e., the part of the propagation track that remains to be taken into account after registry shifts $\underline{s}_i$ have already been incorporated in the diffraction matrices.)

Multiple scattering between the layers is fully taken into account by the inverse-matrix factors of (74). For iteration, the results on the left-hand sides of (74) are substituted into the right-hand sides to produce new left-hand sides, and so on.

The computation time required for the Layer Doubling method scales as N^3, where N is the matrix dimension (i.e. the number of plane waves included in the calculation) and as ln M, where M is the number of layers included.

The Layer Doubling method yields a full reflection matrix for the stack of equal layers. Other layers different from bulk layers can then be added to the surface at will, using the first of the equations (74) again: for efficiency the bulk reflection can be stored and repeatedly used for many different configurations of surface layers (such as different overlayer spacings and registries), a convenient feature in a surface structural search.

8.2 Bulk Treatment: Subroutine SUBREF

In the periodic part of the surface, i.e., the bulk below the first few layers, Layer Doubling is performed by subroutine SUBREF. This subroutine assumes that the interlayer vectors r_{BA} between successive component layers of the bulk are identical (so that the propagators $P_g^{\pm}$ are the same throughout the bulk); it also assumes that each component layer has a symmetry plane parallel to the surface (a planar Bravais-lattice layer always satisfies this condition) so that in (74): $r_i^{+-} = r_i^{-+}$, t_i^{++}, $= t_i^{--}$ for i = A, B; it further assumes that the diffraction matrices for successive component layers of the bulk are either identical or alternate between two sets of values (as in some compound materials, but in simpler cases as well: cf. below).

It should be noted that, even with very simple surfaces [e.g., fcc (100)], the equalities $R^{-+} = R^{+-}$ and $T^{++} = T^{--}$ for the quantities appearing in (74) only rarely hold: it is safer to assume that they never hold (the particular layer stacking arrangement or the particular angle of incidence usually cause these equalities to break down). For this purpose, SUBREF has as input *two* pairs of reflection and transmission matrices, to be given equal values when appropriate, but not identical storage areas (i.e., not identical Fortran names). The output matrices R^{-+}, T^{++}, R^{+-}, T^{--} for a stack of 2^n layers (n = number of doublings that achieved convergence) overwrite the input matrices and are usually not pairwise equal.

A complication arises when the use of symmetry implies registry-dependent symmetrized diffraction matrices [cf. (39), (41)]. According to an above assumption, no more than two registry shifts can be handled by SUBREF. This allows, for example, fcc (100) and bcc (100) surfaces to be handled at normal incidence with the use of the full symmetry (4-fold rotation axis and 2 orthogonal mirror planes), since only 2 alternating registries are involved (ABABA..., in common layer-stacking notation). This holds also for hcp (0001) surfaces at normal incidence (with the use of the 3-fold axis and and the mirror planes). But fcc (111) at normal incidence [with the same symmetries as hcp (0001)] requires 3 registries (ABCABC...). Here the solution, short of not exploiting symmetry at all (or extending Layer Doubling to such cases, which is unwieldy because much additional working space is required), is to use only one of the mirror planes and ignore the 3-fold axis altogether in the entire LEED calculation: it will be seen that this requires only one registry s_i (e.g., $s_i = 0$), coupled with interlayer vectors r'_{BA} that are parallel to the mirror plane: cf. also Sect. 10.4 (the resulting increase in matrix dimensions is compensated for by the fact that with this choice off-normal incidence calculations may be made as well, as long as the incident beam remains parallel to the mirror plane).

No special symmetry information need be given SUBREF beyond the correct input diffraction matrices and interlayer vector, as discussed above.

SUBREF operates as follows (each individual doubling step is carried out by a call to subroutine TLRTA): cf. also Fig. 8.1. The first doubling, starting from 2 symmetrical layers ($r_A^{+-} = r_A^{-+}$ = RA; $t_A^{++} = t_A^{--}$ = TA; $r_B^{+-} = r_B^{-+}$ = RB; $t_B^{++} = t_B^{--}$ = TB; here RA,...,TB are the names used in SUBREF), produces an asymmetrical slab of 2 layers (R^{-+} = RA $\neq$ R^{+-} = RB; T^{++} = TA $\neq$ T^{--} = TB). Two identical such slabs are then combined in the next doubling, producing a further asymmetrical slab consisting of 4 layers; this step is repeated until convergence of the reflection matrices. The number of doublings is limited to LITER (set in the input), but in no case is program execution terminated.

As noted in Sect. 3.3, diffraction matrices in the bulk block-diagonalize according to beam sets when a superlattice is present on the surface. To exploit this feature, SUBREF operates on one block (i.e., one beam set) at a time and must be called separately for each beam set.

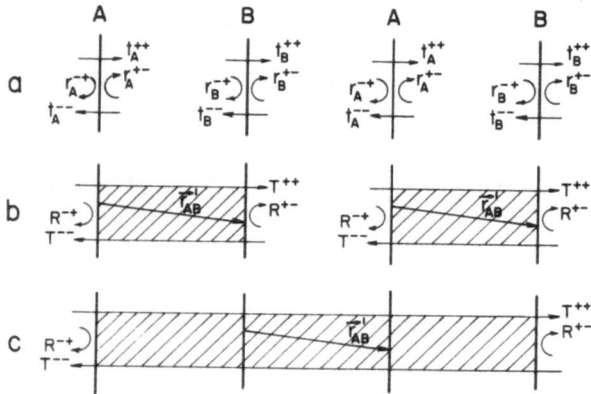

Fig. 8.1 Diagram depicting mechanism of Layer Doubling as implemented in subroutine SUBREF; (a) initial component layers, (b) result of first doubling, (c) result of second doubling. The doubling step from (b) to (c) is iterated until convergence of the reflection matrix R^{-+}

8.3 Surface Treatment: Subroutines ADREF1, DBLG, DBG

In the surface region, the considerations of Sect. 8.2 about layer registry shifts and interlayer vectors apply unchanged.

The simplest surface, the ideal bulk cut with no relaxation must be considered in our programs based on Layer Doubling to be composed of a bulk plus an overlayer (identical to a bulk layer) at the bulk spacing: otherwise no damping outside the outermost nuclei will be included, because SUBREF does not include damping outside of the slab it produces.

Once the substrate reflection matrix has been obtained for the periodic bulk by SUBREF, the surface layers that do not fit into the bulk periodicity perpendicular to the surface have to be included. If a superlattice is present, the block-diagonalized bulk reflection matrix must be built up from its component blocks, vanishing matrix elements filling the remainder of this big matrix: this is simply done by copying the smaller matrices output from SUBREF onto the big matrix (subroutine MATCOP will do this, taking into account the problems of variable dimensions in the main program).

The simplest case is that of a single surface layer: it can be the top layer of the bulk whose position relative to its neighbor differs from the bulk value (contracted or expanded spacing, or shifted registry); or it can be an overlayer (possibly composite). In this case, rather than compute the full matrices R^{-+}, T^{++}, R^{+-}, T^{--} of (74), only the first column of R^{-+} need be computed, since it contains the reflection coefficients of the surface for incidence of the (00) beam, i.e., just the quantities needed to produce the reflected beam intensities. This operation is performed by subroutine ADREF1 (ADREF1 also includes propagation between the outermost nuclear plane and the vacuum edge; this propagation is only necessary to include the damping in that region; the corresponding phase factor disappears when the intensity is computed). The reflected *intensities* (i.e., the final surface reflectivities) are obtained from the reflected *amplitudes* by subroutine RINT, which is to be called from the main program. Subroutine ADREF1 can be called repeatedly to produce reflectivities at different top-layer spacings, for example. (Beware of registry changes, because these may break any assumed symmetry of the system.)

The case of multiple surface layers can be reduced in one scheme to that of a single surface layer by pairing up surface layers by Layer Doubling until only a substrate and a single surface layer (slab) are obtained; then ADREF1 is used as the final step. For example, two surface layers may be combined into one slab to be put on the substrate. Alternatively, in the same situation the deeper of the two surface layers may be combined with the substrate to form a new substrate onto which the topmost surface layer is then added. Which of the two possibilities is chosen depends mainly on which interlayer spacing will be changed most often in a structural search, since this affects the overall computing time. For this purpose two subroutines are available: subroutine DBLG, which combines two layers, producing all reflection and transmission matrices for the combined pair [cf. (74)], and subroutine DBG, which has the same function as DBLG, but produces only the reflection matrix R^{-+} of (74)(DBG is used to put a layer onto a substrate). Note that DBLG, but not DBG, assumes layers with a symmetry plane parallel to the surface: this restriction can be easily lifted by simple changes in the subroutine.

Unfortunately, Layer Doubling, especially with multiple surface layers, tends to require many storage matrices for intermediate results.

Layer Doubling allows the inclusion of a surface potential step in a straightforward manner, by simply considering the step to be a scattering layer with reflection and transmission matrices. No such step is included in our programs, however.

9. Stacking Layers by Renormalized Forward Scattering (RFS) Perturbation

9.1 The Formalism

We assume that the diffraction matrices of the layers (or even slabs) composing the surface have been obtained by any of the methods of Chap. 5-8. The Renormalized Forward Scattering (RFS) perturbation method [15] follows the principle that transmission through any layer should not be described by unperturbed plane waves, but rather by plane waves modified for forward scattering through that layer together with all other plane waves transmitted with various scattering angles. It is the *reflection* by any layer that is considered to be weak and the perturbation is therefore based on an expansion of the total reflectivity of the surface in terms of the number of reflections: the lowest order contains all paths that have been reflected only once, but transmitted any number of times; the next order contains only triple-reflection paths (odd numbers of reflections are needed to bring electrons back out of the surface), and so on.

A convenient way to produce the scattering paths for the lowest order of perturbation is to follow the plane waves generated by the incident beam as they forward-scatter from layer to layer into the surface until damping makes them die out (see Fig. 9.1a); then, starting from the deepest layer reached in penetration, the emerging plane waves, obtained by reflecting the penetrating ones, are forward-scattered outward from layer to layer, picking up reflections from the penetrating waves at each layer; the emerging plane waves constitute the first-order result. The next order of perturbation is obtained by reflecting the emerging plane waves back into the crystal, starting from the top layer, and forward-scattering them inward as in the first order, but now additionally picking up reflections from emerging waves at each layer. Damping again limits the penetration and, as in the first order, the newly reflected emerging plane waves are propagated to the surface, picking up reflections on the way: the second-

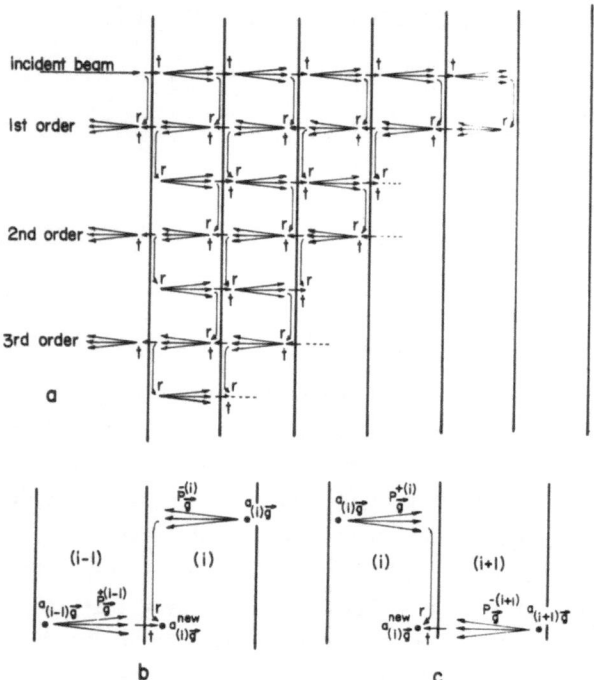

Fig. 9.1 Mechanism of the Renormalized Forward Scattering perturbation method. Vertical lines represent layers. Each triplet of arrows represents the complete set of plane waves that travel from layer to layer; at each layer that set is either transmitted (t) or reflected (r). Panel (a) shows the overall scheme; panels (b) and (c) show the details of each propagation step for penetration into and emergence out of the surface, respectively. Symbols are explained in the text.

order result emerges, to be added to the first order. This procedure is repeated for higher orders until convergence of the reflected amplitudes.

Each transmission through one layer, with the accompanying picking up of reflections, is performed in the following way. The plane wave amplitudes $a_{(i)\underline{g}}$ in the i-th interlayer spacing are computed iteratively using two expressions, one for penetration (cf. Fig.9.1b):

$$a_{(i)\underline{g}}^{new} = \sum_{\underline{g}'} \left[t_{\underline{g}\underline{g}'}^{++} \, P_{\underline{g}'}^{+(i-1)} \, a_{(i-1)\underline{g}'} + r_{\underline{g}\underline{g}'}^{+-} \, P_{\underline{g}'}^{-(i)} \, a_{(i)\underline{g}'} \right] \quad , \qquad (76)$$

and one for emergence (cf. Fig.9.1c):

$$a^{new}_{(i)\underline{g}} = \sum_{\underline{g}'} \left[t^{--}_{\underline{g}\underline{g}'} \, P^{-(i+1)}_{\underline{g}'} \, a_{(i+1)\underline{g}'} + r^{-+}_{\underline{g}\underline{g}'} \, P^{+(i)}_{\underline{g}'} \, a_{(i)\underline{g}'} \right] \quad , \tag{77}$$

where $P^{\pm(i)}_{\underline{g}}$ are plane wave propagators between appropriate reference points on successive layers: cf. (75); $a^{new}_{(i)\underline{g}}$ should be thought of as constantly overwriting $a_{(i)\underline{g}}$. Note that we have arbitrarily chosen to use the amplitudes just past each scattering layer (cf. Figs. 9.1b,c), and not, e.g., half-way between layers. Counting the surface plane (where the inner potential and damping set in) as i = 1, the initial values for the iteration of (76,77) are

$$a_{(i)\underline{g}} = 0 \text{ for all } i,\underline{g}, \text{ except } a_{(1)\underline{0}} = 1. \tag{78}$$

The RFS scheme typically uses 12-15 layers and 3-4 orders of iteration for convergence. The RFS computation requires relatively small amounts of computation time compared with the generation of the layer diffraction matrices.

By its very assumptions RFS fails to converge well when the multiple scattering between any pair of successive layers is very strong (due to strong layer reflections). This happens especially at very low energies (E ~ 10 eV) when the electron damping is small (long mean free path). RFS is also unstable when an interlayer spacing becomes small ($\lesssim 1.0 \text{Å}$), requiring many evanescent plane waves for the description of the wavefield (cf. Sect. 3.2). Under all these conditions Layer Doubling is more reliable than RFS, but for very small spacings, one has to use the matrix inversion method and/ or RSP (cf. Chaps. 6,7).

9.2 Subroutines RFS03, RFS02 and Others

The RFS scheme of calculation is implemented in subroutines RFS03, RFS02, and others. These can handle, for example, a superlattice overlayer, if present, on a bulk-like substrate, with the requirement that it must be possible to describe the substrate layers with diffraction matrices that repeat after 2 layers (RFS02) or after 3 layers (RFS03). This allows the treatment of compounds with alternating layers of simple substrates with ...ABAB... or ...ABCABC... stacking, not only when symmetry is used (requiring registry- and therefore layer-dependent diffraction matrices), but also when no symmetry is used and all layers have the same diffraction matrices. The substrate layers are each assumed to have a plane of symmetry parallel to the surface, while the overlayer(s) may be asymmetrical.

The overlayer may be equal to a substrate layer, corresponding to the case of a clean surface in which the top layer's position can be varied, if desired. The various restrictions in the applicability of these subroutines can be lifted by appropriate simple changes. (Special care must be taken to correctly program the layer spacing counting index I: at any stage of the computation, I is the index of the layer spacing to which the next one-layer propagation by subroutine PRPGAT will lead.) More flexibility is provided, for example, in subroutines RFSO2V and RFS2OV, whose purposes are explained in the subroutine listings of Chap. 11.

Equations (76,77) (depending on the direction of propagation) are evaluated in subroutine PRPGAT, which propagates the plane waves through one layer. This subroutine accepts asymmetrical layers in general: as programmed this feature is mainly used for the overlayer. The convergence for penetration depth is checked inside PRPGAT, which then (by setting IR = 1) signals the RFS subroutine (RFSO3, etc.) to turn around from penetration to emergence. An upper limit for the number of layers penetrated must be input, if only for dimensioning purposes. The convergence test over orders of perturbation is done by the RFS subroutine. These tests never result in the termination of the program execution.

In the presence of a superlattice, the block-diagonalized substrate diffraction matrices are assumed to have been prepared in the form illustrated in Fig. 5.1c (the overlayer-diffraction matrices are not block-diagonalized). To exploit the block-diagonalization, subroutine PRPGAT considers, in the substrate, one beam set at a time, propagating each set independently into the substrate and back out to the overlayer.

As with Layer Doubling (cf. Chap. 8), a surface potential step can be included in the RFS scheme as an additional scattering layer, whose diffraction matrix elements have to be supplied. At very low energies ($\lesssim 10$ eV) RFS may become unstable with a surface potential step, because such a step can produce strong reflections at those energies.

10. Assembling a Program: the Main Program and the Input

10.1 Preparing a Calculation

This section intends to give the user of the programs some guidelines as to which of the various possible calculational methods to use and as to what kind of values to give input parameters in any given situation. We recommend that prospective users of the programs, before embarking on completely new cases, first familiarize themselves with one or more of the supplied programs, learning their organization gradually.

Let us assume that the user is faced with the problem of determining a surface structure, given a set of experimental IV-curves for the surface under consideration. In the choice of experimental data, it should be clear that data taken in conditions of high symmetry (such as at normal incidence or at high-symmetry azimuths) are preferable from the point of view of the computation, since symmetries can substantially reduce the computational effort. One should further remember that a calculation produces the intensities of all emergent beams simultaneously rather than in separate calculations, so that experimental data for different beams should preferably be taken under the same conditions of incidence angle and temperature. The amount of data required for a convincing structural determination can be very roughly described as at least four independent (i.e., not symmetry-related) IV-curves over an energy range from about 25 eV to something like 200 eV, i.e., there should be at least 12-16 major peaks to compare between theory and experiment.

The first task in preparing a calculation is to set up, for the surface under consideration, a list of structural models that one wishes to test, in the hope of eliminating all but one by comparing the computed IV curves with the experimental data. It is recommended that computed IV-curves are produced for as many of the desired model structures as possible in a single calculation, so that partial results valid for various models can be reused (e.g., layer spacings and registries can often be varied quite economically within one calculation, since the computation of the layer diffraction matrices is usually much more time-consuming than that for the stacking of the layers). Note that different surface models may have different symmetries, which requires separate main programs, unless the lowest common symmetry is used.

At this stage one should have clearly in mind the possibility of surface domains. Surface domains may produce a higher symmetry in the LEED intensities than exists within each domain taken by itself, so that lower-symmetry structural models cannot be ruled out a priori. Also, the possibility of domains may impose additional calculations: e.g., four $90°$-rotated domains in general require four calculations done at four different azimuths; however, at normal incidence one single calculation gives results valid for all domains by suitable permutations of the beams. Concerning domains, it is customary to average the calculated intensities rather than to superpose the complex amplitudes reflected by different domains, as one assumes that the domain sizes are larger than the coherence area of the incident electron beam. No such averaging is programmed in our LEED program, since this can easily be done while plotting the results (see the plotting program in App. C).

Given a set of model structures to be tested, the choice between computational methods has to be made (cf. also the discussion of Sect.3.2). The available basic building blocks for a program are:

1. calculation of layer diffraction matrices for:
   a) a single Bravais-lattice layer (Chap.5);
   b) a composite layer by the matrix-inversion method (Chap.6);
   c) a composite layer by the Reverse Scattering Perturbation (RSP) method (Chap.7);
   d) a composite layer by a mixture of the methods of 1b,c (Chap.7);
2. calculation of the reflection by a stack of layers by:
   a) considering the stack as a thick composite layer: cf. 1b, c or d;
   b) the Layer Doubling method (Chap.8);
   c) the Renormalized Forward Scattering (RFS) method (Chap.9).

One will normally combine one of the building blocks 1a-d with one of the building blocks 2b or c. Choice 2a, which is self-sufficient, is to be avoided, if possible. The criteria for choosing (or avoiding) these various computational methods are the following (L: number of phase shifts; N: number of beams after symmetrization; M: number of subplanes or layers; the scaling laws for computation times are given approximately):
1a) fast (time $\propto L^3N^2$), small core size; but spacing to nearest layer must

be $\gtrsim 0.5\text{Å}$ (with choice 2b) or $\gtrsim 0.9\text{Å}$ (with choice 2c);

1b) exact (no perturbation expansion), any subplane spacings; but slow (time $\propto L^3M^3N^2$), large core size if many ($\gtrsim 4$) subplanes present;

1c) somewhat faster (time $\propto L^2M^2N^2$) than 1b, smaller core size than 1b; but can have convergence problems;

1d) can eliminate convergence problems of 1c; otherwise intermediate between 1d and 1c;

2a) as items 1b-d; but normally too many subplanes required for available computers;

2b) good convergence, fast [time $\propto N^3 \ln(M)$]; but interlayer spacings must be $\gtrsim 0.5\text{Å}$, many working matrices required;

2c) very fast (time $\propto N^2M$), small working space, very flexible with respect to surface geometry; but can have convergence problems; interlayer spacings must be $\gtrsim 0.9\text{Å}$.

The convergence problems referred to here are those that result from treating multiple scattering as a perturbation series.

The computationally most favorable situation occurs when methods 1a and 2c (RFS) can be combined: this applies to low-index faces of simple metals, e.g., fcc (111), (100), (110); hcp (0001); bcc (110), (100), and to many simple overlayers (from submonolayers to multilayers, but with at most one atom per unit cell per layer) deposited on such simple metal surfaces, all interlayer spacings being at least 0.9Å.

If RFS does not converge well for the above types of structures, method 1a can be combined with 2b (Layer Doubling). With this combination interlayer spacings should not decrease below about 0.5Å.

If some interlayer spacings are less than about 0.5Å, composite layers should be considered (this includes the case of more than one atom per unit cell per layer). Then method 1c (RSP) should be used for all the closely-spaced layers and the results combined with preferably 2c (RFS) or else 2b (Layer Doubling). RSP is a perturbation method and is therefore liable to have convergence problems. If the convergence problem is limited to only a part of a composite layer (say, for multiple scattering between strongly scattering atoms only), the mixed method 1d can be chosen instead of 1c (note the restrictions mentioned in Chap.7). If the situation is more serious, 1b (matrix inversion) is the answer for composite layers, still combined with 2c (RFS) or 2b (Layer Doubling) to stack all layers.

Once the computational methods have been chosen, the appropriate subroutines can be brought together and the main program adjusted to the particular problem under consideration. Several sample main programs are presented in this chapter, which can be used as such or as starting points to generate main programs for other surface structures.

Because of the almost unbounded variety of conceivable crystalline surface structures, our programs cover only some of the possibilities: for special cases, the user may have to adapt or generalize one or several of the supplied subroutines, which will often be quite straightforward. For example, the RFS subroutines given here allow bulk periodicities (perpendicular to the surface) of 1, 2 or 3 layers and at most two overlayers, while the Layer Doubling subroutines accept bulk periodicities of 1 or 2 layers and any number of overlayers.

The main criterion used in selecting values for the input parameters is that structural determination relies foremost on accurate peak *positions* in IV-curves, secondly on peak *shapes* and *relative* peak *intensities*, and least of all on the *absolute* intensity scale. It follows from this situation that the structural input parameters (and the position of the muffin-tin constant) must be relatively accurate (distances given within about 1%, muffin-tin constant within a few eV), while other physical parameters (optical potential, vibration amplitudes, etc.) need only be known roughly (within about 30% accuracy). Although phase shifts from most standard band structure potentials are adequate (at least for substrates), the user is nevertheless encouraged to choose the best available scattering potentials, maybe comparing calculations done with different potentials.

As pointed out in Sect. 3.2, the input list of beams should be chosen sufficiently large to accommodate all planned variations of the energy and the incidence direction. For the parameter t used by the programs to select only the significant beams at each energy (cf. Sect.3.2; t is called TST and TSTS in the programs), we find that a value of 0.002 is adequate. An increase in computation speed is obtained by changing t to, e.g., 0.008, at the price of reduced accuracy in relative reflected intensities (often not a serious loss), but with little effect on the more important peak positions in the IV-curves.

The inner potential V_0 is normally a poorly known quantity. For a clean surface we often give it a priori a value of 10 eV below the vacuum level, and adjust it a posteriori by simply shifting the energy scale of the theoretical IV-curves to obtain the best fit with experiment. A slight, unimportant, error is made in this a posteriori shifting, because the refraction effects at the vacuum-crystal interface depend on the exact value of the muffin-tin constant, but only low-energy peak intensities are affected by these effects, not the peak positions. With an overlayer, the substrate muffin-tin constant should be changed by the work function change. Inside an overlayer the muffin-tin constant may be chosen to be different from that in the substrate. Sometimes improved agreement with experiment results at the lower energies, but this adds an unknown parameter to be optimized. The most economical and common approach is to give substrate and overlayer identical muffin-tin constants.

The imaginary part V_{0i} of the energy (i.e., the optical potential representing electron damping due to inelastic processes), is not input, but programmed in the main program as an energy-dependent quantity. It is most convenient, and quite adequate, to determine V_{0i} directly from the experiment as the half-width at half-maximum of (non-overlapping) diffraction peaks. For nickel a value of V_{0i} proportional to $E^{1/3}$ is found appropriate; in some other cases a constant value of -4 or -5 eV is better. Different values of V_{0i} can be prescribed in an overlayer and a substrate, but little is gained in doing this: rather, computation time is increased (as with different muffin-tin constants) because the lattice sums done by subroutine FMAT depend on V_0 and V_{0i} and must therefore be done separately for substrate and overlayer in that case. (In a composite layer V_0 and V_{0i} may not vary within the layer.) Again, the common choice is to use the same value of V_{0i} in an overlayer and the substrate.

Thermal vibrations are even less important in structure determinations than damping. In principle they should be taken to be layer-dependent; however their layer dependence is poorly known and difficult to determine accurately. It is recommended that thermal vibrations be made dependent only on the atomic species, and that further layer-dependence be averaged out over the first few surface layers. For example, the surface layers of a clean metal could all be given the same surface-averaged thermal vibration: a reasonable value for a surface enhancement factor for mean square vibration amplitudes lies anywhere between 1 and 2. To consider the anisotropy of thermal vibrations is of little use in structural determination, as is the inclusion of zero-temperature vibrations.

The recommended energy range is 25-200 eV. Below 25 eV, small damping impairs convergence of the perturbation methods and increased sensitivity to the surface potential step decreases the quality of the computed results, while the experimental data are often less reliable. Above 200 eV, one runs into large numbers of beams (especially with superlattices) and of phase shifts (especially with strongly scattering atoms). The largest possible energy step (i.e., the sparsest possible energy grid) is recommended for economy: an energy step about equal to V_{0i} (half-width of peaks) is

physically justified and produces reliable IV-curves when plotted with
third-order polynomial interpolation. The energy loop of the programs allows
the energy step to vary from one energy range to another. In most calculations we use an energy step of 5 eV.

10.2 The Main Program

A generalized flowchart of the main LEED program is given in Fig. 10.1. This
indicates how in general the various methods of calculation can be fitted
together and also which subroutines are needed for each part of a calculation. An equally general main program would be extremely wasteful of computing resources, and therefore probably useless in practice. For any particular calculation only some of the available methods will be selected;
also some of the steps indicated in Fig. 10.1. will often be rearranged,
as circumstances allow, in order to increase efficiency by avoiding duplications and by sharing storage locations. In particular, loops over trial
geometries can be organized efficiently to reuse partial results that do
not vary from one trial geometry to another.

In the following sections a number of representative main programs are
discussed and listed; they apply to various types of surface structures,
they exhibit various combinations of calculational methods, and they exploit various degrees of symmetry. Appropriate input lists are included to
illustrate specific examples, and some output is presented for test purposes. We have a collection of other main programs, available on request.
Only the subroutines contained in Chap. 11 are used. No changes to the listed subroutines are needed. These subroutines in fact suffice for a large
number of conceivable surface structures and circumstances; in particular,
most dimensions are kept variable and can be set in the input and in the
main program. The following list indicates the (minimum) values to be given
the variable array dimensions in the main program (not all listed arrays
appear in each main program). The layer reflection and transmission matrices
(RA1, TA1, etc.) are not included here: their dimensions depend on whether
RFS or Layer Doubling is used; also the number and names of these matrices
vary from program to program.

Fig. 10.1 Generalized flowchart of LEED main program (indicating subroutines called and subroutine nestings, in square brackets)

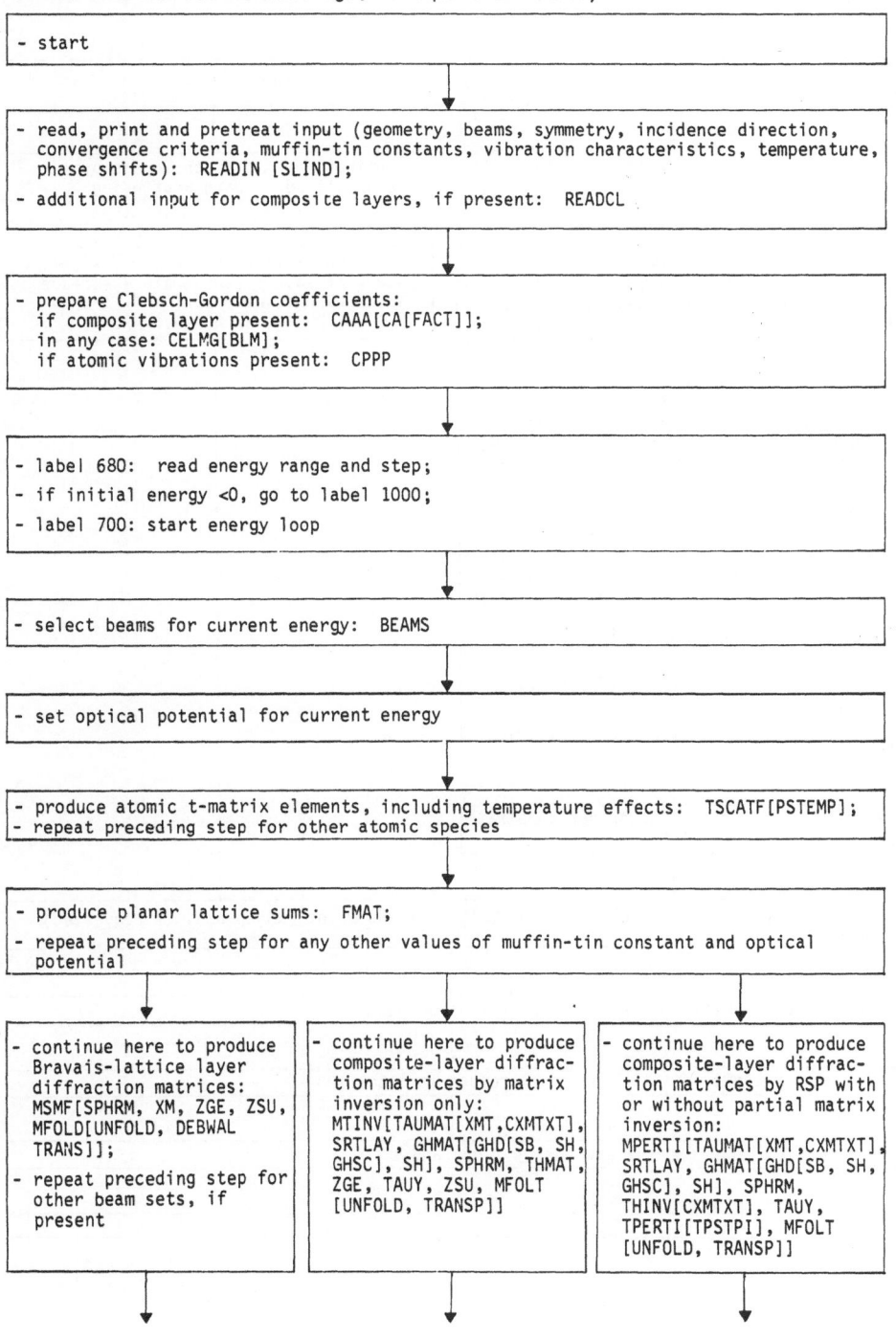

82

Fig. 10.1 (continued)

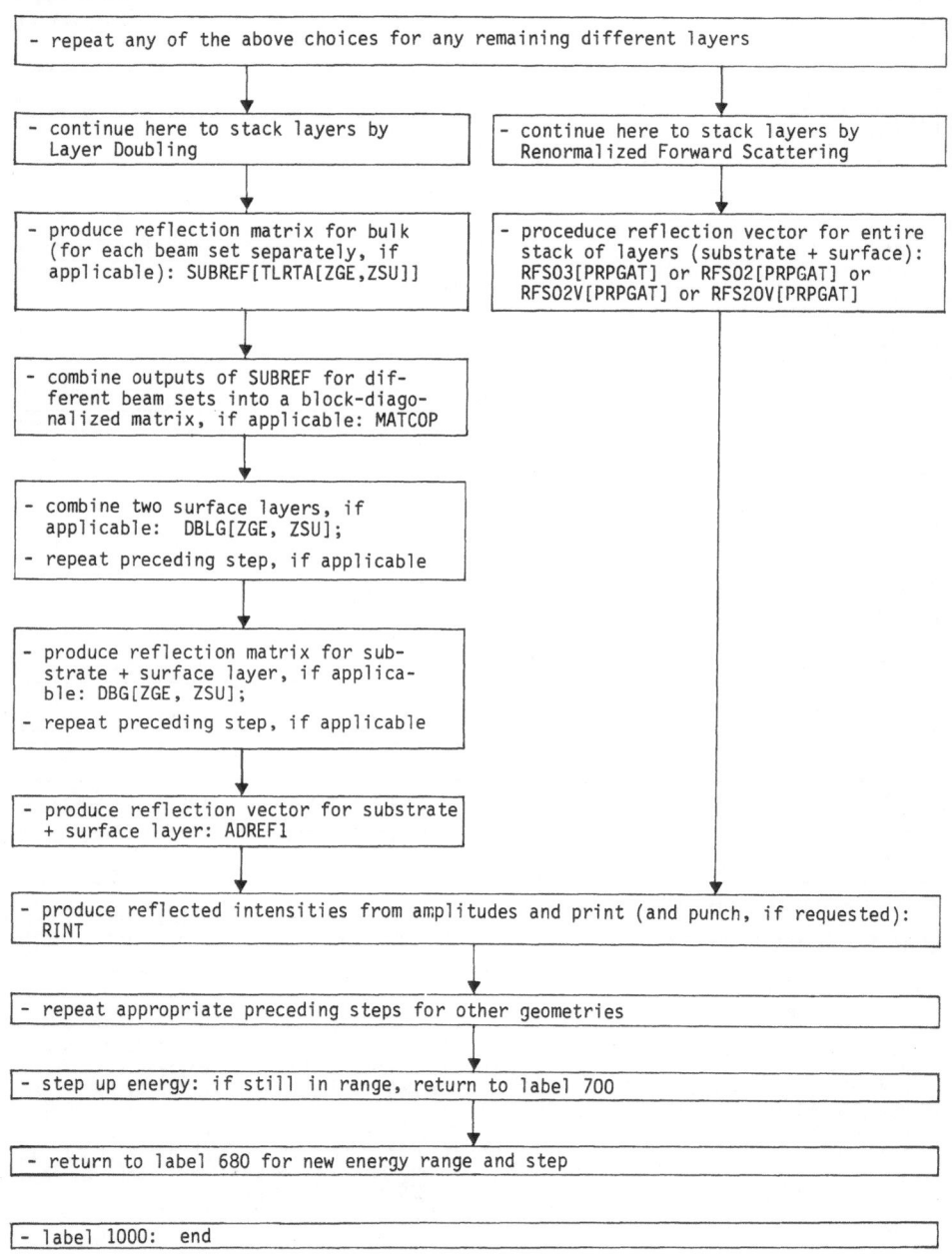

- repeat any of the above choices for any remaining different layers

- continue here to stack layers by
  Layer Doubling

- continue here to stack layers by
  Renormalized Forward Scattering

- produce reflection matrix for bulk
  (for each beam set separately, if
  applicable): SUBREF[TLRTA[ZGE,ZSU]]

- proceduce reflection vector for entire
  stack of layers (substrate + surface):
  RFSO3[PRPGAT] or RFSO2[PRPGAT] or
  RFSO2V[PRPGAT] or RFS2OV[PRPGAT]

- combine outputs of SUBREF for different beam sets into a block-diagonalized matrix, if applicable: MATCOP

- combine two surface layers, if
  applicable: DBLG[ZGE, ZSU];
- repeat preceding step, if applicable

- produce reflection matrix for substrate + surface layer, if applicable: DBG[ZGE, ZSU];
- repeat preceding step, if applicable

- produce reflection vector for substrate
  + surface layer: ADREF1

- produce reflected intensities from amplitudes and print (and punch, if requested):
  RINT

- repeat appropriate preceding steps for other geometries

- step up energy: if still in range, return to label 700

- return to label 680 for new energy range and step

- label 1000: end

ARRAYS WITH VARIABLE DIMENSIONS (EXCLUDING LAYER DIFFRACTION MATRICES)

```
AF      (L1)
AMULT   (KNT)
ANEW    (KNT,ND)
AT      (KNT)
AW      (KNT,2)
CAA     (NCAA)
CAF     (L1)
CLM     (NLM)
CYLM    (KNT,LMMAX)
DRL     (NLAY2,3)
ES      (NPSI)
FAC1    (N)
FAC2    (NN)
FLM     (KLM)
FLMS    (NL,KLM)
FPOS    (NLAY,3)
GH      (LMG,LMMAX)
IPL     (LMNI)
IPLE    (LEV)
IPLO    (LOD)
JJS     (NL,IDEG)
KNB     (KNBS)
KSYM    (2,KNT)
LX      (LMMAX)
LXI     (LMMAX)
LXM     (LMMAX)
LT      (LMMAX)
MGH     (NLAY,NLAY)
NB      (KNBS)
NGEQ    (NLAY2)
NGOL    (NLAY2)
NPU     (NPUN)
NPUC    (NPUN)
NUGH    (NLAY2)
PHSS    (NPSI,NEL*L1)
PK      (KNT,8)   WITH RFS02, RFS03
PK      (KNT,10)  WITH RFS02V, RFS20V
POS     (NLAY,3)
POSS    (NLAY,3)
PPP     (NN1,NN2,NN3)
PQ      (2,KNT)
PQF     (2,KNT)
RG      (3,NLAY,KNT)
SDRL    (NLAY2,3)
SPQ     (2,KNT)
SPQF    (2,KNT)
SYM     (2,KNT)
TAU     (LMT,LEV)
TAUG    (LTAUG)
TAUGM   (LTAUG)
TEST    (NLAY2)
TG      (2,LM2N)
TH      (LMNI,LMNI)   WITH MTINV
TH      (LMNI,LMNI2)  WITH MPERTI
TS      (LMN)
V       (NL,2)
VL      (NL,2)
VT      (LMMAX)
XEV     (LEV,LEV)   WITH MSMF
XEV     (LEV,LEV2)  WITH MTINV, MPERTI
XI      (KNT)
XOD     (LOD,LOD)
YLM     (NN)
YLME    (LEV)
YLMO    (LOD)
```

DEFINITION OF DIMENSIONS OF ABOVE ARRAYS

```
IDEG   = DEGREE OF ROTATION SYMMETRY (IDEG-FOLD AXIS)
KLM    = (2 * LMAX + 1) * (2 * LMAX + 2) / 2
KNBS   = NO. OF BEAM SETS READ IN
KNT    = TOTAL NO. OF BEAMS READ IN
LEV    = (LMAX + 1) * (LMAX + 2) / 2
LEV2   = 2 * LEV
LMAX   = MAXIMUM VALUE OF L USED
LMG    = 2 * NLAY2 * LMMAX
LMMAX  = (LMAX + 1) ** 2
LMN    = NLAY * LMMAX
LMNI   = NINV * LMMAX
LMNI2  = 2 * LMNI
LMT    = NTAU * LMMAX
LM2N   = 2 * LMN
LOD    = LMMAX - LEV
LTAUG  = (NTAU +NINV) * LMMAX
L1     = LMAX + 1 = NO. OF PHASE SHIFTS USED
N      = 2 * LMAX + 1
NCAA   = NO. OF CLEBSCH - GORDON COEFFICIENTS
ND     = NO. OF LAYERS ALLOWED FOR RFS PENETRATION
NEL    = NO. OF CHEMICAL SPECIES IN PHASE SHIFT INPUT
NINV   = NO. OF SUBPLANES IN MATRIX INVERSION OF MPERTI
NL     = NL1 * NL2 = NO. OF SUBLATTICES DUE TO SUPERLATTICE
NLAY   = NO. OF SUBPLANES IN COMPOSITE LAYERS
NLAY2  = NLAY * (NLAY - 1) / 2
NLM    = NO. OF CLEBSCH - GORDON COEFFICIENTS
NN     = N * N
NN1    = NN2 + NN3 - 1
NN2    = LMAX + 1
NN3    = LMAX + 1
NPSI   = NO. OF ENERGIES AT WHICH PHASE SHIFTS ARE READ IN
NPUN   = NO. OF BEAMS IN PUNCH OUTPUT
NTAU   = NO. OF CHEMICAL SPECIES IN COMPOSITE LAYERS
```

10.3 Explanation of Output

10.3.1 Print Output

The print output contains the following information (the exact order of the various items depends somewhat on the individual main program):
- about 2 pages listing most input quantities;
- for each energy value:
  - new page;
  - the current value of the energy in hartrees above the muffin-tin constant, and in eV above the vacuum zero level;
  - the number of beams used, by subset, and their values of $\underline{g}$;
  - the imaginary part of the potential in the substrate and in the overlayer; also the cutoff radii for the planar lattice sums (a.u.);
  - the current temperature (in K);
  - messages announcing the exit from the main subroutines (e.g., "MSMF OK");
  - for each trial geometry: at least one line giving part of the current SURFACE GEOMETRY (in Å);
  - for each call to Layer Doubling (SUBREF): a monitoring of the convergence over number of doublings (X = ...should converge) and printout of the number of doublings; if no convergence is obtained after LITER doublings, an appropriate message is printed; (the number of doublings is checked against LITER *before* the convergence is tested, so that a message of NO CONVergence may appear even when convergence has just occurred);
  - for each call to a RFS subroutine: a monitoring of the convergence of RFS in depth and in perturbation order; for each order are given: IMAX = number of layers penetrated, BNORM = measure of the wavefield magnitude at layer IMAX, ANORM = measure of the total reflectivity accumulated up to this order (ANORM should converge); appropriate warnings are printed when no convergence is obtained;
  - for each call to a matrix inversion and/or RSP subroutine (MTINV, MPERTI): printout of the number of direct-lattice points or reciprocal-lattice points (beams) used in producing the matrices G^{ji}; a monitoring of the convergence of RSP for each subplane (one subplane per print column) and each incident beam $\underline{g}$ in each direction ($\pm x$)(for each iteration only the first and last values of the monitor X are given);

- for each geometry, in addition:
  - a line giving present energy (in hartrees above muffin-tin constant)
    and the incidence direction θ, ϕ (degrees);
  - a set of lines giving beams [(PQ1,PQ2) = $\underline{g}$], REFlected INTensities
    (normalized to 1 for total reflection), and SYM (first symmetry
    code of each beam); only emerging beams are printed;
- after last energy is completed: CORRECT TERMINATION.

10.3.2 Punch Output

The punch output is composed of:
- a card containing a descriptive title for identification of the calculation;
- a card giving the number of punched beams (NPUN);
- NPUN cards, each with the beam name ($\underline{g}$) preceded by its ordinal number in
  the initial input list and followed by its symmetry code KSYM(1,J);
- for each energy:
  - for each geometry:
    - a few cards containing the energy (in eV above vacuum), a geomet-
      rical identifier (e.g., interlayer spacing in Å), followed by the
      beam intensities (normalized to 1 for total reflection), in the
      order given by the above-mentioned NPUN cards containing the beam
      names.

The punch output is designed to be read into a plotting program, such as
the one given in Appendix C.

10.4 Main Program for Clean fcc (111) and hcp (0001) Faces, Including Possible Top-Layer Registry Shifts, Using RFS

We consider the case of a material that might undergo a bulk or surface
structural phase transformation between fcc and hcp lattices. In principle,
any fcc or hcp material might undergo such a transformation, and, in
particular, at an fcc (111) or hcp (0001) surface, the transformation involves only layer registry changes. On either fcc (111) or hcp (0001) surfaces, a registry shift of just the topmost layer can conceivably occur:
this is diagrammatically indicated in comment lines near the end of the
main program listed below and the program considers the four combinations
of registries shown there. A more detailed illustration of a pure fcc (111)

surface is given in Fig. 10.2, showing, in particular, the notation used in the program.

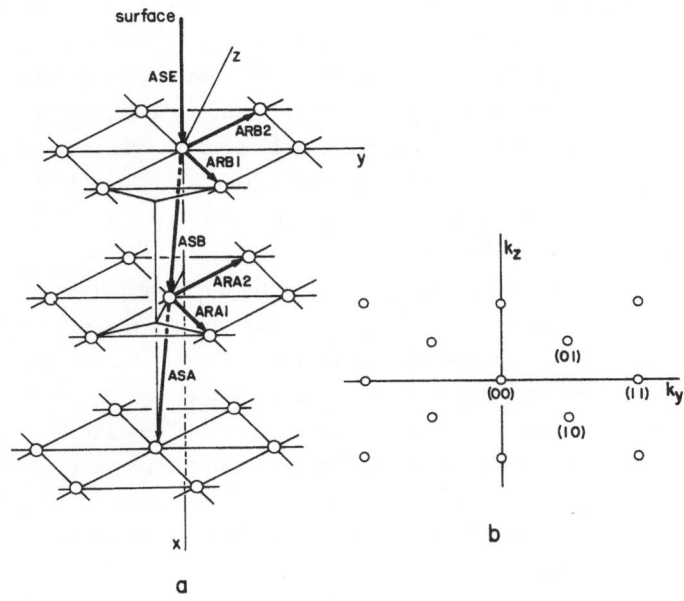

Fig. 10.2 (a) Surface geometry of fcc (111) (surface at top); the (xz)-plane is a mirror plane. (b) Reciprocal space

With each structure one must allow for steps in the surface. With an hcp bulk, mono-atomic steps produce surface structures that are identical but rotated by 180° about the terrace normal, so that the 3-fold surface symmetry of each terrace is lost in favor of 6-fold symmetry obtained by averaging the reflected intensities from 180°-rotated terraces. Thus two calculations are usually required (except at normal incidence, where one calculation and a simple permutation of the beam intensities are sufficient), and these must be subsequently averaged. With an fcc bulk such steps produce identical surface structures with no rotation. However, in this case, there is normally an experimental uncertainty of 180° in the azimuth φ of the crystal orientation, so that calculations for two azimuths differing by 180° are required (except at normal incidence, as above), one of which is to be chosen after comparison with experiment. The program is designed to handle these questions (cf. the comment lines there).

The (xz) plane has been chosen to be a mirror plane: no other symmetry is assumed. The various surface structures have higher symmetry than that of a mirror plane (namely, a 3-fold rotation axis and several mirror planes), but this choice allows additionally off-normal incidence at azimuth $\phi = \pm 90°$, ϕ being measured from the y-axis. Of course, normal incidence ($\theta = 0°$) is accepted by the program as it stands: normal incidence is chosen in the sample run for illustration purposes. In the set of beams input to the program (see the input listing following the main program), all beams to one side of the mirror plane are omitted explicitly, but included implicitly through the code numbers 6 (cf. Fig.4.2) given in the input next to their symmetrical counterparts on the other side of the mirror plane.

Each atomic layer has atoms on the mirror plane, so that a propagation track (cf. Chap.4) can be found that is entirely contained in the mirror plane: cf. Fig. 10.2 (this track is schematically indicated near the end of the main program). Therefore no registry shifts $\underline{s}_j$ (cf. Chap.4, App.A) need be included in the layer diffraction matrices in this case: the registries are entirely taken care of by the interlayer propagators. Furthermore, since all atomic layers have identical internal structures, only one layer reflection matrix (RA1) and one layer transmission matrix (TA1) are required for all layers of the entire crystal.

The atomic layers are well separated, even if the topmost layer is allowed to contract somewhat, and so Renormalized Forward Scattering can be safely used. Subroutine RFS02 is chosen for the hcp bulk structures, since the interlayer vectors alternate (AS1, AS2, AS1, AS2, ...) in a 2-layer periodicity (AS1 and AS2 are generated from the input vector ASA); the topmost layer is treated as an overlayer with variable spacing and registry. Subroutine RFS02 could also handle the fcc bulk structures, with a constant interlayer vector (either AS1, AS1, AS1, ... or AS2, AS2, AS2, ...), i.e., 1-layer periodicity. However, if 3-fold symmetry were exploited, RFS03 would be needed to deal with the 3-layer periodicity induced by the required registry shifts with respect to a symmetry *axis* (...ABCABC...). With that possibility in mind, RFS03 rather than RFS02 is used here with simply 1-layer periodicity. Note that bulk interatomic distances in general change slightly in a phase transformation (and that the c/a ratio in hcp lattices is material-dependent): this may require more than one calculation with different interatomic and interplanar distances.

In this example, the input applies to nickel (with phase shifts from the Wakoh potential) and the imaginary part of the muffin-tin constant is set proportional to $E^{1/3}$.

```
C     MAIN + DATA FOR NI(111), 'NI(0001)', ETC., RFS, (XZ) MIRROR PLANE.
C     MAIN LEED PROGRAM FOR FCC(111) AND HCP(0001), INCLUDING ALSO A TOP-
C     LAYER REGISTRY CHANGE BETWEEN FCC AND HCP. (XZ) PLANE IS ASSUMED
C     TO BE SYMMETRY PLANE (ALLOWING THETA.NE.O IF FI=+-90 DEGREES).
C     RENORMALIZED FORWARD SCATTERING.
      DIMENSION ARA1(2),ARA2(2),RAR1(2),RAR2(2),SS1(2),SS2(2),SS3(2),
     1SS4(2),ASA(3),AS(3),ARB1(2),ARB2(2),RBR1(2),RBR2(2),SO1(2),
     2SO2(2),SO3(2),ASB(3),ASV(3),AS1(3),AS2(3)
C     1=NL, 120=KLM
      COMPLEX VL(1,2),FLMS(1,120),FLM(120)
C     1=NL, 3=IDEG
      DIMENSION V(1,2),JJS(1,3)
C     1=KNBS=NO. OF BEAM SETS READ IN, 18=KNT=NO. OF BEAMS READ IN
      DIMENSION KNB(1),NB(1),SPQF(2,18),SPQ(2,18),PQF(2,18),PQ(2,18)
      INTEGER KSYM(2,18),SYM(2,18)
C     20 MUST BE .GE. NPUN
      DIMENSION NPU(20),NPUC(20)
C     13=NPSI, 16=NEL*(LMAX+1)
      DIMENSION ES(13),PHSS(13,16)
C     7680=NLM FROM DATA STATEMENT, 225=NN, 15=N (FOR CELMG)
C     YLM IS ALSO USED, AS COMPLEX VECTOR OF SMALLER CORE SIZE, IN MSMF,ETC
      DIMENSION NLMS(7),CLM(7680),YLM(225),FAC2(225),FAC1(15)
C     15=NN1, 8=NN2, 8=NN3
      DIMENSION PPP(15,8,8)
C     64=LMMAX
      DIMENSION LX(64),LXI(64),LXM(64),LT(64)
C     8=LMAX+1=L1, 3=NTAU (FOR COMPOSITE LAYERS), 10 MUST BE .GE. L1
      COMPLEX AF(8),CAF(8),TSFO(3,10),TSF(3,10)
      COMPLEX RA1(18,18),TA1(18,18)
C     36=LEV, 28=LOD
      COMPLEX AMULT(18),CYLM(18,64),XEV(36,36),XOD(28,28),
     1YLME(36),YLMO(28)
      DIMENSION IPLE(36),IPLO(28),AT(18)
C     8 FIXED, 20=ND (FOR RFS SUBROUTINES)
      COMPLEX XI(18),PK(18,8),AW(18,2),ANEW(18,20)
      COMMON E3,AK21,AK31,VPI1,TV,EMACH
      COMMON /X4/E,VPI,AK2,AK3,ASA
      COMMON /SL/ARA1, ARA2, ARB1, ARB2, RBR1, RBR2, NL1, NL2
      COMMON /MS/LMAX,EPS,LITER,SO1,SO2,SO3,SS1,SS2,SS3,SS4
      COMMON /ADS/ASL, FR, ASE, VPIS, VPIO, VO, VV
      COMMON /BT/IT, T, TO,DRPER,DRPAR,DRO
      COMMON /TEMP/IT1,IT2,IT3,TI,TF,DT,TO1,DRPER1,DRPAR1,DRO1,DRPER2,
     1DRPAR2,DRO2,DRPER3,DRPAR3,DRO3
C     NLMS IS DIMENSION OF CLM AS FUNCTION OF LMAX
      DATA NLMS(1),NLMS(2),NLMS(3),NLMS(4),NLMS(5),NLMS(6),NLMS(7)/
     10,76,284,809,1925,4032,7680/
  161 FORMAT(3F7.2)
  200 FORMAT(20I3)
  240 FORMAT(8HOIDEG = ,1I3,7H NL1 = ,1I3,7H NL2 = ,1I3)
  370 FORMAT(9H1ENERGY =,F7.4,7H H  OR ,1F7.2,3H EV)
  380 FORMAT(8H VPIS = ,F9.4,8H VPIO = ,F9.4,9H DCUTS = ,F9.4,9H DCUTO =
     1 ,F9.4)
  390 FORMAT(8HOTEMP = ,F9.4)
  410 FORMAT(10H MSMF   OK)
  420 FORMAT(10H SUBREF OK)
  430 FORMAT(20HOSURFACE GEOMETRY : ,3F7.4)
  450 FORMAT(20H CORRECT TERMINATION)
  460 FORMAT(20A4)
  465 FORMAT(1H1,20A4)
C
C
C     READ, WRITE AND PUNCH A DESCRIPTIVE TITLE
      READ (5,460)(CLM(I),I=1,20)
      WRITE(6,465)(CLM(I),I=1,20)
      WRITE(7,460)(CLM(I),I=1,20)
C     EMACH IS MACHINE ACCURACY (USED BY ZGE AND ZSU)
      EMACH=1.0E-6
```

```
C   IDEG: EACH LAYER HAS AN IDEG-FOLD SYMMETRY AXIS
C   NL1, NL2: SUPERLATTICE CHARACTERIZATION  NL1  NL2
C                        P(1*1)                1    1
C                        C(2*2)                2    1
C                        P(2*1)                2    1
C                        P(1*2)                1    2
C                        P(2*2)                2    2
      READ(5,200)IDEG,NL1,NL2
      WRITE(6,240)IDEG,NL1,NL2
      NL=NL1*NL2
C   KNBS= NO.OF BEAM SETS TO BE READ IN (.LE.NL)
C   KNT= TOTAL NO. OF BEAMS TO BE READ IN
C   NPUN= NO. OF BEAMS FOR WHICH INTENSITIES ARE TO BE PUNCHED OUT
      READ(5,200)KNBS,KNT,NPUN
C   NPSI= NO.OF ENERGIES AT WHICH PHASE SHIFTS WILL BE READ IN
      READ(5,200)NPSI
C   READ IN GEOMETRY, PHYSICAL PARAMETERS, CONVERGENCE CRITERIA
      CALL READIN(TVA,RAR1,RAR2,ASA,TVB,ASB,STEP,NSTEP,IDEG,NL,V,VL,
     1JJS,KNBS,KNB,KNT,SPQF,KSYM,SPQ,TST,TSTS,NPUN,NPU,THETA,FI,
     2LMMAX,NPSI,ES,PHSS,L1)
      TO=TO1
      KLM=(2*LMAX+1)*(2*LMAX+2)/2
      LEV=(LMAX+1)*(LMAX+2)/2
      LOD=LMMAX-LEV
  630 N=2*LMAX+1
      NN=N*N
      NLM=NLMS(LMAX)
C   CLM= CLEBSCH-GORDON COEFFICIENTS FOR MATRICES X AND TAU
      CALL CELMG(CLM,NLM,YLM,FAC2,NN,FAC1,N,LMAX)
C   LX,LXI,LT,LXM: PERMUTATIONS OF (L,M)-SEQUENCE
      CALL LXGENT(LX,LXI,LT,LXM,LMAX,LMMAX)
      IF (IT1+IT2+IT3) 640,650,640
  640 NN3=LMAX+1
      NN2=LMAX+1
      NN1=NN2+NN3-1
C   PPP= CLEBSCH-GORDON COEFFICIENTS FOR COMPUTATION OF TEMPERATURE-
C   DEPENDENT PHASE SHIFTS (SKIPPED IF NOT NEEDED)
      CALL  CPPP (PPP, NN1, NN2, NN3)
  650 CONTINUE
C
C
C   READ ENERGY RANGE AND STEP; IF EF AND DE ARE BLANK, PROGRAM COMPUTES
C   FOR ENERGY EI. PROGRAM ALWAYS RETURNS TO THIS LINE TO READ A NEW
C   ENERGY RANGE AND STEP. IF A NEGATIVE EI IS READ, RUN IS TERMINATED
  680 READ (5,161) EI, EF, DE
      IF (EI) 1000, 690, 690
  690 EI=EI/27.18+VV
      EF=EF/27.18+VV
      DE=DE/27.18
      E = EI
      EC = E
      E3 = E
C
C
C   START LOOP OVER ENERGIES IN GIVEN ENERGY RANGE
  700 EEV=(E-VV)*27.18
      WRITE (6,370) E,EEV
C   COMPUTE COMPONENTS OF INCIDENT WAVEVECTOR PARALLEL TO SURFACE
      AK= SQRT(2.0*(E-VV))*SIN(THETA)
      AK2=AK*COS(FI)
      AK3=AK*SIN(FI)
      AK21=AK2
      AK31=AK3
C   SELECT BEAMS APPROPRIATE FOR CURRENT ENERGY
      CALL BEAMS(KNBS,KNB,SPQ,SPQF,KSYM,KNT,NPU,NPUN,AK2,AK3,E,TST,
     1NB,PQ,PQF,SYM,NPUC,MPU,NT,NP)
C   SET OPTICAL POTENTIAL (IMAGINARY PART OF MUFFIN-TIN CONSTANT,
```

```
C   REPRESENTING DAMPING). VPIS FOR SUBSTRATE, VPIO FOR OVERLAYER.
C   HERE VPIS=VPIO PROPORTIONAL TO E**1/3
        VPIS=-(3.8/27.18)*EXP((ALOG((E)/(90.0/27.18+VV)))*0.333333)
        VPIO=VPIS
        DCUTO = SQRT(2.0 * E)
C   SET LIMITING RADII ON LATTICE SUMS, POSSIBLY DIFFERENT FOR SUBSTRATE
C   (DCUTS) AND OVERLAYER (DCUTO)
        DCUTS =  - 5.0 * DCUTO/(AMIN1(VPIS, - 0.05))
        DCUTO =  - 5.0 * DCUTO/(AMIN1(VPIO, - 0.05))
        WRITE (6,380)  VPIS, VPIO, DCUTS, DCUTO
C   FIX TEMPERATURE T. T MAY BE INCREASED STEPWISE IN A LOOP OVER THE
C   INPUT RANGE (TI,TF) WITH INPUT STEP DT
   770 T = TI
   780 WRITE (6,390)   T
C
C
        VPI1 = VPIO
C   PERFORM PLANAR LATTICE SUM IN FMAT. IF MUFFIN-TIN CONSTANTS AND
C   DAMPINGS IN SUBSTRATE AND OVERLAYER ARE EQUAL, DO LATTICE SUMS FOR
C   BOTH TYPES OF LAYERS NOW (NLS=NL IMPLIES: DO ALL SUBLATTICES).
C   OTHERWISE DO ONLY OVERLAYER LATTICE SUM (NLS=1 IMPLIES: DO ONLY THE
C   SUBLATTICE EQUAL TO THE OVERLAYER LATTICE)
        NLS = 1
        IF (ABS(VPIS-VPIO)+ABS(VO) .LE. 1.0E-4)  NLS = NL
        CALL  FMAT (FLMS, V, JJS, NL, NLS, DCUTO, IDEG, LMAX,KLM)
C
C
C   PRODUCE ATOMIC T-MATRIX ELEMENTS
        CALL TSCATF(1,L1,ES,PHSS,NPSI,IT1,E,0.,PPP,NN1,NN2,NN3,DRO1,
       1DRPER1,DRPAR1,TO,T,TSFO,TSF,AF,CAF)
        DRPER=DRPER1
        DRPAR=DRPAR1
        DRO=DRO1
        IT=0
        TV = TVA
        VPI = VPIS
        VPI1 = VPIS
        IF (ABS(VPIS-VPIO)+ABS(VO) .LE. 1.0E-4)   GO TO 930
C   PERFORM PLANAR LATTICE SUMS FOR SUBSTRATE IF NOT DONE BEFORE
        CALL  FMAT (FLMS, V, JJS, NL, NL, DCUTS, IDEG, LMAX,KLM)
C   PRODUCE REFLECTION (RA1) AND TRANSMISSION (TA1) MATRICES FOR
C   ATOMIC LAYERS. ONE REGISTRY: ID=1. NLL=NL=1 SUBLATTICES. SUBSTRATE
C   LAYER SYMMETRIES AND REGISTRIES APPLY: LAY=2. TEMPERATURE-DEPENDENT
C   PHASE SHIFTS: CAF
   930 NAA = NB(1)
        CALL MSMF(RA1,TA1,RA1,TA1,RA1,TA1,RA1,TA1,NAA,NT,NP,AMULT,CYLM,
       1PQ,SYM,NT,FLMS,FLM,V,NL,   0,  0,1,NL,CAF,CAF,L1,LX,LXI,LMMAX,KLM,
       2XEV,XOD,LEV,LOD,YLM,YLME,YLMO,IPLE,IPLO,CLM,NLM,2)
        WRITE (6,410)
C   AS1 AND AS2 BECOME DIFFERENT BULK INTERLAYER VECTORS CORRESPONDING
C   TO DIFFERENT RELATIVE REGISTRIES
        DO 932 J=1,3
        AS1(J)=-ASA(J)
   932 AS2(J)=ASA(J)
        AS1(1)=-AS1(1)
C   LOOP 975 RUNS OVER 4 DIFFERENT REGISTRY CONFIGURATIONS OF THE
C   LAYERS, AS SHOWN IN THE FOLLOWING DIAGRAM IN THE (XZ) PLANE.
C   HERE A,B,C ARE REGISTRIES IN THE LAYER STACKING, * ARE ATOMIC
C   POSITIONS AND ASV, AS1, AS2 ARE THE INTERLAYER VECTORS.
C
C
C
C
C
C
C
C
```

```
C
C   III=    1                2               3                    4
C
C          HCP          HCP+FCC         FCC               FCC+HCP
C
C   A B C A B C A B C A B C A B C A B C A B C A B C A B C A B C A B C A B C
C
C           *                *               *                    *
C          ASV              ASV             ASV                  ASV
C           *                *               *                    *
C          AS1              AS2             AS2                  AS1
C           *                *               *                    *
C          AS2              AS1             AS2                  AS1
C           *                *               *                    *
C          AS1              AS2             AS2                  AS1
C           *                *               *                    *
      DO 975 III=1,4
C ASV BECOMES THE 3-D VECTOR FROM TOP LAYER TO NEXT LAYER AND WILL BE
C VARIED
      DO 937 J=2,3
  937 ASV(J)=ASB(J)
C LOOP 970 RUNS OVER TWO 180-DEGREE-ROTATED ORIENTATIONS OF THE
C SURFACE (FOR FCC(111) THE ABSOLUTE ORIENTATION IS NORMALLY UNKNOWN
C WITH RESPECT TO A 180-DEGREE ROTATION; FOR HCP(0001) STEPS ON THE
C SURFACE PRODUCE 180-DEGREE-ROTATED DOMAINS, WHICH TOGETHER PRODUCE
C AN APPARENT 6-FOLD STRUCTURAL SYMMETRY RATHER THAN 3-FOLD SYMMETRY
C OTHERWISE).
C REMOVE LOOPS 970 AND 965 WHEN TWO 180-DEGREE-ROTATED DOMAINS ARE NOT
C NEEDED (E.G. AT NORMAL INCIDENCE)
      DO 970 IA=1,2
      ASV(1)=ASB(1)
C LOOP 960 VARIES THE TOP-LAYER SPACING ASV(1)
      DO 960 I=1,NSTEP
      DO 940 J=1,3
  940 AS(J)=ASV(J)*0.529
      A=AS(1)
      WRITE (6,430) (AS(J),J=1,3)
      GO TO (941,942,943,944),III
C THIS CALL TO RFSO2 CORRESPONDS TO HCP(0001)
  941 CALL RFSO2(RA1,TA1,RA1,TA1,RA1,TA1,RA1,TA1,NT,NB,KNBS,NP,XI,
     1PQ,PK,AW,ANEW,20,ASV,AS1,AS2)
C COMPUTE REFLECTED INTENSITIES (AT) FROM AMPLITUDES (XI), PRINT AND
C PUNCH (NPNCH=1; PUNCH LAYER SPACING A ALSO FOR LATER IDENTIFICATION)
      CALL RINT(NT,XI,AT,PQ,PQF,SYM,VV,THETA,FI,MPU,NPUC,EEV,A,1)
      GO TO 950
C THIS CALL TO RFSO2 CORRESPONDS TO HCP(0001) + FCC SURFACE
  942 CALL RFSO2(RA1,TA1,RA1,TA1,RA1,TA1,RA1,TA1,NT,NB,KNBS,NP,XI,
     1PQ,PK,AW,ANEW,20,ASV,AS2,AS1)
      CALL RINT(NT,XI,AT,PQ,PQF,SYM,VV,THETA,FI,MPU,NPUC,EEV,A,1)
      GO TO 950
C THIS CALL TO RFSO3 CORRESPONDS TO FCC(111)
  943 CALL RFSO3(RA1,TA1,RA1,TA1,RA1,TA1,RA1,TA1,RA1,TA1,NT,NB,KNBS,
     1NP,XI,PQ,PK,AW,ANEW,20,ASV,AS2)
      CALL RINT(NT,XI,AT,PQ,PQF,SYM,VV,THETA,FI,MPU,NPUC,EEV,A,1)
      GO TO 950
C THIS CALL TO RFSO3 CORRESPONDS TO FCC(111) + HCP SURFACE
  944 CALL RFSO3(RA1,TA1,RA1,TA1,RA1,TA1,RA1,TA1,RA1,TA1,NT,NB,KNBS,
     1NP,XI,PQ,PK,AW,ANEW,20,ASV,AS1)
      CALL RINT(NT,XI,AT,PQ,PQF,SYM,VV,THETA,FI,MPU,NPUC,EEV,A,1)
  950 CONTINUE
  960 ASV(1)=ASV(1)+STEP
C 180-DEGREE ROTATION OF SURFACE OBTAINED HERE
      DO 965 II=2,3
      AS1(II)=-AS1(II)
      AS2(II)=-AS2(II)
  965 ASV(II)=-ASV(II)
  970 CONTINUE
```

```
  975 CONTINUE
      IF (IT1+IT2+IT3) 980,990,980
  980 T = T + DT
      IF (DT.LE.0.0001) GO TO 990
      IF (T-TF)  780, 780, 990
  990 E = E + DE
      EC = E
      E3 = E
      IF (E-EF)  700, 700, 680
 1000 WRITE (6,450)
      STOP
      END
```

---- INPUT DATA FOR PRECEDING PROGRAM FOLLOW ------------

```
NI(111),'(0001)' AND MIXTURES, 1 MIRROR PLANE, RFS
  3   1   1            IDEG NL1 NL2
  1 18 12              KNBS KNT NPUN
 12                    NPSI
 1.2450-2.1564         ARA1
 1.2450 2.1564         ARA2
 0.0000 0.0000         SS1
 0.0000 0.0000         SS2
 0.0000 0.0000         SS3
 0.0000 0.0000         SS4
 2.0330 0.0000-1.4376  ASA
 1.2450-2.1564         ARB1
 1.2450 2.1564         ARB2
 0.0000 0.0000         SO1
 0.0000 0.0000         SO2
 0.0000 0.0000         SO3
 1.8330 0.0000-1.4376  ASB
 0.5000 1.2450 0.1000  FR ASE STEP
  4                    NSTEP
 18                    KNB(1)
 0.0000 0.0000   1   1  BEAMS
 1.0000 0.0000   6   6
-1.0000 1.0000   1   1
 0.0000 1.0000   6   6
 1.0000-1.0000   1   1
 1.0000 1.0000   6   6
-1.0000 2.0000   6   6
 2.0000-1.0000   6   6
 2.0000 0.0000   6   6
-2.0000 2.0000   1   1
 0.0000 2.0000   6   6
 2.0000-2.0000   1   1
 2.0000 1.0000   6   6
-1.0000 3.0000   6   6
 3.0000-2.0000   6   6
 1.0000 2.0000   6   6
-2.0000 3.0000   6   6
 3.0000-1.0000   6   6
 0.0020                        TST
  1   2   3   4   5   6   7   8  9 10 11 12    NPU
   0.00    0.00              THETA FI
   0.00   11.20              VO VV
 0.0010                      EPS
  5                          LITER
  1   0   0                  IT1 IT2 IT3
440.0000    58.7100    1.4000    1.4000    0.0000    THDB1 AM1 FPER1 FPAR1 DRO1
  1.0000     1.0000    1.0000    1.0000    0.0000    THDB2 AM2 FPER2 FPAR2 DRO2
  1.0000     1.0000    1.0000    1.0000    0.0000    THDB3 AM3 FPER3 FPAR3 DRO3
300.0000                                            TI TF DT
  7                          LMAX
  2                          NEL
 .3000
```

```
2.8398 -.0137 -.1876   .0011   .0000   .0000   .0000   .0000

 .4000
2.7639 -.0216 -.3164   .0024   .0001   .0000   .0000   .0000

 .5000
2.6930 -.0322 -.2630   .0048   .0002   .0000   .0000   .0000

 .7500
2.5276 -.0668 -.2322   .0160   .0009   .0000   .0000   .0000

1.0000
2.3803 -.1132 -.2323   .0386   .0028   .0002   .0000   .0000

1.5000
2.1240 -.2247 -.2118   .1202   .0125   .0012   .0001   .0000

2.0000
1.9101 -.3448 -.1758   .2494   .0335   .0042   .0004   .0000

3.0000
1.5727 -.5697 -.1288   .5535   .1138   .0216   .0034   .0004

4.0000
1.3160 -.7574 -.1272   .7884   .2257   .0583   .0125   .0022

5.0000
1.1099 -.9122 -.1445   .9491   .3361   .1107   .0305   .0069

6.0000
 .9366-1.0434 -.1666 1.0690   .4287   .1691   .0571   .0159

7.0000
 .7862-1.1581 -.1902 1.1630   .5045   .2251   .0894   .0297

 40.00 201.00   5.00    EI EF DE
-10.00                  EI EF DE
-------------------------------------------------------
---- FIRST PART OF PRINTER OUTPUT FOLLOWS --------------
----(NEGLECTING MOST BLANK LINES AND PAGE -------------
---- EJECTS AND TRUNCATING LONG LINES)    --------------
-------------------------------------------------------
 NI(111),'(0001)' AND MIXTURES, 1 MIRROR PLANE, RFS
IDEG =   3 NL1 =   1 NL2 =   1
PARAMETERS FOR INTERIOR:
SURF VECS         1.2450 -2.1564
                  1.2450  2.1564
SS     0.0000 0.0000
SS     0.0000 0.0000
SS     0.0000 0.0000
SS     0.0000 0.0000
ASA    2.0330 0.0000-1.4376
PARAMETERS FOR SURFACE:
SURF VECS         1.2450 -2.1564
                  1.2450  2.1564
SO     0.0000 0.0000
SO     0.0000 0.0000
SO     0.0000 0.0000
ASB    1.8330 0.0000-1.4376
FR =   .5000 ASE =  1.2450 STEP =   .1000 NSTEP =   4
         PQ1    PQ2       SYM

    1    0.000  0.000    1  1
    2    1.000  0.000    6  6
    3   -1.000  1.000    1  1
    4    0.000  1.000    6  6
    5    1.000 -1.000    1  1
```

```
  6      1.000   1.000       6   6
  7     -1.000   2.000       6   6
  8      2.000  -1.000       6   6
  9      2.000   0.000       6   6
 10     -2.000   2.000       1   1
 11      0.000   2.000       6   6
 12      2.000  -2.000       1   1
 13      2.000   1.000       6   6
 14     -1.000   3.000       6   6
 15      3.000  -2.000       6   6
 16      1.000   2.000       6   6
 17     -2.000   3.000       6   6
 18      3.000  -1.000       6   6
TST =  .0020
PUNCHED BEAMS        1   2   3   4   5   6   7   8   9  10  11  12
THETA  FI =    0.00    0.00
VO       0.00VV       11.20
EPS =   .0010 LITER =    5
IT1 =   1 IT2 =    0 IT3 =    0
THDB =  440.0000 AM =   58.7100 FPER =      1.4000 FPAR =      1.4000 DRO =
THDB =    1.0000 AM =    1.0000 FPER =      1.0000 FPAR =      1.0000 DRO =
THDB =    1.0000 AM =    1.0000 FPER =      1.0000 FPAR =      1.0000 DRO =
LMAX =    7
        PHASE SHIFTS

E =   .3000  1ST ELEMENT    2.8398  -.0137  -.1876   .0011  0.0000  0.0000
             2ND ELEMENT    0.0000  0.0000  0.0000  0.0000  0.0000  0.0000

E =   .4000  1ST ELEMENT    2.7639  -.0216  -.3164   .0024   .0001  0.0000
             2ND ELEMENT    0.0000  0.0000  0.0000  0.0000  0.0000  0.0000

E =   .5000  1ST ELEMENT    2.6930  -.0322  -.2630   .0048   .0002  0.0000
             2ND ELEMENT    0.0000  0.0000  0.0000  0.0000  0.0000  0.0000

E =   .7500  1ST ELEMENT    2.5276  -.0668  -.2322   .0160   .0009  0.0000
             2ND ELEMENT    0.0000  0.0000  0.0000  0.0000  0.0000  0.0000

E =  1.0000  1ST ELEMENT    2.3803  -.1132  -.2323   .0386   .0028   .0002
             2ND ELEMENT    0.0000  0.0000  0.0000  0.0000  0.0000  0.0000

E =  1.5000  1ST ELEMENT    2.1240  -.2247  -.2118   .1202   .0125   .0012
             2ND ELEMENT    0.0000  0.0000  0.0000  0.0000  0.0000  0.0000

E =  2.0000  1ST ELEMENT    1.9101  -.3448  -.1758   .2494   .0335   .0042
             2ND ELEMENT    0.0000  0.0000  0.0000  0.0000  0.0000  0.0000

E =  3.0000  1ST ELEMENT    1.5727  -.5697  -.1288   .5535   .1138   .0216
             2ND ELEMENT    0.0000  0.0000  0.0000  0.0000  0.0000  0.0000

E =  4.0000  1ST ELEMENT    1.3160  -.7574  -.1272   .7884   .2257   .0583
             2ND ELEMENT    0.0000  0.0000  0.0000  0.0000  0.0000  0.0000

E =  5.0000  1ST ELEMENT    1.1099  -.9122  -.1445   .9491   .3361   .1107
             2ND ELEMENT    0.0000  0.0000  0.0000  0.0000  0.0000  0.0000

E =  6.0000  1ST ELEMENT     .9366 -1.0434  -.1666  1.0690   .4287   .1691
             2ND ELEMENT    0.0000  0.0000  0.0000  0.0000  0.0000  0.0000

E =  7.0000  1ST ELEMENT     .7862 -1.1581  -.1902  1.1630   .5045   .2251
             2ND ELEMENT    0.0000  0.0000  0.0000  0.0000  0.0000  0.0000
ENERGY = 1.8837 H   OR    40.00 EV
   5 BEAMS USED :    5
   0.000 0.000    1.000 0.000  -1.000 1.000    0.000 1.000    1.000-1.000
VPIS =    -.1114 VPIO =    -.1114 DCUTS =   87.1160 DCUTO =   87.1160
TEMP = 300.0000
MSMF   OK
SURFACE GEOMETRY :  1.8330 0.0000-1.4376
```

```
IMAX =  8      BNORM =     .2786E-02
ANORM =    .1176E+00


IMAX =  2      BNORM =     .1912E-02
ANORM =    .1176E+00


E=    1.8837 THETA=    0.0000 FI=    0.0000

    PQ1     PQ2     REF INT      SYM
   0.000   0.000   .18394E-02     1
   1.000   0.000   .90520E-03     6
  -1.000   1.000   .90523E-03     1
   0.000   1.000   .15533E-01     6
   1.000  -1.000   .15533E-01     1
SURFACE GEOMETRY : 1.9330 0.0000-1.4376
IMAX =  8      BNORM =     .2877E-02
ANORM =    .1113E+00


IMAX =  2      BNORM =     .1649E-02
ANORM =    .1113E+00


E=    1.8837 THETA=    0.0000 FI=    0.0000

    PQ1     PQ2     REF INT      SYM
   0.000   0.000   .20836E-02     1
   1.000   0.000   .17917E-03     6
  -1.000   1.000   .17917E-03     1
   0.000   1.000   .15382E-01     6
   1.000  -1.000   .15382E-01     1
SURFACE GEOMETRY : 2.0330 0.0000-1.4376
IMAX =  8      BNORM =     .2970E-02
ANORM =    .1071E+00


IMAX =  2      BNORM =     .1644E-02
ANORM =    .1071E+00


E=    1.8837 THETA=    0.0000 FI=    0.0000

    PQ1     PQ2     REF INT      SYM
   0.000   0.000   .26502E-02     1
   1.000   0.000   .17395E-02     6
  -1.000   1.000   .17394E-02     1
   0.000   1.000   .13447E-01     6
   1.000  -1.000   .13447E-01     1
SURFACE GEOMETRY : 2.1330 0.0000-1.4376
IMAX =  9      BNORM =     .1368E-02
ANORM =    .1054E+00


IMAX =  2      BNORM =     .1791E-02
ANORM =    .1055E+00


E=    1.8837 THETA=    0.0000 FI=    0.0000

    PQ1     PQ2     REF INT      SYM
   0.000   0.000   .34580E-02     1
   1.000   0.000   .50126E-02     6
  -1.000   1.000   .50126E-02     1
   0.000   1.000   .10160E-01     6
   1.000  -1.000   .10160E-01     1
SURFACE GEOMETRY : 1.8330 0.0000 1.4376
IMAX =  8      BNORM =     .2785E-02
ANORM =    .1175E+00


IMAX =  2      BNORM =     .1904E-02
ANORM =    .1176E+00


E=    1.8837 THETA=    0.0000 FI=    0.0000
```

```
    PQ1      PQ2       REF INT      SYM
  0.000    0.000     .18384E-02      1
  1.000    0.000     .15537E-01      6
 -1.000    1.000     .15537E-01      1
  0.000    1.000     .90087E-03      6
  1.000   -1.000     .90090E-03      1
SURFACE GEOMETRY :   1.9330 0.0000 1.4376
IMAX =  8      BNORM =      .2875E-02
ANORM =     .1113E+00

IMAX =  2      BNORM =      .1640E-02
ANORM =     .1113E+00

E=    1.8837 THETA=     0.0000 FI=     0.0000

    PQ1      PQ2       REF INT      SYM
  0.000    0.000     .20826E-02      1
  1.000    0.000     .15385E-01      6
 -1.000    1.000     .15385E-01      1
  0.000    1.000     .17850E-03      6
  1.000   -1.000     .17850E-03      1
SURFACE GEOMETRY :   2.0330 0.0000 1.4376
IMAX =  8      BNORM =      .2969E-02
ANORM =     .1071E+00

IMAX =  2      BNORM =      .1634E-02
ANORM =     .1071E+00

E=    1.8837 THETA=     0.0000 FI=     0.0000

    PQ1      PQ2       REF INT      SYM
  0.000    0.000     .26500E-02      1
  1.000    0.000     .13448E-01      6
 -1.000    1.000     .13448E-01      1
  0.000    1.000     .17430E-02      6
  1.000   -1.000     .17429E-02      1
```

10.5 Main Program for a Small-Atom p(2×1) Overlayer on a fcc (111) Substrate, Using RSP and RFS

We consider adsorption of hydrogen in a half monolayer on Ni(111). The main difficulty of this situation is the small and poorly known radius of chemisorbed hydrogen. This allows the adatoms to take any of a large number of plausible adsorption positions, including positions that imply small interlayer spacings and underlayers. To handle such positions, Reverse Scattering Perturbation can be used: the adlayer and the top substrate layer are then taken to be a composite layer with 3 atoms (one adatom and two substrate atoms) per unit cell. Besides RSP, matrix inversion can equally be used (although less efficiently): the program shows how this is done.

Four possible positions of the adlayer are illustrated in Fig. 10.3 and considered in the calculation presented here. One of these (an underlayer halfway between the two top substrate layers, between which the bulk separation is chosen) does not actually require RSP to be used, because all interlayer spacings are sufficiently large even for RFS in that case.

Since in composite layers *two* reference points (0 and 0', cf. Figs. 10.3, 6.1) are used, one on each side of the composite layer, this implies in the present case of an underlayer that the spacing to the second substrate layer should be reduced by the spacing between the top substrate layer and the underlayer. The coordinates parallel to the surface of both reference points are always the same as those of the local origin chosen in the composite layer (cf. Chap. 6): in this case the local origin 0_U is one of the nickel atom centers [so that FPOS(2,J) = (0,0,0)]. One of the reference points coincides with 0_U, the other is in the plane of the adatoms: in the case of an underlayer $0 = 0_U \neq 0'$, for an overlayer $0 \neq 0_U = 0'$. [In the program, FPOS(1,J) is the adatom position in the unit cell and is overwritten by the hydrogen positions VPOS; FPOS(1,J) is therefore not, in this particular program, actually fixed at the time of input.]

For most plausible adsorption sites on this surface a mirror plane exists, which we can take to be the (xz)-plane, as in Fig. 10.3 (no higher symmetry exists in this case, since the adlayer has a rectangular lattice). This symmetry is exploited in exactly the same way as in Sect. 10.4; the set of extra beams due to the p(2x1) structure does not change this (note that the two beam sets are treated separately in the substrate: block-diagonalization occurs). With this symmetry, off-normal incidence at $\phi = \pm 90°$ is allowed,

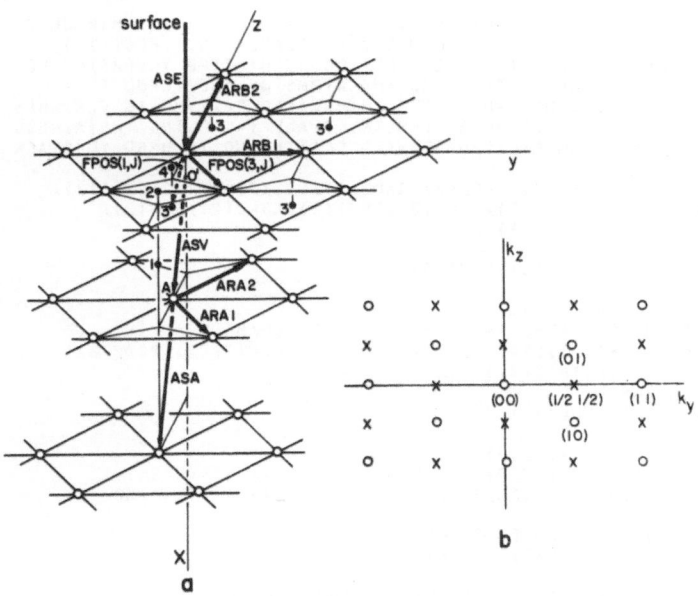

Fig. 10.3 (a) Geometry of fcc(111) with a p(2×1) adlayer; top substrate layer and adlayer are treated as a composite layer. Open and filled circles denote substrate and adatoms, respectively. $0 = 0_u$ is the local origin of coordinates in the composite layer, 0 and 0' are its two reference points in the case of an underlayer. While FPOS(2,J) = (0,0,0), i.e., origin at a substrate atom, 0, FPOS(1,J) points to an adatom, four positions of which are marked 1,2,3,4, corresponding to the positions considered in the program; one of these positions (3) is repeated in the figure with the p(2x1) periodicity. ASV is derived from ASB (which points from 0 to A). The (xz)-plane is a mirror plane. (b) Reciprocal space (o = integral-order beams, x = fractional-order beams)

but only if the one overlayer domain chosen here exists by itself. Domains rotated by ±120° break the assumed symmetry at such off-normal incidence and so require separate, non-symmetrized calculations. At normal incidence this limitation does not arise, and domains can be averaged over after just one calculation.

Note that it is not assumed in this program that the two nickel atoms in the composite layer have bulk positions: they may move together or independently by a simple change of the input, so that a modification of the clean surface due to adsorption may be investigated (cf. also Sect. 10.8). Of course, the assumed symmetry may not be broken.

```
C   MAIN + DATA FOR NI(111)+P(2*1)H, RSP (OR BEEBY) + RFS, (XZ) MIRROR PL
C   MAIN LEED PROGRAM FOR A P(2*1) OVERLAYER/UNDERLAYER ON A FCC(111)
C   SURFACE, FOR A SMALL ADSORBATE, USING RSP OR BEEBY-TYPE INVERSION FOR
C   THE LAYER COMPOSED OF THE ADATOMS AND THE ATOMS OF THE TOPMOST
C   SUBSTRATE LAYER. THE TOPMOST SUBSTRATE INTERLAYER SPACING IS VARIABLE
C   THE (XZ) PLANE IS ASSUMED TO BE A SYMMETRY PLANE (ALLOWING THETA.NE.O
C   IF FI*+-90 DEGREES). RENORMALIZED FORWARD SCATTERING IS USED TO STACK
C   LAYERS.
        DIMENSION ARA1(2),ARA2(2),RAR1(2),RAR2(2),SS1(2),SS2(2),SS3(2),
       1SS4(2),ASA(3),AS(3),ARB1(2),ARB2(2),RBR1(2),RBR2(2),SO1(2),
       2SO2(2),SO3(2),ASB(3),ASV(3)
C   2=NL, 45=KLM
        COMPLEX VL(2,2),FLMS(2,45),FLM(45)
C   2=NL, 2=IDEG
        DIMENSION V(2,2),JJS(2,2)
C   2=KNBS=NO. OF BEAM SETS READ IN, 21=KNT=NO. OF BEAMS READ IN
        DIMENSION KNB(2),NB(2),SPQF(2,21),SPQ(2,21),PQF(2,21),PQ(2,21)
        INTEGER KSYM(2,21),SYM(2,21)
C   20 MUST BE .GE. NPUN
        DIMENSION NPU(20),NPUC(20)
C   12=NPSI, 10=NEL*(LMAX+1)
        DIMENSION ES(12),PHSS(12,10)
C   3(1ST DIM)=NLAY, 20,12,7 FIXED, 759=NCAA FROM DATA STATEMENT
        DIMENSION FPOS(3,3),VPOS(20,3),LPS(12),INV(12),CAA(759),NCA(7),
       1LPSS(12)
C   3(EXCEPT 2ND DIM OF POS AND POSS)=NLAY
        DIMENSION POS(3,3),POSS(3,3),MGH(3,3)
C   3(1ST DIM)=NLAY2
        DIMENSION DRL(3,3),SDRL(3,3),NUGH(3),NGEQ(3),NGOL(3),TEST(3)
C   809=NLM FROM DATA STATEMENT,  81=NN,  9=N (FOR CELMG)
C   YLM IS ALSO USED, AS COMPLEX VECTOR OF SMALLER CORE SIZE, IN MSMF,ETC
        DIMENSION NLMS(7),CLM( 809),YLM( 81),FAC2( 81),FAC1( 9)
C   9=NN1, 5=NN2, 5=NN3
        DIMENSION PPP( 9,5,5)
C   25=LMMAX
        DIMENSION LX(25),LXI(25),LXM(25),LT(25)
C   5=LMAX+1=L1, 3 AND 10 FIXED FOR COMMON /MPT/
        COMPLEX AF(5),CAF(5),TSFO(3,10),TSF(3,10)
        COMPLEX ROP(21,21),TOP(21,21),ROM(21,21),TOM(21,21)
C   12=KNP
        COMPLEX RA1(21,12),TA1(21,12)
C   50=LMT, 150(IN GH)=LMG, 3(2ND DIM IN RG)=NLAY, 75(IN TS)=LMN,
C   150(IN TG)=LM2N, 75(IN TH AND IPL)=LMNI, TH AND IPL ARE NEEDED ONLY
C   IF SUBROUTINE MTINV IS USED
        COMPLEX TAU(50,15),TAUG(50),TAUGM(50),GH(150,25),RG(3,3,21),
       1TS(75),TG(2,150),VT(25)
C       COMPLEX TH(75,75)
C       DIMENSION IPL(75)
C   15=LEV, 30=LEV2, 10=LOD
        COMPLEX AMULT(21),CYLM(21,25),XEV(15,30),XOD(10,10),
       1YLME(15),YLMO(10)
        DIMENSION IPLE(15),IPLO(10),AT(21)
C   8 FIXED, 20=ND (FOR RFS SUBROUTINES)
        COMPLEX XI(21),PK(21,8),AW(21,2),ANEW(21,20)
        COMMON E3,AK21,AK31,VPI1,TV,EMACH
        COMMON /X4/E,VPI,AK2,AK3,ASA
        COMMON  /SL/ARA1, ARA2, ARB1, ARB2, RBR1, RBR2, NL1, NL2
        COMMON /MS/LMAX,EPS,LITER,SO1,SO2,SO3,SS1,SS2,SS3,SS4
        COMMON /ADS/ASL, FR, ASE, VPIS, VPIO, VO, VV
        COMMON /BT/IT,T,TO,DRPER,DRPAR,DRO
        COMMON /TEMP/IT1,IT2,IT3,TI,TF,DT,TO1,DRPER1,DRPAR1,DRO1,DRPER2,
       1DRPAR2,DRO2,DRPER3,DRPAR3,DRO3
        COMMON /MPT/NA,NS,ID,LAY,L1,NTAU,TSTS,TV1,DCUT,EPSN,NPERT,NOPT,
       1NEW,LPS,LPSS,INV,NINV,TSF
C   NCA IS DIMENSION OF CAA AS FUNCTION OF LMAX
        DATA NCA(1),NCA(2),NCA(3),NCA(4),NCA(5),NCA(6),NCA(7)/
       11,70,264,759,1820,3836,7344/
```

```
C   NLMS IS DIMENSION OF CLM AS FUNCTION OF LMAX
        DATA NLMS(1),NLMS(2),NLMS(3),NLMS(4),NLMS(5),NLMS(6),NLMS(7)/
       10,76,284,809,1925,4032,7680/
    161 FORMAT(3F7.2)
    200 FORMAT(20I3)
    240 FORMAT(8HOIDEG = ,1I3,7H NL1 = ,1I3,7H NL2 = ,1I3)
    370 FORMAT(9H1ENERGY =,F7.4,7H H  OR ,1F7.2,3H EV)
    380 FORMAT(8H VPIS = ,F9.4,8H VPIO = ,F9.4,9H DCUTS = ,F9.4,9H DCUTO =
       1 ,F9.4)
    390 FORMAT(8HOTEMP = ,F9.4)
    410 FORMAT(10H MSMF   OK)
    415 FORMAT(10H MPERTI OK)
    416 FORMAT(10H MTINV  OK)
    420 FORMAT(10H SUBREF OK)
    430 FORMAT(20HOSURFACE GEOMETRY : ,3F7.4)
    450 FORMAT(20H CORRECT TERMINATION)
    460 FORMAT(20A4)
    465 FORMAT(1H1,20A4)
C
C
C   READ, WRITE AND PUNCH A DESCRIPTIVE TITLE
        READ (5,460)(CLM(I),I=1,20)
        WRITE(6,465)(CLM(I),I=1,20)
        WRITE(7,460)(CLM(I),I=1,20)
C   EMACH IS MACHINE ACCURACY (USED BY ZGE AND ZSU)
        EMACH=1.0E-6
C   IDEG: EACH LAYER HAS AN IDEG-FOLD SYMMETRY AXIS
C   NL1, NL2: SUPERLATTICE CHARACTERIZATION  NL1  NL2
C                       P(1*1)                 1    1
C                       C(2*2)                 2    1
C                       P(2*1)                 2    1
C                       P(1*2)                 1    2
C                       P(2*2)                 2    2
        READ(5,200)IDEG,NL1,NL2
        WRITE(6,240)IDEG,NL1,NL2
        NL=NL1*NL2
C   KNBS= NO.OF BEAM SETS TO BE READ IN (.LE.NL)
C   KNT= TOTAL NO. OF BEAMS TO BE READ IN
C   NPUN= NO. OF BEAMS FOR WHICH INTENSITIES ARE TO BE PUNCHED OUT
        READ(5,200)KNBS,KNT,NPUN
C   NPSI= NO.OF ENERGIES AT WHICH PHASE SHIFTS WILL BE READ IN
        READ(5,200)NPSI
C   READ IN GEOMETRY, PHYSICAL PARAMETERS, CONVERGENCE CRITERIA
        CALL READIN(TVA,RAR1,RAR2,ASA,TVB,ASB,STEP,NSTEP,IDEG,NL,V,VL,
       1JJS,KNBS,KNB,KNT,SPQF,KSYM,SPQ,TST,TSTS,NPUN,NPU,THETA,FI,
       2LMMAX,NPSI,ES,PHSS,L1)
        TO=TO1
C   NLAY= NO. OF SUBPLANES IN COMPOSITE LAYER (0, IF NO COMP. LAYER)
        READ(5,200)NLAY
        IF (NLAY.EQ.0) GO TO 500
C   READ IN ADDITIONAL DATA FOR COMPOSITE LAYER
        CALL READCL(NLAY,NLAY2,FPOS,VPOS,NVAR,NTAU,LPS,NINV,INV,LMMAX,
       1LMG,LMN,LM2N,LMT,LTAUG,LMNI,LMNI2)
    500 CONTINUE
        KLM=(2*LMAX+1)*(2*LMAX+2)/2
        LEV=(LMAX+1)*(LMAX+2)/2
        LOD=LMMAX-LEV
        IF (NLAY.EQ.0) GO TO 630
        LEV2=2*LEV
        NCAA=NCA(LMAX)
C   CAA= CLEBSCH-GORDON COEFFICIENTS FOR COMPOSITE LAYER
        CALL CAAA(CAA,NCAA,LMMAX)
    630 N=2*LMAX+1
        NN=N*N
        NLM=NLMS(LMAX)
C   CLM= CLEBSCH-GORDON COEFFICIENTS FOR MATRICES X AND TAU
        CALL CELMG(CLM,NLM,YLM,FAC2,NN,FAC1,N,LMAX)
```

```
C   LX,LXI,LT,LXM: PERMUTATIONS OF (L,M)-SEQUENCE
        CALL LXGENT(LX,LXI,LT,LXM,LMAX,LMMAX)
        IF (IT1+IT2+IT3) 640,650,640
   640 NN3=LMAX+1
       NN2=LMAX+1
       NN1=NN2+NN3-1
C   PPP= CLEBSCH-GORDON COEFFICIENTS FOR COMPUTATION OF TEMPERATURE-
C   DEPENDENT PHASE SHIFTS (SKIPPED IF NOT NEEDED)
        CALL  CPPP (PPP, NN1, NN2, NN3)
   650 CONTINUE
C
C
C   READ ENERGY RANGE AND STEP; IF EF AND DE ARE BLANK, PROGRAM COMPUTES
C   FOR ENERGY EI. PROGRAM ALWAYS RETURNS TO THIS LINE TO READ A NEW
C   ENERGY RANGE AND STEP. IF A NEGATIVE EI IS READ, RUN IS TERMINATED
   680 READ (5,161) EI, EF, DE
        IF (EI)  1000, 690, 690
   690 EI=EI/27.18+VV
       EF=EF/27.18+VV
       DE=DE/27.18
       E = EI
       EC = E
       E3 = E
C
C
C   START LOOP OVER ENERGIES IN GIVEN ENERGY RANGE
   700 EEV=(E-VV)*27.18
       WRITE (6,370) E,EEV
C   COMPUTE COMPONENTS OF INCIDENT WAVEVECTOR PARALLEL TO SURFACE
       AK=SQRT(2.0*(E-VV))*SIN(THETA)
       AK2=AK*COS(FI)
       AK3=AK*SIN(FI)
       AK21=AK2
       AK31=AK3
C   SELECT BEAMS APPROPRIATE FOR CURRENT ENERGY
        CALL BEAMS(KNBS,KNB,SPQ,SPQF,KSYM,KNT,NPU,NPUN,AK2,AK3,E,TST,
       1NB,PQ,PQF,SYM,NPUC,MPU,NT,NP)
C   SET OPTICAL POTENTIAL (IMAGINARY PART OF MUFFIN-TIN CONSTANT,
C   REPRESENTING DAMPING). VPIS FOR SUBSTRATE, VPIO FOR OVERLAYER.
C   HERE VPIS=VPIO PROPORTIONAL TO E**1/3
       VPIS=-(3.8/27.18)*EXP((ALOG((E)/(90.0/27.18+VV)))*0.333333)
       VPIO=VPIS
       DCUTO = SQRT(2.0 * E)
C   SET LIMITING RADII ON LATTICE SUMS, POSSIBLY DIFFERENT FOR SUBSTRATE
C   (DCUTS) AND OVERLAYER (DCUTO)
       DCUTS =  - 5.0 * DCUTO/(AMIN1(VPIS, - 0.05))
       DCUTO =  - 5.0 * DCUTO/(AMIN1(VPIO, - 0.05))
       WRITE (6,380)  VPIS, VPIO, DCUTS, DCUTO
C   FIX TEMPERATURE T. T MAY BE INCREASED STEPWISE IN A LOOP OVER THE
C   INPUT RANGE (TI,TF) WITH INPUT STEP DT
   770 T = TI
   780 WRITE (6,390)  T
C
C
       VPI1 = VPIO
C   PERFORM PLANAR LATTICE SUMS IN FMAT. ASSUME SAME MUFFIN-TIN CONSTANTS
C   AND DAMPINGS IN SUBSTRATE AND ADLAYER (MPERTI AND MTINV REQUIRE THIS)
       NLS=NL
       CALL  FMAT (FLMS, V, JJS, NL, NLS, DCUTO, IDEG, LMAX,KLM)
C
C
C   CONSIDER SUBSTRATE LAYERS FIRST
C   PRODUCE ATOMIC T-MATRIX ELEMENTS
        CALL TSCATF(1,L1,ES,PHSS,NPSI,IT1,E,O.,PPP,NN1,NN2,NN3,DRO1,
       1DRPER1,DRPAR1,TO,T,TSFO,TSF,AF,CAF)
       DRPER=DRPER1
       DRPAR=DRPAR1
```

```
      DRO=DRO1
      IT=0
      TV = TVA
      VPI = VPIS
      VPI1 = VPIS
C  PRODUCE REFLECTION (RA1) AND TRANSMISSION (TA1) MATRICES FOR
C  ATOMIC LAYERS. ONE REGISTRY: ID=1. NLL=NL=2 SUBLATTICES. SUBSTRATE
C  LAYER SYMMETRIES AND REGISTRIES APPLY: LAY=2. TEMPERATURE-DEPENDENT
C  PHASE SHIFTS: CAF
C  FIRST BEAM SET FIRST (NAA BEAMS)
  930 NAA = NB(1)
      CALL MSMF(RA1,TA1,RA1,TA1,RA1,TA1,RA1,TA1,NAA,NT,NP,AMULT,CYLM,
     1PQ,SYM,NT,FLMS,FLM,V,NL,   0,   0,1,NL,CAF,CAF,L1,LX,LXI,LMMAX,KLM,
     2XEV,XOD,LEV,LOD,YLM,YLME,YLMO,IPLE,IPLO,CLM,NLM,2)
      WRITE (6,410)
C  SECOND BEAM SET NEXT (NBB BEAMS; NOTE OFFSETS NA=NS=NAA)
      NBB=NB(2)
      CALL MSMF(RA1,TA1,RA1,TA1,RA1,TA1,RA1,TA1,NBB,NT,NP,AMULT,CYLM,
     1PQ,SYM,NT,FLMS,FLM,V,NL,NAA,NAA,1,NL,CAF,CAF,L1,LX,LXI,LMMAX,KLM,
     2XEV,XOD,LEV,LOD,YLM,YLME,YLMO,IPLE,IPLO,CLM,NLM,2)
      WRITE (6,410)
C
C
C  PRODUCE ATOMIC T-MATRIX ELEMENTS FOR ADATOMS
      CALL TSCATF(2,L1,ES,PHSS,NPSI,IT2,E,0.,PPP,NN1,NN2,NN3,DRO2,
     1DRPER2,DRPAR2,TO,T,TSFO,TSF,AF,CAF)
C  SET PARAMETERS FOR COMPOSITE LAYER (OVERLAYER+TOP SUBSTRATE LAYER):
C  ONLY 1 REGISTRY (ID=1), OVERLAYER SYMMETRIES (LAY=1)
      NA=0
      NS=0
      ID=1
      LAY=1
      TV1=TVB
      DCUT=DCUTS
      EPSN=0.01
      NPERT=5
      NOPT=1
      NEW=1
C
C
C  LOOP 960 RUNS OVER ADLAYER POSITIONS VPOS IN THE COMPOSITE LAYER
      DO 960 ISTEP=1,NVAR
      DO 935 J=1,3
      AS(J)=VPOS(ISTEP,J)*0.529
  935 POS(1,J)=VPOS(ISTEP,J)
      WRITE(6,430)(AS(J),J=1,3)
      DO 936 I=2,NLAY
      DO 936 J=1,3
  936 POS(I,J)=FPOS(I,J)
C
C  COMPUTE REFLECTION AND TRANSMISSION MATRICES FOR THE COMPOSITE
C  LAYER BY RSP (NO PARTIAL INVERSION)
      CALL MPERTI(ROP,TOP,ROM,TOM,ROP,TOP,ROM,TOM,ROP,TOP,ROM,TOM,
     1NT,NT,NT,AMULT,CYLM,PQ,SYM,NT,FLMS,NL,LX,LXI,LT,LXM,LMMAX,
     2KLM,XEV,LEV,LEV2,TAU,LMT,TAUG,TAUGM,LTAUG,CLM,NLM,POS,POSS,
     3MGH,NLAY,DRL,SDRL,NUGH,NGEQ,NGOL,NLAY2,TEST,
     4GH,LMG,RG,TS,LMN,TG,LM2N,VT,CAA,NCAA,TH,1,1)
      WRITE(6,415)
C  DO SAME WITH BEEBY-TYPE INVERSION ONLY INSTEAD OF RSP, IF DESIRED
C  (MAKE MATRICES TH AND IPL AVAILABLE IN DIMENSION STATEMENTS).
C  ALSO CHANGE IN INPUT: NINV=3.
C     CALL MTINV(ROP,TOP,ROM,TOM,ROP,TOP,ROM,TOM,ROP,TOP,ROM,TOM,
C    1NT,NT,NT,AMULT,CYLM,PQ,SYM,NT,FLMS,NL,LX,LXI,LT,LXM,LMMAX,
C    2KLM,XEV,LEV,LEV2,TAU,LMT,TAUG,TAUGM,      CLM,NLM,POS,POSS,
C    3MGH,NLAY,DRL,SDRL,NUGH,NGEQ,NGOL,NLAY2,TEST,
C    4GH,LMG,RG,TS,LMN,TG,LM2N,VT,CAA,NCAA,TH,LMNI,IPL)
C     WRITE(6,416)
```

```
C
C
C   NOW STACK LAYERS
C   ASV BECOMES VECTOR BETWEEN COMPOSITE LAYER AND 2ND SUBSTRATE LAYER,
C   AND CAN BE VARIED BY STEP IN LOOP 969
      DO 938 J=1,3
  938 ASV(J)=ASB(J)
C   IF ADLAYER IS UNDERLAYER, REDUCE SPACING BETWEEN COMPOSITE LAYER AND
C   SECOND SUBSTRATE LAYER ACCORDINGLY (ASV(1) INITIALLY IS SPACING
C   BETWEEN TOP SUBSTRATE LAYERS)
      IF (VPOS(ISTEP,1).GT.FPOS(2,1)) ASV(1)=ASV(1)-(VPOS(ISTEP,1)-
     1FPOS(2,1))
      DO 969 ISUB=1,NSTEP
      DO 940 J=1,3
  940 AS(J)=ASV(J)*0.529
      WRITE(6,430)(AS(J),J=1,3)
C
C   COMPUTE REFLECTED BEAM AMPLITUDES BY RFS
      CALL RFSO2(ROP,TOP,ROM,TOM,RA1,TA1,RA1,TA1,NT,NB,KNBS,NP,XI,
     1PQ,PK,AW,ANEW,20,ASV,ASA,ASA)
C   COMPUTE REFLECTED INTENSITIES (AT) FROM AMPLITUDES (XI), PRINT AND
C   PUNCH (NPNCH=1; PUNCH A ALSO FOR LATER IDENTIFICATION)
      A=FLOAT(ISTEP)
      CALL RINT(NT,XI,AT,PQ,PQF,SYM,VV,THETA,FI,MPU,NPUC,EEV,A,1)
  969 ASV(1)=ASV(1)+STEP
C
C
C   REUSE UNCHANGED QUANTITIES IN MPERTI OR MTINV IN NEXT GEOMETRY OF THE
C   COMPOSITE LAYER
  960 NEW=-1
      IF (IT1+IT2+IT3) 980,990,980
  980 T = T + DT
      IF (DT.LE.0.0001) GO TO 990
      IF (T-TF)  780, 780, 990
  990 E = E + DE
      EC = E
      E3 = E
      IF (E-EF)  700, 700, 680
 1000 WRITE (6,450)
      STOP
      END
```

---- INPUT DATA FOR PRECEDING PROGRAM FOLLOW ------------

```
NI(111)+P(2*1)H, RSP (OR BEEBY) + RFS, 1 MIRROR PLANE, 5 PH. SH.
  2  2  1            IDEG NL1 NL2
  2 21 19            KNBS KNT NPUN
 12                  NPSI
 1.2450-2.1564       ARA1
 1.2450 2.1564       ARA2
 0.0000 0.0000       SS1
 0.0000 0.0000       SS2
 0.0000 0.0000       SS3
 0.0000 0.0000       SS4
 2.0330 0.0000 1.4376 ASA
 2.4900 0.0000       ARB1
 0.0000 4.3128       ARB2
 0.0000 0.0000       SO1
 0.0000 0.0000       SO2
 0.0000 0.0000       SO3
 2.0330 0.0000 1.4376 ASB
 0.5000 1.2450 0.0500 FR ASE STEP
  1                  NSTEP
 12                  KNB(1)
 0.0000 0.0000  1  1 BEAMS
 1.0000 0.0000  6  6
-1.0000 1.0000  1  1
```

```
0.0000 1.0000   6   6
1.0000-1.0000   1   1
1.0C00 1.0000   6   6
-1.0000 2.0000   6   6
2.0000-1.0000   6   6
2.0000 0.0000   6   6
-2.0000 2.0000   1   1
0.0000 2.0000   6   6
2.0000-2.0000   1   1
    9                      KNB(2)
-0.5000 0.5000   1   1
0.5000-0.5000   1   1
0.5000 0.5000   6   6
-1.5C00 1.5000   1   1
-0.5000 1.5000   6   6
0.5000 1.5000   6   6
1.5000 0.5000   6   6
1.5C00-0.5000   6   6
1.5000-1.5000   1   1
0.0020                      TST
    1   2   3   4   5   6   7   8   9  10  11  12  13  14  15  16  17  20  21      NPU
    0.00     0.00          THETA FI
    0.00    11.20          VO VV
0.0010                      EPS
    5                       LITER
    1   1   0               IT1 IT2 IT3
440.0000      58.7100      1.4000      1.4000      0.0000       THDB1 AM1 FPER1 FPAR1 DRO1
3357.9827      1.0080      2.0000      2.0000      0.0000       THDB2 AM2 FPER2 FPAR2 DRO2
    1.0000      1.0000      1.0000      1.0000      0.0000       THDB3 AM3 FPER3 FPAR3 DRO3
110.0000                                                        TI TF DT
    4                       LMAX
    2                       NEL
    .3000
2.8398 -.0137 -.1876  .0011  .0000  .0000  .0000  .0000   NI
1.3979 0.0211 0.0003 0.0000 0.0000 0.0000 0.0000
    .4000
2.7639 -.0216 -.3164  .0024  .0001  .0000  .0000  .0000   NI
1.3321 0.0320 0.0006 0.0000 0.0000 0.0000 0.0000
    .5000
2.6930 -.0322 -.2630  .0048  .0002  .0000  .0000  .0000   NI
1.2785 0.0440 0.0010 0.0000 0.0000 0.0000 0.0000
    .7500
2.5276 -.0668 -.2322  .0160  .0009  .0000  .0000  .0000   NI
1.1743 0.0772 0.0025 0.0001 0.0000 0.0000 0.0000
1.0000
2.3803 -.1132 -.2323  .0386  .0028  .0002  .0000  .0000   NI
1.0944 0.1125 0.0050 0.0001 0.0000 0.0000 0.0000
1.5000
2.1240 -.2247 -.2118  .1202  .0125  .0012  .0001  .0000   NI
0.9739 0.1818 0.0125 0.0006 0.0000 0.0000 0.0000
2.0000
1.9101 -.3448 -.1758  .2494  .0335  .0042  .0004  .0000   NI
0.8847 0.2416 0.0233 0.0014 0.0001 0.0000 0.0000
3.0000
1.5727 -.5697 -.1288  .5535  .1138  .0216  .0034  .0004   NI
0.7610 0.3215 0.0527 0.0049 0.0003 0.0000 0.0000
4.0C00
1.3160 -.7574 -.1272  .7884  .2257  .0583  .0125  .0022   NI
0.6829 0.3567 0.0877 0.0113 0.0009 0.0001 0.0000
5.0000
1.1099 -.9122 -.1445  .9491  .3361  .1107  .0305  .0069   NI
0.6334 0.3644 0.1227 0.0207 0.0022 0.0002 0.0000
6.0000
 .9366-1.0434 -.1666 1.0690  .4287  .1691  .0571  .0159   NI
0.6C21 0.3577 0.1534 0.0328 0.0043 0.0004 0.0000
7.0000
 .7862-1.1581 -.1902 1.1630  .5045  .2251  .0894  .0297   NI
```

```
 0.5819 0.3447 0.1778 0.0469 0.0074 0.0008 0.0001
 3                         NLAY
 0.0000 0.0000 0.0000     FPOS(1,J) NOT USED
 0.0000 0.0000 0.0000     FPOS(2,J)
 0.0000 1.2450-2.1564     FPOS(3,J)
 4                         NVAR
 1.0166 0.0000-1.4400     VPOS(1,J)
-0.2000 0.0000-1.4400     VPOS(2,J)
 0.5080 0.0000-2.8800     VPOS(3,J)
-0.2000 0.0000-2.8800     VPOS(4,J)
 2                         NTAU
 2  1  1                   LPS
 0                         NINV
 0   0   0                 INV
 40.00 151.00    5.00      EI EF DE
-10.00                     EI EF DE
----------------------------------------------------------------
---- FIRST PART OF PRINTER OUTPUT FOLLOWS ----------------
----(NEGLECTING MOST BLANK LINES AND PAGE ----------------
---- EJECTS AND TRUNCATING LONG LINES)     ----------------
----------------------------------------------------------------
 NI(111)+P(2*1)H, RSP (OR BEEBY) + RFS, 1 MIRROR PLANE, 5 PH. SH.
 IDEG =  2 NL1 =  2 NL2 =  1
 PARAMETERS FOR INTERIOR:
 SURF VECS        1.2450 -2.1564
                  1.2450  2.1564
 SS     0.0000 0.0000
 SS     0.0000 0.0000
 SS     0.0000 0.0000
 SS     0.0000 0.0000
 ASA    2.0330 0.0000 1.4376
 PARAMETERS FOR SURFACE:
 SURF VECS        2.4900  0.0000
                  0.0000  4.3128
 SO     0.0000 0.0000
 SO     0.0000 0.0000
 SO     0.0000 0.0000
 ASB    2.0330 0.0000 1.4376
 FR =  .5000 ASE = 1.2450 STEP =   .0500 NSTEP =   1
          PQ1      PQ2       SYM

    1     0.000    0.000    1  1
    2     1.000    0.000    6  6
    3    -1.000    1.000    1  1
    4     0.000    1.000    6  6
    5     1.000   -1.000    1  1
    6     1.000    1.000    6  6
    7    -1.000    2.000    6  6
    8     2.000   -1.000    6  6
    9     2.000    0.000    6  6
   10    -2.000    2.000    1  1
   11     0.000    2.000    6  6
   12     2.000   -2.000    1  1

   13     -.500     .500    1  1
   14      .500    -.500    1  1
   15      .500     .500    6  6
   16    -1.500    1.500    1  1
   17     -.500    1.500    6  6
   18      .500    1.500    6  6
   19     1.500     .500    6  6
   20     1.500    -.500    6  6
   21     1.500   -1.500    1  1
 TST =  .0020
 PUNCHED BEAMS    1  2  3  4  5  6  7  8  9 10 11 12 13 14 15 16 17 20 2
 THETA  FI =    0.00    0.00
 VO       0.00VV       11.20
```

```
EPS =   .0010  LITER =    5
IT1 =    1 IT2 =    1 IT3 =    0
THDB =  440.0000 AM =   58.7100 FPER =    1.4000 FPAR =    1.4000 DRO =
THDB = 3357.9827 AM =    1.0080 FPER =    2.0000 FPAR =    2.0000 DRO =
THDB =    1.0000 AM =    1.0000 FPER =    1.0000 FPAR =    1.0000 DRO =
LMAX =   4
     PHASE SHIFTS

E =   .3000  1ST ELEMENT    2.8398  -.0137  -.1876   .0011  0.0000
             2ND ELEMENT    1.3979   .0211   .0003  0.0000  0.0000

E =   .4000  1ST ELEMENT    2.7639  -.0216  -.3164   .0024   .0001
             2ND ELEMENT    1.3321   .0320   .0006  0.0000  0.0000

E =   .5000  1ST ELEMENT    2.6930  -.0322  -.2630   .0048   .0002
             2ND ELEMENT    1.2785   .0440   .0010  0.0000  0.0000

E =   .7500  1ST ELEMENT    2.5276  -.0668  -.2322   .0160   .0009
             2ND ELEMENT    1.1743   .0772   .0025   .0001  0.0000

E =  1.0000  1ST ELEMENT    2.3803  -.1132  -.2323   .0386   .0028
             2ND ELEMENT    1.0944   .1125   .0050   .0001  0.0000

E =  1.5000  1ST ELEMENT    2.1240  -.2247  -.2118   .1202   .0125
             2ND ELEMENT     .9739   .1818   .0125   .0006  0.0000

E =  2.0000  1ST ELEMENT    1.9101  -.3448  -.1758   .2494   .0335
             2ND ELEMENT     .8847   .2416   .0233   .0014   .0001

E =  3.0000  1ST ELEMENT    1.5727  -.5697  -.1288   .5535   .1138
             2ND ELEMENT     .7610   .3215   .0527   .0049   .0003

E =  4.0000  1ST ELEMENT    1.3160  -.7574  -.1272   .7884   .2257
             2ND ELEMENT     .6829   .3567   .0877   .0113   .0009

E =  5.0000  1ST ELEMENT    1.1099  -.9122  -.1445   .9491   .3361
             2ND ELEMENT     .6334   .3644   .1227   .0207   .0022

E =  6.0000  1ST ELEMENT     .9366 -1.0434  -.1666  1.0690   .4287
             2ND ELEMENT     .6021   .3577   .1534   .0328   .0043

E =  7.0000  1ST ELEMENT     .7862 -1.1581  -.1902  1.1630   .5045
             2ND ELEMENT     .5819   .3447   .1778   .0469   .0074
COMPOSITE LAYER VECTOR :  0.0000 0.0000 0.0000
COMPOSITE LAYER VECTOR :  0.0000 0.0000 0.0000
COMPOSITE LAYER VECTOR :  0.0000 1.2450-2.1564
VARYING POSITION IN COMPOSITE LAYER :  1.0166 0.0000-1.4400
VARYING POSITION IN COMPOSITE LAYER :  -.2000 0.0000-1.4400
VARYING POSITION IN COMPOSITE LAYER :   .5080 0.0000-2.8800
VARYING POSITION IN COMPOSITE LAYER :  -.2000 0.0000-2.8800
NTAU =   2
PHASE SHIFT ASSIGNMENT IN COMPOSITE LAYER   2  1  1
  0 SUBPLANES IN BEEBY INVERSION :   0  0  0
NO. OF ELEMENTS IN CAAA :    759
ENERGY = 1.8837 H  OR   40.00 EV
  12 BEAMS USED :    5   7
   0.000 0.000    1.000 0.000   -1.000 1.000    0.000 1.000    1.000-1.000
  -.500 1.500    1.500 -.500    1.500-1.500
VPIS =    -.1114 VPIO =    -.1114 DCUTS =   87.1160 DCUTO =   87.1160
TEMP =  110.0000
MSMF   OK
MSMF   OK
SURFACE GEOMETRY :  1.0166 0.0000-1.4400
NO. OF LATT. PTS. FOR GH :   616
NO. OF LATT. PTS. FOR GH :   616
NO. OF BEAMS FOR GH :    41
X IN TPERTI =     .19742E+00    .19742E+00    .92227E-01
```

```
X IN TPERTI =    .19289E+00    .20031E+00    .11110E+00
NO CONV IN TPERTI AFTER    5 ITERATIONS
X IN TPERTI =    .19742E+00    .19742E+00    .92227E-01
X IN TPERTI =    .19352E+00    .18844E+00    .10118E+00
   CONV IN TPERTI AFTER    4 ITERATIONS
X IN TPERTI =    .32288E+00    .32288E+00    .11663E+00
X IN TPERTI =    .32950E+00    .32831E+00    .10225E+00
   CONV IN TPERTI AFTER    5 ITERATIONS
X IN TPERTI =    .32288E+00    .32288E+00    .11663E+00
X IN TPERTI =    .32073E+00    .33388E+00    .10984E+00
   CONV IN TPERTI AFTER    4 ITERATIONS
X IN TPERTI =    .31767E+00    .31767E+00    .11940E+00
X IN TPERTI =    .32778E+00    .32955E+00    .10487E+00
   CONV IN TPERTI AFTER    5 ITERATIONS
X IN TPERTI =    .31767E+00    .31767E+00    .11940E+00
X IN TPERTI =    .35048E+00    .30918E+00    .11039E+00
   CONV IN TPERTI AFTER    5 ITERATIONS
X IN TPERTI =    .32288E+00    .32288E+00    .11663E+00
X IN TPERTI =    .32930E+00    .32049E+00    .10229E+00
   CONV IN TPERTI AFTER    5 ITERATIONS
X IN TPERTI =    .32288E+00    .32288E+00    .11663E+00
X IN TPERTI =    .31393E+00    .29124E+00    .12321E+00
   CONV IN TPERTI AFTER    4 ITERATIONS
X IN TPERTI =    .31767E+00    .31767E+00    .11940E+00
X IN TPERTI =    .32975E+00    .31915E+00    .10491E+00
   CONV IN TPERTI AFTER    5 ITERATIONS
X IN TPERTI =    .31767E+00    .31767E+00    .11940E+00
X IN TPERTI =    .31947E+00    .28182E+00    .12688E+00
   CONV IN TPERTI AFTER    4 ITERATIONS
X IN TPERTI =    .28682E+00    .28682E+00    .11011E+00
X IN TPERTI =    .28483E+00    .28671E+00    .11998E+00
   CONV IN TPERTI AFTER    5 ITERATIONS
X IN TPERTI =    .28682E+00    .28682E+00    .11011E+00
X IN TPERTI =    .29678E+00    .28060E+00    .95219E-01
   CONV IN TPERTI AFTER    5 ITERATIONS
X IN TPERTI =    .28682E+00    .28682E+00    .11011E+00
X IN TPERTI =    .28942E+00    .28040E+00    .98745E-01
NO CONV IN TPERTI AFTER    5 ITERATIONS
X IN TPERTI =    .28682E+00    .28682E+00    .11011E+00
X IN TPERTI =    .28206E+00    .26233E+00    .11126E+00
   CONV IN TPERTI AFTER    4 ITERATIONS
X IN TPERTI =    .32800E+00    .32800E+00    .11477E+00
X IN TPERTI =    .32460E+00    .33002E+00    .80985E-01
NO CONV IN TPERTI AFTER    5 ITERATIONS
X IN TPERTI =    .32800E+00    .32800E+00    .11477E+00
X IN TPERTI =    .30578E+00    .33408E+00    .10366E+00
   CONV IN TPERTI AFTER    4 ITERATIONS
X IN TPERTI =    .78131E+00    .78131E+00    .15024E+00
X IN TPERTI =    .78659E+00    .78513E+00    .99735E+00
NO CONV IN TPERTI AFTER    5 ITERATIONS
X IN TPERTI =    .78131E+00    .78131E+00    .15024E+00
X IN TPERTI =    .13377E+01    .11155E+01    .15493E+00
   CONV IN TPERTI AFTER    4 ITERATIONS
X IN TPERTI =    .46587E+00    .46587E+00    .12486E+00
X IN TPERTI =    .44922E+00    .45002E+00    .17243E+00
NO CONV IN TPERTI AFTER    5 ITERATIONS
X IN TPERTI =    .46587E+00    .46587E+00    .12486E+00
X IN TPERTI =    .54924E+00    .55929E+00    .12519E+00
   CONV IN TPERTI AFTER    4 ITERATIONS
X IN TPERTI =    .46587E+00    .46587E+00    .12486E+00
X IN TPERTI =    .44611E+00    .44413E+00    .14885E+00
NO CONV IN TPERTI AFTER    5 ITERATIONS
X IN TPERTI =    .46587E+00    .46587E+00    .12486E+00
X IN TPERTI =    .48998E+00    .56797E+00    .12438E+00
   CONV IN TPERTI AFTER    4 ITERATIONS
X IN TPERTI =    .78131E+00    .78131E+00    .15024E+00
X IN TPERTI =    .78165E+00    .78331E+00    .61331E+00
```

```
NO CONV IN TPERTI AFTER     5 ITERATIONS
X IN TPERTI =       .78131E+00    .78131E+00     .15024E+00
X IN TPERTI =       .99533E+00    .94266E+00     .15050E+00
   CONV IN TPERTI AFTER     4 ITERATIONS
MPERTI OK
SURFACE GEOMETRY :  1.0164 0.0000 1.4376
IMAX = 12     BNORM =     .2039E-02
ANORM =     .8267E-01

IMAX = 4      BNORM =     .1009E-02
ANORM =     .8301E-01

IMAX = 2      BNORM =     .2651E-03
ANORM =     .8302E-01

E=   1.8837 THETA=    0.0000 FI=    0.0000

   PQ1     PQ2       REF INT      SYM
  0.000   0.000   .49899E-03       1
  1.000   0.000   .10339E-01       6
 -1.000   1.000   .10399E-01       1
  0.000   1.000   .13756E-02       6
  1.000  -1.000   .14041E-02       1
  -.500    .500   .27782E-04       1
   .500   -.500   .30052E-03       1
   .500    .500   .25011E-04       6
SURFACE GEOMETRY :   -.2000 0.0000-1.4400
NO. OF LATT. PTS. FOR GH :    620
NO. OF LATT. PTS. FOR GH :    620
NO. OF LATT. PTS. FOR GH :    622
NO. OF LATT. PTS. FOR GH :    622
X IN TPERTI =       .92227E-01    .19742E+00     .19742E+00
X IN TPERTI =       .11818E+00    .23655E+00     .23390E+00
   CONV IN TPERTI AFTER     5 ITERATIONS
X IN TPERTI =       .92227E-01    .19742E+00     .19742E+00
X IN TPERTI =       .12020E+00    .22760E+00     .21330E+00
   CONV IN TPERTI AFTER     5 ITERATIONS
X IN TPERTI =       .11663E+00    .32288E+00     .32288E+00
X IN TPERTI =       .13106E+00    .32806E+00     .32229E+00
   CONV IN TPERTI AFTER     5 ITERATIONS
X IN TPERTI =       .11663E+00    .32288E+00     .32288E+00
X IN TPERTI =       .11393E+00    .32779E+00     .30967E+00
   CONV IN TPERTI AFTER     4 ITERATIONS
X IN TPERTI =       .11940E+00    .31767E+00     .31767E+00
X IN TPERTI =       .13566E+00    .34000E+00     .30712E+00
   CONV IN TPERTI AFTER     5 ITERATIONS
X IN TPERTI =       .11940E+00    .31767E+00     .31767E+00
X IN TPERTI =       .11970E+00    .32056E+00     .32251E+00
NO CONV IN TPERTI AFTER     5 ITERATIONS
X IN TPERTI =       .11663E+00    .32288E+00     .32288E+00
X IN TPERTI =       .92734E-01    .30901E+00     .32597E+00
NO CONV IN TPERTI AFTER     5 ITERATIONS
X IN TPERTI =       .11663E+00    .32288E+00     .32288E+00
X IN TPERTI =       .91161E-01    .32582E+00     .33242E+00
NO CONV IN TPERTI AFTER     5 ITERATIONS
X IN TPERTI =       .11940E+00    .31767E+00     .31767E+00
X IN TPERTI =       .97972E-01    .31889E+00     .31657E+00
NO CONV IN TPERTI AFTER     5 ITERATIONS
X IN TPERTI =       .11940E+00    .31767E+00     .31767E+00
X IN TPERTI =       .99073E-01    .33200E+00     .32753E+00
NO CONV IN TPERTI AFTER     5 ITERATIONS
X IN TPERTI =       .11011E+00    .28682E+00     .28682E+00
X IN TPERTI =       .13053E+00    .26985E+00     .28936E+00
   CONV IN TPERTI AFTER     5 ITERATIONS
X IN TPERTI =       .11011E+00    .28682E+00     .28682E+00
X IN TPERTI =       .14931E+00    .27669E+00     .27311E+00
NO CONV IN TPERTI AFTER     5 ITERATIONS
```

```
X IN TPERTI =        .11011E+00    .28682E+00    .28682E+00
X IN TPERTI =        .12371E+00    .30616E+00    .28231E+00
   CONV IN TPERTI AFTER    5 ITERATIONS
X IN TPERTI =        .11011E+00    .28682E+00    .28682E+00
X IN TPERTI =        .12245E+00    .29798E+00    .29099E+00
   CCNV IN TPERTI AFTER    5 ITERATIONS
X IN TPERTI =        .11477E+00    .32800E+00    .32800E+00
X IN TPERTI =        .13598E+00    .30796E+00    .31033E+00
NO CONV IN TPERTI AFTER    5 ITERATIONS
X IN TPERTI =        .11477E+00    .32800E+00    .32800E+00
X IN TPERTI =        .13271E+00    .32043E+00    .30891E+00
NO CONV IN TPERTI AFTER    5 ITERATIONS
X IN TPERTI =        .15024E+00    .78131E+00    .78131E+00
X IN TPERTI =        .16235E+00    .87029E+00    .76832E+00
   CCNV IN TPERTI AFTER    5 ITERATIONS
X IN TPERTI =        .15024E+00    .78131E+00    .78131E+00
X IN TPERTI =        .20852E+00    .81066E+00    .77560E+00
NO CONV IN TPERTI AFTER    5 ITERATIONS
X IN TPERTI =        .12486E+00    .46587E+00    .46587E+00
X IN TPERTI =        .10633E+00    .48175E+00    .45824E+00
NO CONV IN TPERTI AFTER    5 ITERATIONS
X IN TPERTI =        .12486E+00    .46587E+00    .46587E+00
X IN TPERTI =        .94376E-01    .45080E+00    .44812E+00
NO CONV IN TPERTI AFTER    5 ITERATIONS
X IN TPERTI =        .12486E+00    .46587E+00    .46587E+00
X IN TPERTI =        .11836E+00    .44122E+00    .48589E+00
NO CONV IN TPERTI AFTER    5 ITERATIONS
X IN TPERTI =        .12486E+00    .46587E+00    .46587E+00
X IN TPERTI =        .11449E+00    .43675E+00    .45401E+00
NO CONV IN TPERTI AFTER    5 ITERATIONS
X IN TPERTI =        .15024E+00    .78131E+00    .78131E+00
X IN TPERTI =        .19862E+00    .80500E+00    .75692E+00
NO CONV IN TPERTI AFTER    5 ITERATIONS
X IN TPERTI =        .15024E+00    .78131E+00    .78131E+00
X IN TPERTI =        .30437E+00    .80052E+00    .74147E+00
NO CONV IN TPERTI AFTER    5 ITERATIONS
MPERTI OK
SURFACE GEOMETRY :   2.0330 0.0000 1.4376
IMAX = 12      BNORM =     .2097E-02
ANORM =     .7736E-01

IMAX =  3      BNORM =     .1795E-02
ANORM =     .7758E-01

IMAX =  2      BNORM =     .8832E-04
ANORM =     .7759E-01

E=     1.8837 THETA=     0.0000 FI=     0.0000

   PQ1      PQ2      REF INT      SYM
  0.000    0.000    .31923E-02     1
  1.000    0.000    .63228E-02     6
 -1.000    1.000    .64306E-02     1
  0.000    1.000    .30378E-02     6
  1.000   -1.000    .34254E-02     1
  -.500     .500    .45687E-03     1
   .500    -.500    .15910E-02     1
   .500     .500    .13155E-02     6
SURFACE GEOMETRY :      .5080 0.0000-2.8800
NO. OF LATT. PTS. FOR GH :    620
NO. OF LATT. PTS. FOR GH :    620
NO. OF LATT. PTS. FOR GH :    622
NO. OF LATT. PTS. FOR GH :    622
X IN TPERTI =        .19742E+00    .19742E+00    .92227E-01
X IN TPERTI =        .19888E+00    .18324E+00    .11057E+00
NO CONV IN TPERTI AFTER    5 ITERATIONS
X IN TPERTI =        .19742E+00    .19742E+00    .92227E-01
```

```
X IN TPERTI =      .20240E+00    .20685E+00    .82584E-01
NO CONV IN TPERTI AFTER     5 ITERATIONS
X IN TPERTI =      .32288E+00    .32288E+00    .11663E+00
X IN TPERTI =      .33764E+00    .32921E+00    .96074E-01
   CONV IN TPERTI AFTER     5 ITERATIONS
X IN TPERTI =      .32288E+00    .32288E+00    .11663E+00
X IN TPERTI =      .30802E+00    .31179E+00    .11173E+00
   CONV IN TPERTI AFTER     5 ITERATIONS
```

10.6 Main Program for a Clean fcc or bcc (100) Surface, Using RFS

The clean surfaces of fcc and bcc (100) are simple in that no reconstruction of the surface is plausible unless the surface unit cell is strongly enlarged [such reconstructions do occur, e.g. on Pt(100), and W(100)]: therefore, basically, only a relaxation of the topmost interlayer spacing need be searched for when no superstructure is observed. In order to allow every incidence direction, no symmetry is exploited here, although the surface structure has high symmetry (this symmetry is used in Sect. 10.7). So the interlayer vectors should be chosen to connect atom centers in successive layers, no registry shifts being needed. All interlayer vectors may be taken equal, see Fig. 10.4, except the topmost one whose x-component should vary.

RFS can be used for efficiency and will work also if the atoms under consideration are stronger scatterers than nickel, as in the case of tungsten (bcc), for example.

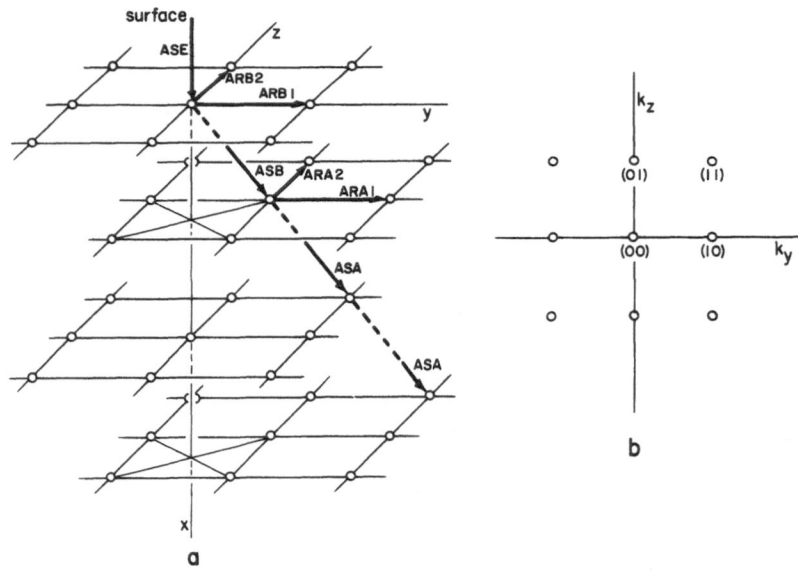

Fig. 10.4 (a) Geometry of a fcc or bcc(100) surface, no symmetry being considered. (b) Reciprocal space

```
C    MAIN + DATA FOR NI(100), RFS, NO SYMMETRY.
C    MAIN LEED PROGRAM FOR A CLEAN FCC OR BCC (100) SURFACE, USING NO
C    SYMMETRY, BY RENORMALIZED FORWARD SCATTERING. THE TOPMOST INTERLAYER
C    SPACING IS VARIABLE
      DIMENSION ARA1(2),ARA2(2),RAR1(2),RAR2(2),SS1(2),SS2(2),SS3(2),
     1SS4(2),ASA(3),AS(3),ARB1(2),ARB2(2),RBR1(2),RBR2(2),SO1(2),
     2SO2(2),SO3(2),ASB(3),ASV(3)
C    1=NL, 120=KLM
      COMPLEX VL(1,2),FLMS(1,120),FLM(120)
C    1(1ST DIMENSION)=NL=NL1*NL2, 4(2ND DIMENSION)=IDEG
      DIMENSION V(1,2),JJS(1,4)
C    1=KNBS=NO. OF BEAM SETS READ IN, 57=KNT=NO. OF BEAMS READ IN
      DIMENSION KNB(1),N8(1),SPQF(2,57),SPQ(2,57),PQF(2,57),PQ(2,57)
      INTEGER KSYM(2,57),SYM(2,57)
C    20 MUST BE .GE. NPUN
      DIMENSION NPU(20),NPUC(20)
C    12=NPSI, 16=NEL*(LMAX+1)
      DIMENSION ES(12),PHSS(12,16)
C    7680=NLM FROM DATA STATEMENT, 225=NN, 15=N (FOR CELMG)
C    YLM IS ALSO USED, AS COMPLEX VECTOR OF SMALLER CORE SIZE, IN MSMF,ETC
      DIMENSION NLMS(7),CLM(7680),YLM(225),FAC2(225),FAC1(15)
C    15=NN1, 8=NN2, 8=NN3
      DIMENSION PPP(15,8,8)
C    64=LMMAX
      DIMENSION LX(64),LXI(64),LXM(64),LT(64)
C    8=LMAX+1=L1, 3 AND 10 FIXED FOR COMMON /MPT/ WHEN PRESENT
      COMPLEX AF(8),CAF(8),TSFO(3,10),TSF(3,10)
      COMPLEX RA1(57,57),TA1(57,57)
C    36=LEV, 28=LOD
      COMPLEX AMULT(57),CYLM(57,64),XEV(36,36),XOD(28,28),
     1YLME(36),YLMO(28)
      DIMENSION IPLE(36),IPLO(28),AT(57)
C    8 FIXED (FOR RFSO2), 20=ND (FOR RFS SUBROUTINES)
      COMPLEX XI(57),PK(57,8),AW(57,2),ANEW(57,20)
      COMMON E3,AK21,AK31,VPI1,TV,EMACH
      COMMON /X4/E,VPI,AK2,AK3,ASA
      COMMON /SL/ARA1, ARA2, ARB1, ARB2, RBR1, RBR2, NL1, NL2
      COMMON /MS/LMAX,EPS,LITER,SO1,SO2,SO3,SS1,SS2,SS3,SS4
      COMMON /ADS/ASL, FR, ASE, VPIS, VPIO, VO, VV
      COMMON /BT/IT, T,TO,DRPER,DRPAR,DRO
      COMMON /TEMP/IT1,IT2,IT3,TI,TF,DT,TO1,DRPER1,DRPAR1,DRO1,DRPER2,
     1DRPAR2,DRO2,DRPER3,DRPAR3,DRO3
C    NLMS IS DIMENSION OF CLM AS FUNCTION OF LMAX
      DATA NLMS(1),NLMS(2),NLMS(3),NLMS(4),NLMS(5),NLMS(6),NLMS(7)/
     110,76,284,809,1925,4032,7680/
  161 FORMAT(3F7.2)
  200 FORMAT(20I3)
  240 FORMAT(8HOIDEG = ,1I3,7H NL1 = ,1I3,7H NL2 = ,1I3)
  370 FORMAT(9H1ENERGY =,F7.4,7H H  OR ,1F7.2,3H EV)
  380 FORMAT(8H VPIS = ,F9.4,8H VPIO = ,F9.4,9H DCUTS = ,F9.4,9H DCUTO =
     1 ,F9.4)
  390 FORMAT(8HOTEMP = ,F9.4)
  410 FORMAT(10H MSMF   OK)
  420 FORMAT(10H SUBREF OK)
  430 FORMAT(20HOSURFACE GEOMETRY : ,3F7.4)
  450 FORMAT(20H CORRECT TERMINATION)
  460 FORMAT(20A4)
  465 FORMAT(1H1,20A4)
C
C
C    READ, WRITE AND PUNCH A DESCRIPTIVE TITLE
      READ (5,460)(CLM(I),I=1,20)
      WRITE(6,465)(CLM(I),I=1,20)
      WRITE(7,460)(CLM(I),I=1,20)
C    EMACH IS MACHINE ACCURACY (USED BY ZGE AND ZSU)
      EMACH=1.0E-6
C    IDEG: EACH LAYER HAS AN IDEG-FOLD SYMMETRY AXIS
```

```
C   NL1, NL2: SUPERLATTICE CHARACTERIZATION  NL1   NL2
C                       P(1*1)                 1     1
C                       C(2*2)                 2     1
C                       P(2*1)                 2     1
C                       P(1*2)                 1     2
C                       P(2*2)                 2     2
      READ(5,200)IDEG,NL1,NL2
      WRITE(6,240)IDEG,NL1,NL2
      NL=NL1*NL2
C   KNBS= NO.OF BEAM SETS TO BE READ IN (.LE.NL)
C   KNT= TOTAL NO. OF BEAMS TO BE READ IN
C   NPUN= NO. OF BEAMS FOR WHICH INTENSITIES ARE TO BE PUNCHED OUT
      READ(5,200)KNBS,KNT,NPUN
C   NPSI= NO.OF ENERGIES AT WHICH PHASE SHIFTS WILL BE READ IN
      READ(5,200)NPSI
C   READ IN GEOMETRY, PHYSICAL PARAMETERS, CONVERGENCE CRITERIA
      CALL READIN(TVA,RAR1,RAR2,ASA,TVB,ASB,STEP,NSTEP,IDEG,NL,V,VL,
     1JJS,KNBS,KNB,KNT,SPQF,KSYM,SPQ,TST,TSTS,NPUN,NPU,THETA,FI,
     2LMMAX,NPSI,ES,PHSS,L1)
      TO=TO1
      KLM=(2*LMAX+1)*(2*LMAX+2)/2
      LEV=(LMAX+1)*(LMAX+2)/2
      LOD=LMMAX-LEV
  630 N=2*LMAX+1
      NN=N*N
      NLM=NLMS(LMAX)
C   CLM= CLEBSCH-GORDON COEFFICIENTS FOR MATRICES X AND TAU
      CALL CELMG(CLM,NLM,YLM,FAC2,NN,FAC1,N,LMAX)
C   LX,LXI,LT,LXM: PERMUTATIONS OF (L,M)-SEQUENCE
      CALL LXGENT(LX,LXI,LT,LXM,LMAX,LMMAX)
      IF (IT1+IT2+IT3) 640,650,640
  640 NN3=LMAX+1
      NN2=LMAX+1
      NN1=NN2+NN3-1
C   PPP= CLEBSCH-GORDON COEFFICIENTS FOR COMPUTATION OF TEMPERATURE-
C   DEPENDENT PHASE SHIFTS (SKIPPED IF NOT NEEDED)
      CALL  CPPP (PPP, NN1, NN2, NN3)
  650 CONTINUE
C
C
C   READ ENERGY RANGE AND STEP; IF EF AND DE ARE BLANK, PROGRAM COMPUTES
C   FOR ENERGY EI. PROGRAM ALWAYS RETURNS TO THIS LINE TO READ A NEW
C   ENERGY RANGE AND STEP. IF A NEGATIVE EI IS READ, RUN IS TERMINATED
  680 READ (5,161)  EI, EF, DE
      IF (EI)  1000, 690, 690
  690 EI=EI/27.18+VV
      EF=EF/27.18+VV
      DE=DE/27.18
      E = EI
      EC = E
      E3 = E
C
C
C   START LOOP OVER ENERGIES IN GIVEN ENERGY RANGE
  700 EEV=(E-VV)*27.18
      WRITE (6,370)  E,EEV
C   COMPUTE COMPONENTS OF INCIDENT WAVEVECTOR PARALLEL TO SURFACE
      AK=SQRT(2.0*(E-VV))*SIN(THETA)
      AK2=AK*COS(FI)
      AK3=AK*SIN(FI)
      AK21=AK2
      AK31=AK3
C   SELECT BEAMS APPROPRIATE FOR CURRENT ENERGY
      CALL BEAMS(KNBS,KNB,SPQ,SPQF,KSYM,KNT,NPU,NPUN,AK2,AK3,E,TST,
     1NB,PQ,PQF,SYM,NPUC,MPU,NT,NP)
C   SET OPTICAL POTENTIAL (IMAGINARY PART OF MUFFIN-TIN CONSTANT,
C   REPRESENTING DAMPING). VPIS FOR SUBSTRATE, VPIO FOR OVERLAYER.
```

```
C   HERE VPIS=VPIO PROPORTIONAL TO E**1/3
        VPIS=-(3.8/27.18)*EXP((ALOG((E)/(90.0/27.18+VV)))*0.333333)
        VPIO=VPIS
        DCUTO = SQRT(2.0 * E)
C   SET LIMITING RADII ON LATTICE SUMS, POSSIBLY DIFFERENT FOR SUBSTRATE
C   (DCUTS) AND OVERLAYER (DCUTO)
        DCUTS = - 5.0 * DCUTO/(AMIN1(VPIS, - 0.05))
        DCUTO = - 5.0 * DCUTO/(AMIN1(VPIO, - 0.05))
        WRITE (6,380)  VPIS, VPIO, DCUTS, DCUTO
C   FIX TEMPERATURE T. T MAY BE INCREASED STEPWISE IN A LOOP OVER THE
C   INPUT RANGE (TI,TF) WITH INPUT STEP DT
  770 T = TI
  780 WRITE (6,390)  T
C
C
C   CONSIDER ONE ATOMIC LAYER
        E = EC
        E3 = E
C   PRODUCE ATOMIC T-MATRIX ELEMENTS
        CALL TSCATF(1,L1,ES,PHSS,NPSI,IT1,E,0.,PPP,NN1,NN2,NN3,DRO1,
       1DRPER1,DRPAR1,TO,T,TSFO,TSF,AF,CAF)
        DRPER=DRPER1
        DRPAR=DRPAR1
        DRO=DRO1
        IT=0
        TV = TVA
        VPI = VPIS
        VPI1 = VPIS
C   PERFORM PLANAR LATTICE SUMS
        CALL  FMAT (FLMS, V, JJS, NL, NL, DCUTS, IDEG, LMAX,KLM)
C   PRODUCE REFLECTION (RA1) AND TRANSMISSION (TA1) MATRICES FOR
C   SUBSTRATE LAYERS. ONE REGISTRY: ID=1. SUBSTRATE LAYER SYMMETRIES AND
C   REGISTRIES APPLY: LAY=2
        CALL MSMF(RA1,TA1,RA1,TA1,RA1,TA1,RA1,TA1,NT,NT,NT,AMULT,CYLM,
       1PQ,SYM,NT,FLMS,FLM,V,NL,  0,  0,1,NL,CAF,CAF,L1,LX,LXI,LMMAX,KLM,
       2XEV,XOD,LEV,LOD,YLM,YLME,YLMO,IPLE,IPLO,CLM,NLM,2)
        WRITE (6,410)
C
C
C   NOW STACK LAYERS.
C   ASV BECOMES THE 3-D VECTOR FROM TOP LAYER TO SECOND LAYER
C   AND WILL BE VARIED
        DO 937 J=1,3
  937 ASV(J)=ASB(J)
C   LOOP 960 RUNS OVER GEOMETRIES: HERE THE TOP LAYER SPACING ASV(1) IS
C   VARIED
        DO 960 I = 1, NSTEP
        DO 940 J=1,3
  940 AS(J)=ASV(J)*0.529
        A=AS(1)
        WRITE (6,430) (AS(J),J=1,3)
C   RFS IS USED HERE TO PRODUCE REFLECTED BEAM AMPLITUDES XI
        CALL RFS02(RA1,TA1,RA1,TA1,RA1,TA1,RA1,TA1,NT,NB,KNBS,NP,XI,PQ,
       1PK,AW,ANEW,20,ASV,ASA,ASA)
C   COMPUTE REFLECTED INTENSITIES (AT) FROM AMPLITUDES (XI), PRINT AND
C   PUNCH (NPNCH=1; PUNCH LAYER SPACING A ALSO FOR LATER IDENTIFICATION)
        CALL RINT(NT,XI,AT,PQ,PQF,SYM,VV,THETA,FI,MPU,NPUC,EEV,A,1)
  960 ASV(1)=ASV(1)+STEP
        IF (IT1+IT2+IT3) 980,990,980
  980 T = T + DT
        IF (DT.LE.0.0001) GO TO 990
        IF (T-TF) 780, 780, 990
  990 E = E + DE
        EC = E
        E3 = E
        IF (E-EF)  700, 700, 680
 1000 WRITE (6,450)
```

```
      STOP
      END
------------------------------------------------------------
---- INPUT DATA FOR PRECEDING PROGRAM FOLLOW ------------
------------------------------------------------------------
 NI(100), RFS, NO SYMMETRY
   4  1  1               IDEG NL1 NL2
   1 57 13               KNBS KNT NPUN
  12                     NPSI
  2.4900 0.0000          ARA1
  0.0000 2.4900          ARA2
  0.0000 0.0000          SS1
  0.0000 0.0000          SS2
  0.0000 0.0000          SS3
  0.0000 0.0000          SS4
  1.7650 1.2450 1.2450   ASA
  2.4900 0.0000          ARB1
  0.0000 2.4900          ARB2
  0.0000 0.0000          SO1
  0.0000 0.0000          SO2
  0.0000 0.0000          SO3
  1.6150 1.2450 1.2450   ASB
  0.5000 1.2450 0.0500   FR ASE STEP
   5                     NSTEP
  57                     KNB(1)
  0.0000 0.0000   1  1   BEAMS
  1.0000 0.0000   1  1
 -1.0000 0.0000   1  1
  0.0000 1.0000   1  1
  0.0000-1.0000   1  1
  1.0000 1.0000   1  1
 -1.0000 1.0000   1  1
 -1.0000-1.0000   1  1
  1.0000-1.0000   1  1
  2.0000 0.0000   1  1
 -2.0000 0.0000   1  1
  0.0000 2.0000   1  1
  0.0000-2.0000   1  1
  2.0000 1.0000   1  1
 -2.0000 1.0000   1  1
 -2.0000-1.0000   1  1
  2.0000-1.0000   1  1
  1.0000 2.0000   1  1
 -1.0000 2.0000   1  1
 -1.0000-2.0000   1  1
  1.0000-2.0000   1  1
  2.0000 2.0000   1  1
 -2.0000 2.0000   1  1
 -2.0000-2.0000   1  1
  2.0000-2.0000   1  1
  3.0000 0.0000   1  1
 -3.0000 0.0000   1  1
  0.0000 3.0000   1  1
  0.0000-3.0000   1  1
  3.0000 1.0000   1  1
 -3.0000 1.0000   1  1
 -3.0000-1.0000   1  1
  3.0000-1.0000   1  1
  1.0000 3.0000   1  1
 -1.0000 3.0000   1  1
 -1.0000-3.0000   1  1
  1.0000-3.0000   1  1
  3.0000 2.0000   1  1
 -3.0000 2.0000   1  1
 -3.0000-2.0000   1  1
  3.0000-2.0000   1  1
  2.0000 3.0000   1  1
```

```
-2.0000 3.0000   1  1
-2.0000-3.0000   1  1
 2.0000-3.0000   1  1
 4.0000 0.0000   1  1
-4.0000 0.0000   1  1
 0.0000 4.0000   1  1
 0.0000-4.0000   1  1
 4.0000 1.0000   1  1
-4.0000 1.0000   1  1
-4.0000-1.0000   1  1
 4.0000-1.0000   1  1
 1.0000 4.0000   1  1
-1.0000 4.0000   1  1
-1.0000-4.0000   1  1
 1.0000-4.0000   1  1
 0.0020                 TST
   1  2  3  4  5  6  7  8  9 10 11 12 13     NPU
    0.00    0.00        THETA FI
    0.00   11.20        VO VV
 0.0010                 EPS
    5                   LITER
    1  0  0             IT1 IT2 IT3
 440.0000    58.7100    1.4000   1.4000   0.0000    THDB1 AM1 FPER1 FPAR1 DRO1
   1.0000     1.0000    1.0000   1.0000   0.0000    THDB2 AM2 FPER2 FPAR2 DRO2
   1.0000     1.0000    1.0000   1.0000   0.0000    THDB3 AM3 FPER3 FPAR3 DRO3
 300.0000                                           TI TF DT
    7                   LMAX
    1                   NEL
  .3000
 2.8398 -.0137 -.1876   .0011   .0000   .0000   .0000   .0000   NI
  .4000
 2.7639 -.0216 -.3164   .0024   .0001   .0000   .0000   .0000   NI
  .5000
 2.6930 -.0322 -.2630   .0048   .0002   .0000   .0000   .0000   NI
  .7500
 2.5276 -.0668 -.2322   .0160   .0009   .0000   .0000   .0000   NI
 1.0000
 2.3803 -.1132 -.2323   .0386   .0028   .0002   .0000   .0000   NI
 1.5000
 2.1240 -.2247 -.2118   .1202   .0125   .0012   .0001   .0000   NI
 2.0000
 1.9101 -.3448 -.1758   .2494   .0335   .0042   .0004   .0000   NI
 3.0000
 1.5727 -.5697 -.1288   .5535   .1138   .0216   .0034   .0004   NI
 4.0000
 1.3160 -.7574 -.1272   .7884   .2257   .0583   .0125   .0022   NI
 5.0000
 1.1099 -.9122 -.1445   .9491   .3361   .1107   .0305   .0069   NI
 6.0000
  .9366-1.0434 -.1666  1.0690   .4287   .1691   .0571   .0159   NI
 7.0000
  .7862-1.1581 -.1902  1.1630   .5045   .2251   .0894   .0297   NI
  40.00 201.00    5.00     EI EF DE
 -10.00                     EI EF DE
------------------------------------------------------------
---- FIRST PART OF PRINTER OUTPUT FOLLOWS ---------------
----(NEGLECTING MOST BLANK LINES AND PAGE --------------
---- EJECTS AND TRUNCATING LONG LINES)     --------------
------------------------------------------------------------
NI(100), RFS, NO SYMMETRY
IDEG =   4 NL1 =   1 NL2 =    1
PARAMETERS FOR INTERIOR:
SURF VECS        2.4900   0.0000
                 0.0000   2.4900
SS    0.0000 0.0000
SS    0.0000 0.0000
SS    0.0000 0.0000
```

```
SS      0.0000 0.0000
ASA     1.7650 1.2450 1.2450
PARAMETERS FOR SURFACE:
SURF VECS        2.4900  0.0000
                 0.0000  2.4900
SO      0.0000 0.0000
SO      0.0000 0.0000
SO      0.0000 0.0000
ASB     1.6150 1.2450 1.2450
FR =    .5000 ASE =  1.2450 STEP =    .0500 NSTEP =    5
           PQ1      PQ2        SYM

     1      0.000    0.000      1  1
     2      1.000    0.000      1  1
     3     -1.000    0.000      1  1
     4      0.000    1.000      1  1
     5      0.000   -1.000      1  1
     6      1.000    1.000      1  1
     7     -1.000    1.000      1  1
     8     -1.000   -1.000      1  1
     9      1.000   -1.000      1  1
    10      2.000    0.000      1  1
    11     -2.000    0.000      1  1
    12      0.000    2.000      1  1
    13      0.000   -2.000      1  1
    14      2.000    1.000      1  1
    15     -2.000    1.000      1  1
    16     -2.000   -1.000      1  1
    17      2.000   -1.000      1  1
    18      1.000    2.000      1  1
    19     -1.000    2.000      1  1
    20     -1.000   -2.000      1  1
    21      1.000   -2.000      1  1
    22      2.000    2.000      1  1
    23     -2.000    2.000      1  1
    24     -2.000   -2.000      1  1
    25      2.000   -2.000      1  1
    26      3.000    0.000      1  1
    27     -3.000    0.000      1  1
    28      0.000    3.000      1  1
    29      0.000   -3.000      1  1
    30      3.000    1.000      1  1
    31     -3.000    1.000      1  1
    32     -3.000   -1.000      1  1
    33      3.000   -1.000      1  1
    34      1.000    3.000      1  1
    35     -1.000    3.000      1  1
    36     -1.000   -3.000      1  1
    37      1.000   -3.000      1  1
    38      3.000    2.000      1  1
    39     -3.000    2.000      1  1
    40     -3.000   -2.000      1  1
    41      3.000   -2.000      1  1
    42      2.000    3.000      1  1
    43     -2.000    3.000      1  1
    44     -2.000   -3.000      1  1
    45      2.000   -3.000      1  1
    46      4.000    0.000      1  1
    47     -4.000    0.000      1  1
    48      0.000    4.000      1  1
    49      0.000   -4.000      1  1
    50      4.000    1.000      1  1
    51     -4.000    1.000      1  1
    52     -4.000   -1.000      1  1
    53      4.000   -1.000      1  1
    54      1.000    4.000      1  1
    55     -1.000    4.000      1  1
```

```
    56    -1.000 -4.000      1  1
    57     1.000 -4.000      1  1
TST =   .0020
PUNCHED BEAMS     1   2   3   4   5   6   7   8   9  10  11  12  13
THETA  FI =    0.00    0.00
VO.       0.00VV        11.20
EPS =   .0010 LITER =    5
IT1 =   1 IT2 =    0 IT3 =    0
THDB =  440.0000 AM =   58.7100 FPER =    1.4000 FPAR =    1.4000 DRO =
THDB =    1.0000 AM =    1.0000 FPER =    1.0000 FPAR =    1.0000 DRO =
THDB =    1.0000 AM =    1.0000 FPER =    1.0000 FPAR =    1.0000 DRO =
LMAX =    7
        PHASE SHIFTS

E =    .3000  1ST ELEMENT     2.8398   -.0137   -.1876    .0011  0.0000  0.0000

E =    .4000  1ST ELEMENT     2.7639   -.0216   -.3164    .0024   .0001  0.0000

E =    .5000  1ST ELEMENT     2.6930   -.0322   -.2630    .0048   .0002  0.0000

E =    .7500  1ST ELEMENT     2.5276   -.0668   -.2322    .0160   .0009  0.0000

E =   1.0000  1ST ELEMENT     2.3803   -.1132   -.2323    .0386   .0028   .0002

E =   1.5000  1ST ELEMENT     2.1240   -.2247   -.2118    .1202   .0125   .0012

E =   2.0000  1ST ELEMENT     1.9101   -.3448   -.1758    .2494   .0335   .0042

E =   3.0000  1ST ELEMENT     1.5727   -.5697   -.1288    .5535   .1138   .0216

E =   4.0000  1ST ELEMENT     1.3160   -.7574   -.1272    .7884   .2257   .0583

E =   5.0000  1ST ELEMENT     1.1099   -.9122   -.1445    .9491   .3361   .1107

E =   6.0000  1ST ELEMENT      .9366  -1.0434   -.1666   1.0690   .4287   .1691

E =   7.0000  1ST ELEMENT      .7862  -1.1581   -.1902   1.1630   .5045   .2251
ENERGY = 1.8837 H   OR    40.00 EV
  13 BEAMS USED :  13
   0.000 0.000    1.000 0.000   -1.000 0.000    0.000 1.000    0.000-1.000
   2.000 0.000   -2.000 0.000    0.000 2.000    0.000-2.000
VPIS =   -.1114 VPIO =    -.1114 DCUTS =    87.1160 DCUTO =    87.1160
TEMP =  300.0000
MSMF    OK
SURFACE GEOMETRY :  1.6150 1.2450 1.2450
IMAX = 10     BNORM =    .1701E-02
ANORM =    .6593E-01

IMAX =  9     BNORM =    .2628E-02
ANORM =    .6885E-01

IMAX =  2     BNORM =    .2834E-02
ANORM =    .6888E-01

E=    1.8837 THETA=    0.0000 FI=    0.0000

   PQ1    PQ2     REF INT     SYM
 0.000  0.000   .57036E-02    1
 1.000  0.000   .10155E-02    1
-1.000  0.000   .10155E-02    1
 0.000  1.000   .10155E-02    1
 0.000 -1.000   .10155E-02    1
SURFACE GEOMETRY :  1.6650 1.2450 1.2450
IMAX = 10     BNORM =    .1645E-02
ANORM =    .6976E-01

IMAX =  9     BNORM =    .2614E-02
```

```
ANORM =    .7266E-01

IMAX = 2    BNORM =    .2613E-02
ANORM =    .7269E-01

E=    1.8837 THETA=    0.0000 FI=    0.0000

   PQ1    PQ2    REF INT    SYM
 0.000  0.000  .86486E-02    1
 1.000  0.000  .98162E-03    1
-1.000  0.000  .98162E-03    1
 0.000  1.000  .98162E-03    1
 0.000 -1.000  .98162E-03    1
SURFACE GEOMETRY :   1.7150 1.2450 1.2450
IMAX = 10    BNORM =    .1592E-02
ANORM =    .7413E-01

IMAX = 9    BNORM =    .2581E-02
ANORM =    .7698E-01

IMAX = 2    BNORM =    .2404E-02
ANORM =    .7701E-01

E=    1.8837 THETA=    0.0000 FI=    0.0000

   PQ1    PQ2    REF INT    SYM
 0.000  0.000  .10874E-01    1
 1.000  0.000  .91350E-03    1
-1.000  0.000  .91350E-03    1
 0.000  1.000  .91350E-03    1
 0.000 -1.000  .91350E-03    1
```

10.7 Main Program for a p(2×2) Overlayer on an fcc or bcc (100) Substrate,
Using Layer Doubling

We consider an fcc or bcc (100) substrate (treated in Sect. 10.6) with a
p(2×2) overlayer: see Fig. 10.5. The fourfold increase in the area of the
surface unit cell multiplies the number of beams by four, by generating
three extra beam sets. As a result, matrix dimensions become large. If the
adsorption site has some symmetry, that symmetry can be exploited. If we
limit ourselves to hollow and top adsorption sites and to normal incidence,
the full symmetry of the clean surface is maintained: a fourfold rotation
axis and four mirror planes (a bridge site would leave two orthogonal mirror
planes, and two 90°-rotated overlayer domains would have to be considered).

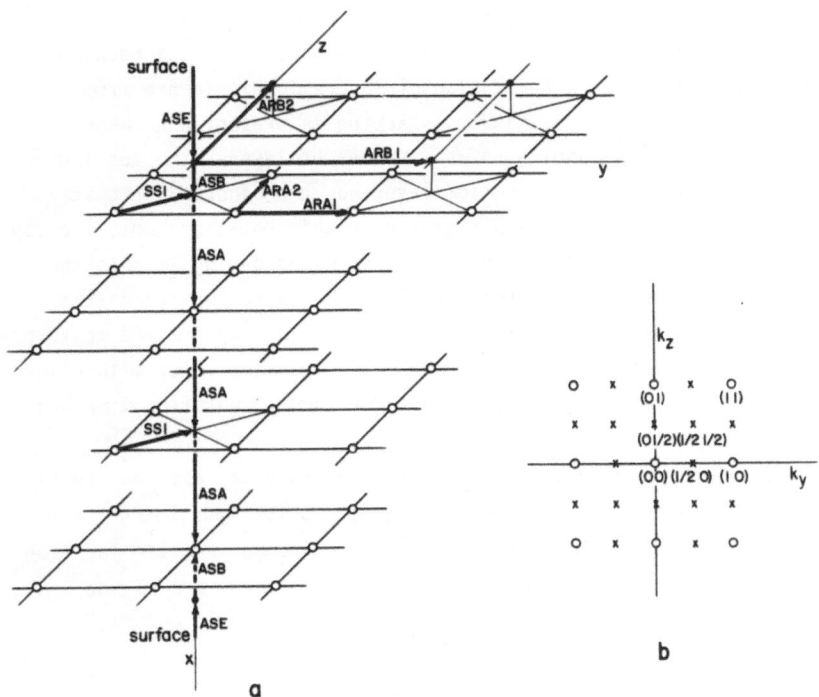

Fig. 10.5 (a) Geometry of a p(2×2) overlayer in hollow site (top of figure)
and top site (bottom of figure) on an fcc or bcc(100) surface. The 2nd, 4th,
etc. substrate layers have a registry SS2 = (0,0). (b) Reciprocal space
(o = integral-order beams, x = fractional-order beams)

This symmetry gives rise to groups of 8, of 4, or of just 1 symmetrical beams. Furthermore, two of the extra beam sets are mutually symmetrical: they are therefore symmetrized together as explained in Sect. 4.2, which gives rise to different symmetry code numbers in that combined beam set for the overlayer and for substrate layers. (Of course, in the absence of symmetry, all beams in all beam sets would have to be input, including the fourth beam set which can be omitted here.) Note the reduction in matrix dimensions compared with Sect. 10.6.

A symmetry axis is chosen that passes through atom centers in the 2nd, 4th, 6th, etc., substrate layers and through an overlayer atom center if it is positioned at the hollow site. In the substrate two layer registries are needed [SS1 = (1/2, 1/2) and SS2 = (0,0)], while only one is needed in the overlayer [SO1 = (0,0)](see below for obtaining the top site).

Layer Doubling is used, because in the system chosen here, Ni(100) + p(2×2)0, the overlayer can easily come too close to the substrate for the faster RFS method to remain convergent. Layer Doubling for the substrate produces reflection and transmission matrices (split into smaller matrices because of block-diagonalization) for a thick slab of layers for incidence from either side of the slab. Since ...ABAB... stacking is present and Layer Doubling produces a slab containing an *even* number of layers, the top layer has, say, the A registry while the bottom layer has the other (B) registry. By putting the overlayer (which has B registry in this notation) onto the top layer one obtains the hollow adsorption site, and by adsorbing it onto the bottom layer one obtains the top site: see Fig. 10.5 (a reflection matrix does not depend on the direction of the x-axis and so we may imagine that we flip the substrate slab upside down, thereby exposing a top layer with B instead of A registry). This simple procedure saves much computation time and storage.

Here the substrate and overlayer muffin-tin constants are assumed identical (V0 = 0.0): to change that, all that is needed is to input the desired non-zero difference V0 and choose, through the input number FR, the location of the change in the muffin-tin constant. The program takes care of the rest.

```
C    MAIN + DATA FOR NI(100)+P(2*2)O, LAYER DOUBLING, FULL SYMMETRY.
C    MAIN LEED PROGRAM FOR A P(2*2) OVERLAYER ON FCC OR BCC (100), USING
C    FULL SYMMETRY OF NORMAL INCIDENCE (4-FOLD AXIS OF ROTATION AND 4
C    MIRROR PLANES), BY LAYER DOUBLING.
     DIMENSION ARA1(2),ARA2(2),RAR1(2),RAR2(2),SS1(2),SS2(2),SS3(2),
    1SS4(2),ASA(3),AS(3),ARB1(2),ARB2(2),RBR1(2),RBR2(2),SO1(2),
    2SO2(2),SO3(2),ASB(3),ASV(3)
C    4=NL, 120=KLM
     COMPLEX VL(4,2),FLMS(4,120),FLM(120)
C    4(1ST DIMENSION)=NL=NL1*NL2, 4(2ND DIMENSION)=IDEG
     DIMENSION V(4,2),JJS(4,4)
C    3=KNBS=NO. OF BEAM SETS READ IN, 39=KNT=NO. OF BEAMS READ IN
     DIMENSION KNB(3),NB(3),SPQF(2,39),SPQ(2,39),PQF(2,39),PQ(2,39)
     INTEGER KSYM(2,39),SYM(2,39)
C    20 MUST BE .GE. NPUN
     DIMENSION NPU(20),NPUC(20)
C    12=NPSI, 16=NEL*(LMAX+1)
     DIMENSION ES(12),PHSS(12,16)
C    7680=NLM FROM DATA STATEMENT, 225=NN, 15=N (FOR CELMG)
C    YLM IS ALSO USED, AS COMPLEX VECTOR OF SMALLER CORE SIZE, IN MSMF,ETC
     DIMENSION NLMS(7),CLM(7680),YLM(225),FAC2(225),FAC1(15)
C    15=NN1, 8=NN2, 8=NN3
     DIMENSION PPP(15,8,8)
C    64=LMMAX
     DIMENSION LX(64),LXI(64),LXM(64),LT(64)
C    8=LMAX+1=L1, 3 AND 10 FIXED FOR COMMON /MPT/
     COMPLEX AF(8),CAF(8),TSFO(3,10),TSF(3,10)
     COMPLEX ROV(39,39),TOV(39,39)
C    13=KNB(1)
     COMPLEX RA1(13,13),TA1(13,13),RA2(13,13),TA2(13,13)
C    9=KNB(2)
     COMPLEX RB1(9,9),TB1(9,9),RB2(9,9),TB2(9,9)
C    17=KNB(3)
     COMPLEX RC1(17,17),TC1(17,17),RC2(17,17),TC2(17,17)
C    17=KNP
     COMPLEX PP(2,17),XS(17),U(39),RS(39,39),S1(39,39),S2(17,17),
    1S3(17,17),S4(17,17)
     DIMENSION JNT(39)
C    36=LEV, 28=LOD
     COMPLEX AMULT(39),CYLM(39,64),XEV(36,36),XOD(28,28),
    1YLME(36),YLMO(28)
     DIMENSION IPLE(36),IPLO(28),AT(39)
     COMPLEX XI(39)
     COMMON E3,AK21,AK31,VPI1,TV,EMACH
     COMMON /X4/E,VPI,AK2,AK3,ASA
     COMMON /SL/ARA1, ARA2, ARB1, ARB2, RBR1, RBR2, NL1, NL2
     COMMON /MS/LMAX,EPS,LITER,SO1,SO2,SO3,SS1, SS2,SS3,SS4
     COMMON /ADS/ASL, FR, ASE, VPIS, VPIO, VO, VV
     COMMON /BT/IT,T,TO,DRPER,DRPAR,DRO
     COMMON /TEMP/IT1,IT2,IT3,TI,TF,DT,TO1,DRPER1,DRPAR1,DRO1,DRPER2,
    1DRPAR2,DRO2,DRPER3,DRPAR3,DRO3
C    NLMS IS DIMENSION OF CLM AS FUNCTION OF LMAX
     DATA NLMS(1),NLMS(2),NLMS(3),NLMS(4),NLMS(5),NLMS(6),NLMS(7)/
    10,76,284,809,1925,4032,7680/
 161 FORMAT(3F7.2)
 200 FORMAT(20I3)
 240 FORMAT(8HOIDEG = ,1I3,7H NL1 = ,1I3,7H NL2 = ,1I3)
 370 FORMAT(9H1ENERGY =,F7.4,7H H  OR ,1F7.2,3H EV)
 380 FORMAT(8H VPIS = ,F9.4,8H VPIO = ,F9.4,9H DCUTS = ,F9.4,9H DCUTO =
    1 ,F9.4)
 390 FORMAT(8HOTEMP = ,F9.4)
 410 FORMAT(10H MSMF   OK)
 420 FORMAT(10H SUBREF OK)
 430 FORMAT(20HOSURFACE GEOMETRY : ,3F7.4)
 450 FORMAT(20H CORRECT TERMINATION)
 460 FORMAT(20A4)
 465 FORMAT(1H1,20A4)
```

```
C
C
C    READ, WRITE AND PUNCH A DESCRIPTIVE TITLE
        READ (5,460)(CLM(I),I=1,20)
        WRITE(6,465)(CLM(I),I=1,20)
        WRITE(7,460)(CLM(I),I=1,20)
C    EMACH IS MACHINE ACCURACY (USED BY ZGE AND ZSU)
        EMACH=1.0E-6
C    IDEG: EACH LAYER HAS AN IDEG-FOLD SYMMETRY AXIS
C    NL1, NL2: SUPERLATTICE CHARACTERIZATION   NL1   NL2
C                         P(1*1)                1     1
C                         C(2*2)                2     1
C                         P(2*1)                2     1
C                         P(1*2)                1     2
C                         P(2*2)                2     2
        READ(5,200)IDEG,NL1,NL2
        WRITE(6,240)IDEG,NL1,NL2
        NL=NL1*NL2
C    KNBS= NO.OF BEAM SETS TO BE READ IN (.LE.NL)
C    KNT= TOTAL NO. OF BEAMS TO BE READ IN
C    NPUN= NO. OF BEAMS FOR WHICH INTENSITIES ARE TO BE PUNCHED OUT
        READ(5,200)KNBS,KNT,NPUN
C    NPSI= NO.OF ENERGIES AT WHICH PHASE SHIFTS WILL BE READ IN
        READ(5,200)NPSI
C    READ IN GEOMETRY, PHYSICAL PARAMETERS, CONVERGENCE CRITERIA
        CALL READIN(TVA,RAR1,RAR2,ASA,TVB,ASB,STEP,NSTEP,IDEG,NL,V,VL,
       1JJS,KNBS,KNB,KNT,SPQF,KSYM,SPQ,TST,TSTS,NPUN,NPU,THETA,FI,
       2LMMAX,NPSI,ES,PHSS,L1)
        TO=TO1
        KLM=(2*LMAX+1)*(2*LMAX+2)/2
        LEV=(LMAX+1)*(LMAX+2)/2
        LOD=LMMAX-LEV
630     N=2*LMAX+1
        NN=N*N
        NLM=NLMS(LMAX)
C    CLM= CLEBSCH-GORDON COEFFICIENTS FOR MATRICES X AND TAU
        CALL CELMG(CLM,NLM,YLM,FAC2,NN,FAC1,N,LMAX)
C    LX,LXI,LT,LXM: PERMUTATIONS OF (L,M)-SEQUENCE
        CALL LXGENT(LX,LXI,LT,LXM,LMAX,LMMAX)
        IF (IT1+IT2+IT3) 640,650,640
640     NN3=LMAX+1
        NN2=LMAX+1
        NN1=NN2+NN3-1
C    PPP= CLEBSCH-GORDON COEFFICIENTS FOR COMPUTATION OF TEMPERATURE-
C    DEPENDENT PHASE SHIFTS (SKIPPED IF NOT NEEDED)
        CALL   CPPP (PPP, NN1, NN2, NN3)
650     CONTINUE
C
C
C    READ ENERGY RANGE AND STEP; IF EF AND DE ARE BLANK, PROGRAM COMPUTES
C    FOR ENERGY EI. PROGRAM ALWAYS RETURNS TO THIS LINE TO READ A NEW
C    ENERGY RANGE AND STEP. IF A NEGATIVE EI IS READ, RUN IS TERMINATED
680     READ (5,161) EI, EF, DE
        IF (EI)  1000, 690, 690
690     EI=EI/27.18+VV
        EF=EF/27.18+VV
        DE=DE/27.18
        E = EI
        EC = E
        E3 = E
C
C
C    START LOOP OVER ENERGIES IN GIVEN ENERGY RANGE
700     EEV=(E-VV)*27.18
        WRITE (6,370)  E,EEV
C    COMPUTE COMPONENTS OF INCIDENT WAVEVECTOR PARALLEL TO SURFACE
        AK=SQRT(2.0*(E-VV))*SIN(THETA)
```

```
      AK2=AK*COS(FI)
      AK3=AK*SIN(FI)
      AK21=AK2
      AK31=AK3
C SELECT BEAMS APPROPRIATE FOR CURRENT ENERGY
      CALL BEAMS(KNBS,KNB,SPQ,SPQF,KSYM,KNT,NPU,NPUN,AK2,AK3,E,TST,
     1NB,PQ,PQF,SYM,NPUC,MPU,NT,NP)
C SET OPTICAL POTENTIAL (IMAGINARY PART OF MUFFIN-TIN CONSTANT,
C REPRESENTING DAMPING). VPIS FOR SUBSTRATE, VPIO FOR OVERLAYER.
C HERE VPIS=VPIO PROPORTIONAL TO E**1/3
      VPIS=-(3.8/27.18)*EXP((ALOG((E)/(90.0/27.18+VV)))*0.333333)
      VPIO=VPIS
      DCUTO = SQRT(2.0 * E)
C SET LIMITING RADII ON LATTICE SUMS, POSSIBLY DIFFERENT FOR SUBSTRATE
C (DCUTS) AND OVERLAYER (DCUTO)
      DCUTS =  - 5.0 * DCUTO/(AMIN1(VPIS, - 0.05))
      DCUTO =  - 5.0 * DCUTO/(AMIN1(VPIO, - 0.05))
      WRITE (6,380)  VPIS, VPIO, DCUTS, DCUTO
C FIX TEMPERATURE T. T MAY BE INCREASED STEPWISE IN A LOOP OVER THE
C INPUT RANGE (TI,TF) WITH INPUT STEP DT
  770 T = TI
  780 WRITE (6,390)  T
C
C
C START WITH OVERLAYER: E NOW BECOMES ENERGY IN OVERLAYER
      E = EC - VO
      E3 = E
C PRODUCE ATOMIC T-MATRIX ELEMENTS FROM PHASE SHIFTS, CORRECTED FOR
C THERMAL VIBRATION (CAF AND TSF) OR NOT (AF AND TSFO). AF AND CAF
C ARE USED FOR BRAVAIS-LATTICE LAYERS. TSFO AND TSF ARE USED FOR
C COMPOSITE LAYERS.
C THIS CALL IS FOR OVERLAYER (IEL=2 IN INPUT SEQUENCE OF PHASE SHIFTS)
      CALL TSCATF(2,L1,ES,PHSS,NPSI,IT2,E,0.,PPP,NN1,NN2,NN3,DRO2,
     1DRPER2,DRPAR2,TO,T,TSFO,TSF,AF,CAF)
C CHOOSE PARAMETERS RELEVANT TO OVERLAYER
      DRPER=DRPER2
      DRPAR=DRPAR2
      DRO=DRO2
C IT=(0),1 MEANS INCLUDE (NO) EXPLICIT DEBYE-WALLER FACTORS IN MSMF
      IT=0
      TV = TVB
      VPI = VPIO
      VPI1 = VPIO
C PERFORM PLANAR LATTICE SUM IN FMAT. IF MUFFIN-TIN CONSTANTS AND
C DAMPINGS IN SUBSTRATE AND OVERLAYER ARE EQUAL, DO LATTICE SUMS FOR
C BOTH TYPES OF LAYERS NOW (NLS=NL IMPLIES: DO ALL SUBLATTICES).
C OTHERWISE DO ONLY OVERLAYER LATTICE SUM (NLS=1 IMPLIES: DO ONLY THE
C SUBLATTICE EQUAL TO THE OVERLAYER LATTICE)
      NLS = 1
      IF (ABS(VPIS-VPIO)+ABS(VO) .LE. 1.0E-4)  NLS = NL
      CALL   FMAT (FLMS, V, JJS, NL, NLS, DCUTO, IDEG, LMAX,KLM)
C PRODUCE REFLECTION (ROV) AND TRANSMISSION (TOV) MATRICES FOR THE
C OVERLAYER. ONLY ONE REGISTRY: ID=1. ONLY ONE SUBLATTICE: NLL=1.
C OVERLAYER SYMMETRIES AND REGISTRIES APPLY: LAY=1. TEMPERATURE-
C DEPENDENT PHASE SHIFTS USED: CAF
      CALL MSMF(ROV,TOV,ROV,TOV,ROV,TOV,ROV,TOV,NT,NT,NT,AMULT,CYLM,
     1PQ,SYM,NT,FLMS,FLM,V,NL,0,0,1,1,CAF,CAF,L1,LX,LXI,LMMAX,KLM,
     2XEV,XOD,LEV,LOD,YLM,YLME,YLMO,IPLE,IPLO,CLM,NLM,1)
      WRITE (6,410)
C
C
C CONSIDER SUBSTRATE NEXT. CHOOSE CORRECT ENERGY
      E = EC
      E3 = E
C PRODUCE ATOMIC T-MATRIX ELEMENTS FOR SUBSTRATE
      CALL TSCATF(1,L1,ES,PHSS,NPSI,IT1,E,0.,PPP,NN1,NN2,NN3,DRO1,
     1DRPER1,DRPAR1,TO,T,TSFO,TSF,AF,CAF)
```

```
            DRPER=DRPER1
            DRPAR=DRPAR1
            DRO=DRO1
            IT=0
            TV = TVA
            VPI = VPIS
            VPI1 = VPIS
            IF (ABS(VPIS-VPIO)+ABS(VO) .LE. 1.0E-4)  GO TO 930
C   PERFORM PLANAR LATTICE SUMS FOR SUBSTRATE IF NOT DONE BEFORE
            CALL   FMAT (FLMS, V, JJS, NL, NL, DCUTS, IDEG, LMAX,KLM)
C   PRODUCE REFLECTION (RA1,RA2) AND TRANSMISSION (TA1,TA2) MATRICES
C   FOR SUBSTRATE LAYERS. TWO REGISTRIES: ID=2 (RA1,TA1 FOR REGISTRY
C   SS1, RA2,TA2 FOR REGISTRY SS2). NLL=NL SUBLATTICES. SUBSTRATE LAYER
C   SYMMETRIES AND REGISTRIES APPLY: LAY=2.
C   FIRST BEAM SET FIRST (NAA BEAMS)
      930 NAA = NB(1)
            CALL MSMF(RA1,TA1,RA2,TA2,RA1,TA1,RA1,TA1,NAA,NAA,NAA,AMULT,CYLM,
           1PQ,SYM,NT,FLMS,FLM,V,NL,    0,   0,2,NL,CAF,CAF,L1,LX,LXI,LMMAX,KLM,
           2XEV,XOD,LEV,LOD,YLM,YLME,YLMO,IPLE,IPLO,CLM,NLM,2)
            WRITE (6,410)
C   LAYER DOUBLING IS DONE FOR FIRST BEAM SET
            CALL SUBREF(RA1,TA1,RA2,TA2,NAA,S1,S2,S3,S4,PP,XS,JNT,NP,0,NT,PQ)
            WRITE(6,420)
C   SECOND BEAM SET NEXT (NBB BEAMS; NOTE OFFSETS NA=NAA, NS=0)
            NBB = NB(2)
            CALL MSMF(RB1,TB1,RB2,TB2,RB1,TB1,RB1,TB1,NBB,NBB,NBB,AMULT,CYLM,
           1PQ,SYM,NT,FLMS,FLM,V,NL,NAA,   0,2,NL,CAF,CAF,L1,LX,LXI,LMMAX,KLM,
           2XEV,XOD,LEV,LOD,YLM,YLME,YLMO,IPLE,IPLO,CLM,NLM,2)
            WRITE (6,410)
C   LAYER DOUBLING IS DONE FOR SECOND BEAM SET
            N=NAA
            CALL SUBREF(RB1,TB1,RB2,TB2,NBB,S1,S2,S3,S4,PP,XS,JNT,NP,N,NT,PQ)
            WRITE(6,420)
C   THIRD (AND FOURTH, BY SYMMETRY) BEAM SET LAST (NCC BEAMS; NOTE
C   OFFSETS NA=N=NAA+NBB, NS=0)
            NCC=NB(3)
            N=NAA+NBB
            CALL MSMF(RC1,TC1,RC2,TC2,RC1,TC1,RC1,TC1,NCC,NCC,NCC,AMULT,CYLM,
           1PQ,SYM,NT,FLMS,FLM,V,NL,   N,   0,2,NL,CAF,CAF,L1,LX,LXI,LMMAX,KLM,
           2XEV,XOD,LEV,LOD,YLM,YLME,YLMO,IPLE,IPLO,CLM,NLM,2)
            WRITE(6,410)
C   LAYER DOUBLING IS DONE FOR THIRD (AND FOURTH) BEAM SET
            CALL SUBREF(RC1,TC1,RC2,TC2,NCC,S1,S2,S3,S4,PP,XS,JNT,NP,N,NT,PQ)
            WRITE(6,420)
C
C
C   NOW ADD OVERLAYER
C   WITH THE ASSUMED SURFACE SYMMETRIES (4-FOLD AXIS AND 2 ORTHOGONAL
C   MIRROR PLANES) TWO OVERLAYER REGISTRIES ARE POSSIBLE: HOLLOW SITE
C   (IO=1) AND ON-TOP SITE (IO=2), BOTH OF WHICH WILL BE SELECTED HERE.
C   THE REGISTRY OF THE OVERLAYER IS HERE SO1=(0,0) (FROM INPUT), THAT OF
C   THE TOPMOST SUBSTRATE LAYER IS (.5,.5), THAT OF THE BACKMOST LAYER OF
C   THE SLAB TREATED BY LAYER DOUBLING IS (0,0) (BECAUSE LAYER DOUBLING
C   PUTS AN EVEN NUMBER OF ALTERNATING LAYERS TOGETHER, THE BACKMOST
C   LAYER HAS THE SAME REGISTRY AS THE SECOND LAYER)
            IO=1
      932 CONTINUE
C   COPY SUBSTRATE REFLECTION MATRICES ONTO A LARGER BLOCK-DIAGONALIZED
C   MATRIX RS
            GO TO (934,935),IO
      934 CALL MATCOP(RA1,NAA,RB1,NBB,RC1,NCC,RC1,NCC,RS,NT,3)
            GO TO 936
      935 CALL MATCOP(RA2,NAA,RB2,NBB,RC2,NCC,RC2,NCC,RS,NT,3)
      936 CONTINUE
C   ASV BECOMES THE 3-D VECTOR FROM OVERLAYER TO TOP SUBSTRATE LAYER
C   AND WILL BE VARIED
            DO 937 J=1,3
```

```
  937 ASV(J)=ASB(J)
C   LOOP 960 RUNS OVER GEOMETRIES: HERE THE OVERLAYER SPACING ASV(1) IS
C   VARIED
      DO 960 I = 1, NSTEP
      DO 940 J=1,3
  940 AS(J)=ASV(J)*0.529
      A=AS(1)
      WRITE (6,430) (AS(J),J=1,3)
C   THE OVERLAYER IS ADDED BY LAYER DOUBLING, PRODUCING REFLECTED BEAM
C   AMPLITUDES.
C   SOME STORAGE IS SAVED BY USING S2 AND S3 HERE, ALTHOUGH THEY HAVE
C   IMPROPER (BUT SUFFICIENT) DIMENSIONS
      CALL ADREF1(ROV,TOV,ROV,TOV,RS,S1,S2,S3,JNT,XI,PQ,NT,EMACH,ASV)
C   COMPUTE REFLECTED INTENSITIES (AT) FROM AMPLITUDES (XI), PRINT AND
C   PUNCH (NPNCH=1; PUNCH LAYER SPACING A ALSO FOR LATER IDENTIFICATION)
      CALL RINT(NT,XI,AT,PQ,PQF,SYM,VV,THETA,FI,MPU,NPUC,EEV,A,1)
  960 ASV(1)=ASV(1)+STEP
      IF (IO.EQ.2) GO TO 969
      IO=2
      GO TO 932
  969 CONTINUE
      IF (IT1+IT2+IT3) 980,990,980
  980 T = T + DT
      IF (DT.LE.0.0001) GO TO 990
      IF (T-TF)  780, 780, 990
  990 E = E + DE
      EC = E
      E3 = E
      IF (E-EF)  700, 700, 680
 1000 WRITE (6,450)
      STOP
      END
----------------------------------------------------------------
---- INPUT DATA FOR PRECEDING PROGRAM FOLLOW -----------
----------------------------------------------------------------
NI(100)+P(2*2)O, LAYER DOUBLING, FULL SYMMETRY
  4  2  2              IDEG NL1 NL2
  3 39  6              KNBS KNT NPUN
 12                    NPSI
2.4900 0.0000          ARA1
0.0000 2.4900          ARA2
0.5000 0.5000          SS1
0.0000 0.0000          SS2
0.0000 0.0000          SS3
0.0000 0.0000          SS4
1.7650 0.0000 0.0000   ASA
4.9800 0.0000          ARB1
0.0000 4.9800          ARB2
0.0000 0.0000          SO1
0.0000 0.0000          SO2
0.0000 0.0000          SO3
0.9000 0.0000 0.0000   ASB
0.6290 0.7300 0.2000   FR ASE STEP
  6                    NSTEP
 13                    KNB(1)
0.0000 0.0000   1  1   BEAMS
1.0000 0.0000   7  7
1.0000 1.0000   7  7
2.0000 0.0000   7  7
2.0000 1.0000  10 10
2.0000 2.0000   7  7
3.0000 0.0000   7  7
3.0000 1.0000  10 10
3.0000 2.0000  10 10
4.0000 0.0000   7  7
4.0000 1.0000  10 10
3.0000 3.0000   7  7
```

```
4.0000 2.0000 10 10
   9                      KNB(2)
0.5000 0.5000  7  7
1.5000 0.5000 10 10
1.5000 1.5000  7  7
2.5000 0.5000 10 10
2.5000 1.5000 10 10
3.5000 0.5000 10 10
2.5000 2.5000  7  7
3.5000 1.5000 10 10
3.5000 2.5000 10 10
  17                      KNB(3)
0.0000 0.5000  7  2
1.0000 0.5000 10  9
0.0000 1.5000  7  2
1.0000 1.5000 10  9
2.0000 0.5000 10  9
2.0000 1.5000 10  9
0.0000 2.5000  7  2
1.0000 2.5000 10  9
2.0000 2.5000 10  9
3.0000 0.5000 10  9
3.0000 1.5000 10  9
0.0000 3.5000  7  2
1.0000 3.5000 10  9
3.0000 2.5000 10  9
4.0000 0.5000 10  9
2.0000 3.5000 10  9
4.0000 1.5000 10  9
0.0020                    TST
   1   2   3 14 23 24     NPU
   0.00    0.00           THETA FI
   0.00   11.20           VO VV
0.0010                    EPS
   5                      LITER
   1   1   0              IT1 IT2 IT3
440.0000    58.7100    1.4000   1.4000   0.0000    THDB1 AM1 FPER1 FPAR1 DRO1
843.0000    15.9994    2.0000   2.0000   0.0000    THDB2 AM2 FPER2 FPAR2 DRO2
  1.0000     1.0000    1.0000   1.0000   0.0000    THDB3 AM3 FPER3 FPAR3 DRO3
300.0000                                           TI TF DT
   7                      LMAX
   2                      NEL
 .3000
2.8398 -.0137 -.1876   .0011   .0000   .0000   .0000   .0000
1.9481 2.5909  .0113   .0003   .0000   .0000   .0000   .0000
 .4000
2.7639 -.0216 -.3164   .0024   .0001   .0000   .0000   .0000
1.8026 2.5183  .0223   .0006   .0000   .0000   .0000   .0000
 .5000
2.6930 -.0322 -.2630   .0048   .0002   .0000   .0000   .0000
1.6839 2.4586  .0374   .0016   .0000   .0000   .0000   .0000
 .7500
2.5276 -.0668 -.2322   .0160   .0009   .0000   .0000   .0000
1.4577 2.3455  .0920   .0053   .0003   .0000   .0000   .0000
1.0000
2.3803 -.1132 -.2323   .0386   .0028   .0002   .0000   .0000
1.2921 2.2663  .1681   .0129   .0009   .0000   .0000   .0000
1.5000
2.1240 -.2247 -.2118   .1202   .0125   .0012   .0001   .0000
1.0568 2.1658  .3484   .0408   .0041   .0003   .0000   .0000
2.0000
1.9101 -.3448 -.1758   .2494   .0335   .0042   .0004   .0000
 .8888 2.1039  .5102   .0842   .0116   .0013   .0000   .0000
3.0000
1.5727 -.5697 -.1288   .5535   .1138   .0216   .0034   .0004
 .6434 2.0175  .6993   .1948   .0424   .0072   .0009   .0000
4.0000
```

```
1.3160 -.7574 -.1272  .7884  .2257  .0583  .0125  .0022
 .4577 1.9446  .7876  .2934  .0902  .0210  .0038  .0006
5.0000
1.1099 -.9122 -.1445  .9491  .3361  .1107  .0305  .0069
 .3079 1.8821  .8413  .3625  .1420  .0437  .0104  .0019
6.0000
 .9366-1.0434 -.1666 1.0690  .4287  .1691  .0571  .0159
 .1847 1.8312  .8781  .4109  .1876  .0713  .0210  .0047
7.0000
 .7862-1.1581 -.1902 1.1630  .5045  .2251  .0894  .0297
 .0616 1.7803  .9148  .4593  .2331  .0990  .0317  .0075
 40.00 201.00   5.00    EI EF DE
-10.00                  EI EF DE
```

---- FIRST PART OF PRINTER OUTPUT FOLLOWS ---------------
----(NEGLECTING MOST BLANK LINES AND PAGE ---------------
---- EJECTS AND TRUNCATING LONG LINES)   ---------------

```
NI(100)+P(2*2)O, LAYER DOUBLING, FULL SYMMETRY
IDEG =   4 NL1 =   2 NL2 =   2
PARAMETERS FOR INTERIOR:
SURF VECS       2.4900  0.0000
                0.0000  2.4900
SS      .5000   .5000
SS     0.0000  0.0000
SS     0.0000  0.0000
SS     0.0000  0.0000
ASA     1.7650 0.0000 0.0000
PARAMETERS FOR SURFACE:
SURF VECS       4.9800  0.0000
                0.0000  4.9800
SO     0.0000  0.0000
SO     0.0000  0.0000
SO     0.0000  0.0000
ASB     .9000 0.0000 0.0000
FR =   .6290 ASE =    .7300 STEP =   .2000 NSTEP =    6
           PQ1     PQ2        SYM

     1     0.000   0.000      1  1
     2     1.000   0.000      7  7
     3     1.000   1.000      7  7
     4     2.000   0.000      7  7
     5     2.000   1.000     10 10
     6     2.000   2.000      7  7
     7     3.000   0.000      7  7
     8     3.000   1.000     10 10
     9     3.000   2.000     10 10
    10     4.000   0.000      7  7
    11     4.000   1.000     10 10
    12     3.000   3.000      7  7
    13     4.000   2.000     10 10

    14      .500    .500      7  7
    15     1.500    .500     10 10
    16     1.500   1.500      7  7
    17     2.500    .500     10 10
    18     2.500   1.500     10 10
    19     3.500    .500     10 10
    20     2.500   2.500      7  7
    21     3.500   1.500     10 10
    22     3.500   2.500     10 10

    23     0.000    .500      7  2
    24     1.000    .500     10  9
    25     0.000   1.500      7  2
    26     1.000   1.500     10  9
    27     2.000    .500     10  9
```

```
 28     2.000  1.500      10   9
 29     0.000  2.500       7   2
 30     1.000  2.500      10   9
 31     2.000  2.500      10   9
 32     3.000   .500      10   9
 33     3.000  1.500      10   9
 34     0.000  3.500       7   2
 35     1.000  3.500      10   9
 36     3.000  2.500      10   9
 37     4.000   .500      10   9
 38     2.000  3.500      10   9
 39     4.000  1.500      10   9
TST =   .0020
PUNCHED BEAMS        1   2   3  14  23  24
THETA  FI =      0.00     0.00
VO       0.00VV         11.20
EPS =   .0010  LITER =    5
IT1 =   1 IT2 =   1 IT3 =   0
THDB =  440.0000 AM =   58.7100 FPER =    1.4000 FPAR =    1.4000 DRO =
THDB =  843.0000 AM =   15.9994 FPER =    2.0000 FPAR =    2.0000 DRO =
THDB =    1.0000 AM =    1.0000 FPER =    1.0000 FPAR =    1.0000 DRO =
LMAX =   7
        PHASE SHIFTS


E =   .3000  1ST ELEMENT      2.8398  -.0137  -.1876   .0011  0.0000  0.0000
             2ND ELEMENT      1.9481  2.5909   .0113   .0003  0.0000  0.0000

E =   .4000  1ST ELEMENT      2.7639  -.0216  -.3164   .0024   .0001  0.0000
             2ND ELEMENT      1.8026  2.5183   .0223   .0006  0.0000  0.0000

E =   .5000  1ST ELEMENT      2.6930  -.0322  -.2630   .0048   .0002  0.0000
             2ND ELEMENT      1.6839  2.4586   .0374   .0016  0.0000  0.0000

E =   .7500  1ST ELEMENT      2.5276  -.0668  -.2322   .0160   .0009  0.0000
             2ND ELEMENT      1.4577  2.3455   .0920   .0053   .0003  0.0000

E =  1.0000  1ST ELEMENT      2.3803  -.1132  -.2323   .0386   .0028   .0002
             2ND ELEMENT      1.2921  2.2663   .1681   .0129   .0009  0.0000

E =  1.5000  1ST ELEMENT      2.1240  -.2247  -.2118   .1202   .0125   .0012
             2ND ELEMENT      1.0568  2.1658   .3484   .0408   .0041   .0003

E =  2.0000  1ST ELEMENT      1.9101  -.3448  -.1758   .2494   .0335   .0042
             2ND ELEMENT       .8888  2.1039   .5102   .0842   .0116   .0013

E =  3.0000  1ST ELEMENT      1.5727  -.5697  -.1288   .5535   .1138   .0216
             2ND ELEMENT       .6434  2.0175   .6993   .1948   .0424   .0072

E =  4.0000  1ST ELEMENT      1.3160  -.7574  -.1272   .7884   .2257   .0583
             2ND ELEMENT       .4577  1.9446   .7876   .2934   .0902   .0210

E =  5.0000  1ST ELEMENT      1.1099  -.9122  -.1445   .9491   .3361   .1107
             2ND ELEMENT       .3079  1.8821   .8413   .3625   .1420   .0437

E =  6.0000  1ST ELEMENT       .9366 -1.0434  -.1666  1.0690   .4287   .1691
             2ND ELEMENT       .1847  1.8312   .8781   .4109   .1876   .0713

E =  7.0000  1ST ELEMENT       .7862 -1.1581  -.1902  1.1630   .5045   .2251
             2ND ELEMENT       .0616  1.7803   .9148   .4593   .2331   .0990
ENERGY = 1.8837 H  OR    40.00 EV
  21 BEAMS USED :    7    5    9
   0.000 0.000    1.000 0.000    1.000 1.000    2.000 0.000    2.000 1.000
   1.500 1.500    2.500  .500    2.500 1.500    0.000  .500    1.000  .500
   0.000 2.500    1.000 2.500    3.000  .500
VPIS =    -.1114 VPIO =     -.1114 DCUTS =    87.1160 DCUTO =    87.1160
TEMP =   300.0000
MSMF    OK
```

```
MSMF    OK
X =        .1405E+01
X =        .1414E+01
X =        .1425E+01
X =        .1426E+01
    4   ITER IN SUBREF
SUBREF OK
MSMF    OK
X =        .2024E+00
X =        .1843E+00
X =        .1835E+00
X =        .1822E+00
X =        .1822E+00
NO CONV IN SUBREF AFTER  5  ITER, X =        .1822E+00
SUBREF OK
MSMF    OK
X =        .8200E+00
X =        .7753E+00
X =        .7643E+00
X =        .7656E+00
X =        .7656E+00
NO CONV IN SUBREF AFTER  5  ITER, X =        .7656E+00
SUBREF OK
SURFACE GEOMETRY :    .9000 0.0000 0.0000
E=    1.8837 THETA=    0.0000 FI=    0.0000

   PQ1      PQ2      REF INT      SYM
  0.000    0.000    .16642E-02      1
  1.000    0.000    .21325E-03      7
   .500     .500    .17053E-02      7
  0.000     .500    .20641E-02      7
  1.000     .500    .16551E-02     10
SURFACE GEOMETRY :   1.1000 0.0000 0.0000
E=    1.8837 THETA=    0.0000 FI=    0.0000

   PQ1      PQ2      REF INT      SYM
  0.000    0.000    .72632E-02      1
  1.000    0.000    .14570E-02      7
   .500     .500    .96796E-03      7
  0.000     .500    .83157E-03      7
  1.000     .500    .51368E-03     10
SURFACE GEOMETRY :   1.3000 0.0000 0.0000
E=    1.8837 THETA=    0.0000 FI=    0.0000

   PQ1      PQ2      REF INT      SYM
  0.000    0.000    .10836E-01      1
  1.000    0.000    .25769E-02      7
   .500     .500    .95906E-03      7
  0.000     .500    .10737E-02      7
  1.000     .500    .22231E-04     10
```

10.8 Main Program for a c(2×2) Overlayer on an fcc or bcc (100) Substrate, with Rumpling of the Topmost Substrate Layer, Using Matrix Inversion and RFS

In this section we consider a program that can investigate a rumpling in the topmost substrate layer due to a half-monolayer overlayer adsorbed on an fcc or bcc (100) substrate. The case of W(100) + c(2×2)H is chosen. For example, if the adatoms choose the top site, cf. Fig. 10.6, half the top-layer substrate atoms may be pulled outward somewhat, creating a slightly rumpled layer with a larger unit cell. This rumpled layer cannot normally be treated as two separate layers, but should be considered to be a composite layer with two atoms per unit cell; this assumes (as we do here) that the overlayer atoms are far enough from the rumpled layer, so that they do not have to be taken as part of the composite layer. As a result, one has to effectively consider two overlayers on a substrate. Using RFS, subroutine RFS20V (not to be confused with RFS02V) should be used for this case. Be-

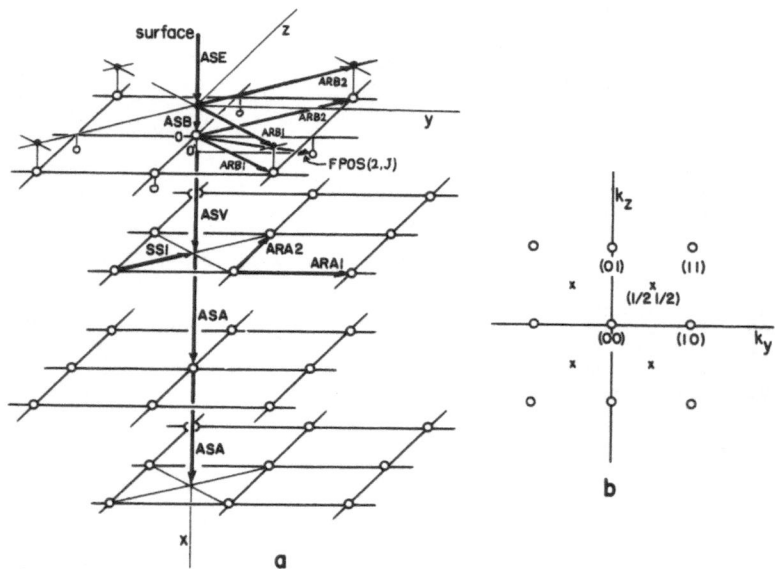

Fig. 10.6 (a) Geometry of a c(2×2) top-site overlayer creating a rumpling of the topmost substrate layer, which is treated as a composite layer. 0 is the local origin of coordinates in the composite layer, where FPOS(1,J) = (0,0,0). ASV is measured from 0' and its component perpendicular to the surface can be varied. (b) Reciprocal space (o = integral-order beams, x = fractional-order beams)

cause of strong multiple scattering in tungsten, the RSP method converges poorly for the rumpled layer and matrix inversion is used here instead.

Full symmetry of normal incidence is used, as in Sect. 10.7, where the same symmetry applies. Two registries are therefore needed for the substrate layers (which RFS20V is designed to handle), while the adlayer and the rumpled layer need only one registry each.

In order to keep the computation effort for the rumpled layer down, only 5 phase shifts are used: with tungsten this can lead to poor results at energies above about 100 eV, where more phase shifts should be used.

```
C     MAIN + DATA FOR W(100)+C(2*2)H, WITH W RUMPLING(BEEBY),RFS,FULL SYMM.
C     MAIN LEED PROGRAM FOR A C(2*2) OVERLAYER ON FCC OR BCC (100), USING
C     FULL SYMMETRY OF NORMAL INCIDENCE (4-FOLD AXIS OF ROTATION AND 4
C     MIRROR PLANES), BY RENORMALIZED FORWARD SCATTERING.
C     THE TOP SUBSTRATE LAYER MAY BE RUMPLED (BEEBY-TYPE INVERSION USED).
C     BOTH THE OVERLAYER AND THE TOP SUBSTRATE LAYER SPACINGS ARE VARIABLE.
      DIMENSION ARA1(2),ARA2(2),RAR1(2),RAR2(2),SS1(2),SS2(2),SS3(2),
     1SS4(2),ASA(3),AS(3),ARB1(2),ARB2(2),RBR1(2),RBR2(2),SO1(2),
     2SO2(2),SO3(2),ASB(3),ASV(3),ASW(3)
C     2=NL, 45=KLM
      COMPLEX VL(2,2),FLMS(2, 45),FLM( 45)
C     2(1ST DIMENSION)=NL=NL1*NL2, 4(2ND DIMENSION)=IDEG
      DIMENSION V(2,2),JJS(2,4)
C     2=KNBS=NO. OF BEAM SETS READ IN, 30=KNT=NO. OF BEAMS READ IN
      DIMENSION KNB(2),NB(2),SPQF(2,30),SPQ(2,30),PQF(2,30),PQ(2,30)
      INTEGER KSYM(2,30),SYM(2,30)
C     20 MUST BE .GE. NPUN
      DIMENSION NPU(20),NPUC(20)
C     12=NPSI, 10=NEL*(LMAX+1)
      DIMENSION ES(12),PHSS(12,10)
C     2=NLAY, VPOS NOT USED, 12 FIXED, 759 FROM DATA STATEMENT.
      DIMENSION FPOS(2,3),VPOS(1,3),LPS(12),INV(12),CAA(759),NCA(7),
     1LPSS(12)
C     2=NLAY, 1=NLAY2
      DIMENSION POSS(2,3),MGH(2,2),DRL(1,3),SDRL(1,3),NUGH(1),NGEQ(1),
     1NGOL(1),TEST(1)
C     809=NLM FROM DATA STATEMENT,  81=NN,  9=N (FOR CELMG)
C     YLM IS ALSO USED, AS COMPLEX VECTOR OF SMALLER CORE SIZE, IN MSMF,ETC
      DIMENSION NLMS(7),CLM( 809),YLM( 81),FAC2( 81),FAC1( 9)
C     9=NN1, 5=NN2, 5=NN3
      DIMENSION PPP( 9,5,5)
C     25=LMMAX
      DIMENSION LX(25),LXI(25),LXM(25),LT(25)
C     5=LMAX+1=L1, 3 AND 10 FIXED FOR COMMON /MPT/
      COMPLEX AF(5),CAF(5),TSFO(3,10),TSF(3,10)
      COMPLEX ROV(30,30),TOV(30,30)
      COMPLEX RTP(30,30),TTP(30,30),RTM(30,30),TTM(30,30)
C     17=KNP
      COMPLEX RA1(30,17),TA1(30,17),RA2(30,17),TA2(30,17)
C     25=LMT, 15=LEV, 50(IN GH)=LMG, 2=NLAY, 50(IN TS)=LMN, 100=LM2N
C     50(IN TH, IPL)=LMNI
      COMPLEX TAU(25,15),TAUG(25),TAUGM(25),GH(50,25),RG(3,2,30),
     1TS(50),TG(2,100),VT(25),TH(50,50)
      DIMENSION IPL(50)
C     15=LEV, 10=LOD
      COMPLEX AMULT(30),CYLM(30,25),XEV(15,30),XOD(10,10),
     1YLME(15),YLMO(10)
      DIMENSION IPLE(15),IPLO(10),AT(30)
C     10 FIXED (FOR RFS2OV), 20=ND (FOR RFS SUBROUTINES)
      COMPLEX XI(30),PK(30,10),AW(30,2),ANEW(30,20)
      COMMON E3,AK21,AK31,VPI1,TV,EMACH
      COMMON /X4/E,VPI,AK2,AK3,ASA
      COMMON  /SL/ARA1, ARA2, ARB1, ARB2, RBR1, RBR2, NL1, NL2
      COMMON /MS/LMAX,EPS,LITER,SO1,SO2,SO3,SS1,SS2,SS3,SS4
      COMMON /ADS/ASL, FR, ASE, VPIS, VPIO, VO, VV
      COMMON /BT/IT,T,TO,DRPER,DRPAR,DRO
      COMMON /TEMP/IT1,IT2,IT3,TI,TF,DT,TO1,DRPER1,DRPAR1,DRO1,DRPER2,
     1DRPAR2,DRO2,DRPER3,DRPAR3,DRO3
      COMMON /MPT/NA,NS,ID,LAY,L1,NTAU,TSTS,TV1,DCUT,EPSN,NPERT,NOPT,
     1NEW,LPS,LPSS,INV,NINV,TSF
C     NCA IS DIMENSION OF CAA AS FUNCTION OF LMAX
      DATA NCA(1),NCA(2),NCA(3),NCA(4),NCA(5),NCA(6),NCA(7)/
     11,70,264,759,1820,3836,7344/
C     NLMS IS DIMENSION OF CLM AS FUNCTION OF LMAX
      DATA NLMS(1),NLMS(2),NLMS(3),NLMS(4),NLMS(5),NLMS(6),NLMS(7)/
     10,76,284,809,1925,4032,7680/
  161 FORMAT(3F7.2)
```

```
  200 FORMAT(20I3)
  240 FORMAT(8HOIDEG = ,1I3,7H NL1 = ,1I3,7H NL2 = ,1I3)
  370 FORMAT(9H1ENERGY =,F7.4,7H H  OR ,1F7.2,3H EV)
  380 FORMAT(8H VPIS = ,F9.4,8H VPIO = ,F9.4,9H DCUTS = ,F9.4,9H DCUTO =
     1 ,F9.4)
  390 FORMAT(8HOTEMP = ,F9.4)
  410 FORMAT(10H MSMF   OK)
  415 FORMAT(10H MPERTI OK)
  416 FORMAT(10H MTINV  OK)
  420 FORMAT(10H SUBREF OK)
  430 FORMAT(20HOSURFACE GEOMETRY : ,3F7.4)
  450 FORMAT(20H CORRECT TERMINATION)
  460 FORMAT(20A4)
  465 FORMAT(1H1,20A4)
C
C
C   READ, WRITE AND PUNCH A DESCRIPTIVE TITLE
        READ (5,460)(CLM(I),I=1,20)
        WRITE(6,465)(CLM(I),I=1,20)
        WRITE(7,460)(CLM(I),I=1,20)
C   EMACH IS MACHINE ACCURACY (USED BY ZGE AND ZSU)
        EMACH=1.0E-6
C   IDEG: EACH LAYER HAS AN IDEG-FOLD SYMMETRY AXIS
C   NL1, NL2: SUPERLATTICE CHARACTERIZATION  NL1  NL2
C                        P(1*1)               1    1
C                        C(2*2)               2    1
C                        P(2*1)               2    1
C                        P(1*2)               1    2
C                        P(2*2)               2    2
        READ(5,200)IDEG,NL1,NL2
        WRITE(6,240)IDEG,NL1,NL2
        NL=NL1*NL2
C   KNBS= NO.OF BEAM SETS TO BE READ IN (.LE.NL)
C   KNT= TOTAL NO. OF BEAMS TO BE READ IN
C   NPUN= NO. OF BEAMS FOR WHICH INTENSITIES ARE TO BE PUNCHED OUT
        READ(5,200)KNBS,KNT,NPUN
C   NPSI= NO.OF ENERGIES AT WHICH PHASE SHIFTS WILL BE READ IN
        READ(5,200)NPSI
C   READ IN GEOMETRY, PHYSICAL PARAMETERS, CONVERGENCE CRITERIA
        CALL READIN(TVA,RAR1,RAR2,ASA,TVB,ASB,STEP,NSTEP,IDEG,NL,V,VL,
     1JJS,KNBS,KNB,KNT,SPQF,KSYM,SPQ,TST,TSTS,NPUN,NPU,THETA,FI,
     2LMMAX,NPSI,ES,PHSS,L1)
        TO=TO1
C   NLAY=NO. OF SUBPLANES IN COMPOSITE LAYER (0,IF NO COMP. LAYER)
        READ(5,200)NLAY
        IF (NLAY.EQ.0) GO TO 500
C   READ IN ADDITIONAL DATA FOR COMPOSITE LAYER
        CALL READCL(NLAY,NLAY2,FPOS,VPOS,NVAR,NTAU,LPS,NINV,INV,LMMAX,
     1LMG,LMN,LM2N,LMT,LTAUG,LMNI,LMNI2)
  500 CONTINUE
        KLM=(2*LMAX+1)*(2*LMAX+2)/2
        LEV=(LMAX+1)*(LMAX+2)/2
        LOD=LMMAX-LEV
        IF (NLAY.EQ.0) GO TO 630
        LEV2=2*LEV
        NCAA=NCA(LMAX)
C   CAA= CLEBSCH-GORDON COEFFICIENTS FOR COMPOSITE LAYER
        CALL CAAA(CAA,NCAA,LMMAX)
  630 N=2*LMAX+1
        NN=N*N
        NLM=NLMS(LMAX)
C   CLM= CLEBSCH-GORDON COEFFICIENTS FOR MATRICES X AND TAU
        CALL CELMG(CLM,NLM,YLM,FAC2,NN,FAC1,N,LMAX)
C   LX,LXI,LT,LXM: PERMUTATIONS OF (L,M)-SEQUENCE
        CALL LXGENT(LX,LXI,LT,LXM,LMAX,LMMAX)
        IF (IT1+IT2+IT3) 640,650,640
  640 NN3=LMAX+1
```

```
      NN2=LMAX+1
      NN1=NN2+NN3-1
C  PPP= CLEBSCH-GORDON COEFFICIENTS FOR COMPUTATION OF TEMPERATURE-
C  DEPENDENT PHASE SHIFTS (SKIPPED IF NOT NEEDED)
      CALL  CPPP (PPP, NN1, NN2, NN3)
  650 CONTINUE
C
C
C  READ ENERGY RANGE AND STEP; IF EF AND DE ARE BLANK, PROGRAM COMPUTES
C  FOR ENERGY EI. PROGRAM ALWAYS RETURNS TO THIS LINE TO READ A NEW
C  ENERGY RANGE AND STEP. IF A NEGATIVE EI IS READ, RUN IS TERMINATED
  680 READ (5,161) EI, EF, DE
      IF (EI)  1000, 690, 690
  690 EI=EI/27.18+VV
      EF=EF/27.18+VV
      DE=DE/27.18
      E = EI
      EC = E
      E3 = E
C
C
C  START LOOP OVER ENERGIES IN GIVEN ENERGY RANGE
  700 EEV=(E-VV)*27.18
      WRITE (6,370)  E,EEV
C  COMPUTE COMPONENTS OF INCIDENT WAVEVECTOR PARALLEL TO SURFACE
      AK=SQRT(2.0*(E-VV))*SIN(THETA)
      AK2=AK*COS(FI)
      AK3=AK*SIN(FI)
      AK21=AK2
      AK31=AK3
C  SELECT BEAMS APPROPRIATE FOR CURRENT ENERGY
      CALL BEAMS(KNBS,KNB,SPQ,SPQF,KSYM,KNT,NPU,NPUN,AK2,AK3,E,TST,
     1NB,PQ,PQF,SYM,NPUC,MPU,NT,NP)
C  SET OPTICAL POTENTIAL (IMAGINARY PART OF MUFFIN-TIN CONSTANT,
C  REPRESENTING DAMPING). VPIS FOR SUBSTRATE, VPIO FOR OVERLAYER.
      VPIS=-5.0/27.18
      VPIO=VPIS
      DCUTO = SQRT(2.0 * E)
C  SET LIMITING RADII ON LATTICE SUMS, POSSIBLY DIFFERENT FOR SUBSTRATE
C  (DCUTS) AND OVERLAYER (DCUTO)
      DCUTS =  - 5.0 * DCUTO/(AMIN1(VPIS, - 0.05))
      DCUTO =  - 5.0 * DCUTO/(AMIN1(VPIO, - 0.05))
      WRITE (6,380)  VPIS, VPIO, DCUTS, DCUTO
C  FIX TEMPERATURE T. T MAY BE INCREASED STEPWISE IN A LOOP OVER THE
C  INPUT RANGE (TI,TF) WITH INPUT STEP DT
  770 T = TI
  780 WRITE (6,390)   T
C
C
C  START WITH OVERLAYER: E NOW BECOMES ENERGY IN OVERLAYER
      E = EC - VO
      E3 = E
C  PRODUCE ATOMIC T-MATRIX ELEMENTS FROM PHASE SHIFTS, CORRECTED FOR
C  THERMAL VIBRATION (CAF AND TSF) OR NOT (AF AND TSFO). AF AND CAF
C  ARE USED FOR BRAVAIS-LATTICE LAYERS. TSFO AND TSF ARE USED FOR
C  COMPOSITE LAYERS.
C  THIS CALL IS FOR OVERLAYER (IEL=2 IN INPUT SEQUENCE OF PHASE SHIFTS)
      CALL TSCATF(2,L1,ES,PHSS,NPSI,IT2,E,0.,PPP,NN1,NN2,NN3,DRO2,
     1DRPER2,DRPAR2,TO,T,TSFO,TSF,AF,CAF)
C  CHOOSE PARAMETERS RELEVANT TO OVERLAYER
      DRPER=DRPER2
      DRPAR=DRPAR2
      DRO=DRO2
C  IT=(0),1 MEANS INCLUDE (NO) EXPLICIT DEBYE-WALLER FACTORS IN MSMF
      IT=0
      TV = TVB
      VPI = VPIO
```

```
          VPI1 = VPIO
C   PERFORM PLANAR LATTICE SUM IN FMAT. IF MUFFIN-TIN CONSTANTS AND
C   DAMPINGS IN SUBSTRATE AND OVERLAYER ARE EQUAL, DO LATTICE SUMS FOR
C   BOTH TYPES OF LAYERS NOW (NLS=NL IMPLIES: DO ALL SUBLATTICES).
C   OTHERWISE DO ONLY OVERLAYER LATTICE SUM (NLS=1 IMPLIES: DO ONLY THE
C   SUBLATTICE EQUAL TO THE OVERLAYER LATTICE)
          NLS = 1
          IF (ABS(VPIS-VPIO)+ABS(VO) .LE. 1.0E-4)  NLS = NL
          CALL  FMAT (FLMS, V, JJS, NL, NLS, DCUTO, IDEG, LMAX,KLM)
C   PRODUCE REFLECTION (ROV) AND TRANSMISSION (TOV) MATRICES FOR THE
C   OVERLAYER. ONLY ONE REGISTRY: ID=1. ONLY ONE SUBLATTICE: NLL=1.
C   OVERLAYER SYMMETRIES AND REGISTRIES APPLY: LAY=1. TEMPERATURE-
C   DEPENDENT PHASE SHIFTS USED: CAF
          CALL MSMF(ROV,TOV,ROV,TOV,ROV,TOV,ROV,TOV,NT,NT,NT,AMULT,CYLM,
         1PQ,SYM,NT,FLMS,FLM,V,NL,0,0,1,1,CAF,CAF,L1,LX,LXI,LMMAX,KLM,
         2XEV,XOD,LEV,LOD,YLM,YLME,YLMO,IPLE,IPLO,CLM,NLM,1)
          WRITE (6,410)
C
C
C   CONSIDER SUBSTRATE NEXT. CHOOSE CORRECT ENERGY
          E = EC
          E3 = E
C   PRODUCE ATOMIC T-MATRIX ELEMENTS FOR SUBSTRATE
          CALL TSCATF(1,L1,ES,PHSS,NPSI,IT1,E,0.,PPP,NN1,NN2,NN3,DRO1,
         1DRPER1,DRPAR1,TO,T,TSFO,TSF,AF,CAF)
          VPI = VPIS
          VPI1 = VPIS
          IF (ABS(VPIS-VPIO)+ABS(VO) .LE. 1.0E-4)  GO TO 930
C   PERFORM PLANAR LATTICE SUMS FOR SUBSTRATE IF NOT DONE BEFORE
          CALL  FMAT (FLMS, V, JJS, NL, NL, DCUTS, IDEG, LMAX,KLM)
      930 CONTINUE
C   SET PARAMETERS FOR RUMPLED TOP SUBSTRATE LAYER (LAY=1 FOR OVERLAYER
C   SYMMETRIES AND REGISTRY)
          NA=0
          NS=0
          ID=1
          LAY=1
          TV1=TVB
          DCUT=DCUTS
          EPSN=0.01
          NPERT=5
          NOPT=1
          NEW=1
          VPI1=VPIS
C   COMPUTE REFLECTION AND TRANSMISSION MATRICES FOR RUMPLED LAYER
          CALL MTINV(RTP,TTP,RTM,TTM,RTP,TTP,RTM,TTM,RTP,TTP,RTM,TTM,
         1NT,NT,NT,AMULT,CYLM,PQ,SYM,NT,FLMS,NL,LX,LXI,LT,LXM,LMMAX,
         2KLM,XEV,LEV,LEV2,TAU,LMT,TAUG,TAUGM,CLM,NLM,FPOS,POSS,
         3MGH,NLAY,DRL,SDRL,NUGH,NGEQ,NGOL,NLAY2,TEST,
         4GH,LMG,RG,TS,LMN,TG,LM2N,VT,CAA,NCAA,TH,LMNI,IPL)
          WRITE(6,416)
C   SET PARAMETERS FOR UNRUMPLED BULK LAYERS
          DRPER=DRPER1
          DRPAR=DRPAR1
          DRO=DRO1
          IT=0
          TV = TVA
C   PRODUCE REFLECTION (RA1,RA2) AND TRANSMISSION (TA1,TA2) MATRICES
C   FOR SUBSTRATE LAYERS. TWO REGISTRIES: ID=2 (RA1,TA1 FOR REGISTRY
C   SS1, RA2,TA2 FOR REGISTRY SS2). NLL=NL SUBLATTICES. SUBSTRATE LAYER
C   SYMMETRIES AND REGISTRIES APPLY: LAY=2.
C   FIRST BEAM SET FIRST (NAA BEAMS)
          NAA = NB(1)
          CALL MSMF(RA1,TA1,RA2,TA2,RA1,TA1,RA1,TA1,NAA,NT,NP,AMULT,CYLM,
         1PQ,SYM,NT,FLMS,FLM,V,NL,  0,  0,2,NL,CAF,CAF,L1,LX,LXI,LMMAX,KLM,
         2XEV,XOD,LEV,LOD,YLM,YLME,YLMO,IPLE,IPLO,CLM,NLM,2)
          WRITE (6,410)
```

```
C   SECOND BEAM SET NEXT (NBB BEAMS; NOTE OFFSETS NA=NS=NAA)
      NBB = NB(2)
      CALL MSMF(RA1,TA1,RA2,TA2,RA1,TA1,RA1,TA1,NBB,NT,NP,AMULT,CYLM,
     1PQ,SYM,NT,FLMS,FLM,V,NL,NAA,NAA,2,NL,CAF,CAF,L1,LX,LXI,LMMAX,KLM,
     2XEV,XOD,LEV,LOD,YLM,YLME,YLMO,IPLE,IPLO,CLM,NLM,2)
      WRITE (6,410)
C
C
C   NOW STACK LAYERS.
C   ASW BECOMES VECTOR BETWEEN 1ST AND 2ND SUBSTRATE LAYERS, AND WILL
C   BE VARIED BY SSTEP IN LOOP 969
      DO 936 J=2,3
  936 ASW(J)=ASA(J)
      ASW(1)=ASA(1)-FPOS(2,1)
      SSTEP=0.05/0.529
      DO 969 ISUB=1,3
      DO 935 J=1,3
  935 AS(J)=ASW(J)*0.529
      WRITE(6,430)(AS(J),J=1,3)
C   ASV BECOMES THE 3-D VECTOR FROM OVERLAYER TO TOP SUBSTRATE LAYER
C   AND WILL BE VARIED
      DO 937 J=1,3
  937 ASV(J)=ASB(J)
C   LOOP 960 RUNS OVER GEOMETRIES: HERE THE OVERLAYER SPACING ASV(1) IS
C   VARIED
      DO 960 I = 1, NSTEP
      DO 940 J=1,3
  940 AS(J)=ASV(J)*0.529
      A=AS(1)
      WRITE (6,430) (AS(J),J=1,3)
C   COMPUTE REFLECTED BEAM AMPLITUDES (XI) BY RFS
      CALL RFS2OV(ROV,TOV,ROV,TOV,RTP,TTP,RTM,TTM,RA1,TA1,RA2,TA2,
     1NT,NB,KNBS,NP,XI,PQ,PK,AW,ANEW,20,ASV,ASW,ASA,ASA)
C   COMPUTE REFLECTED INTENSITIES (AT) FROM AMPLITUDES (XI), PRINT AND
C   PUNCH (NPNCH=1; PUNCH LAYER SPACING A ALSO FOR LATER IDENTIFICATION)
      CALL RINT(NT,XI,AT,PQ,PQF,SYM,VV,THETA,FI,MPU,NPUC,EEV,A,1)
  960 ASV(1)=ASV(1)+STEP
      ASW(1)=ASW(1)-SSTEP
  969 CONTINUE
      IF (IT1+IT2+IT3) 980,990,980
  980 T = T + DT
      IF (DT.LE.0.0001) GO TO 990
      IF (T-TF)   780, 780, 990
  990 E = E + DE
      EC = E
      E3 = E
      IF (E-EF)   700, 700, 680
 1000 WRITE (6,450)
      STOP
      END
```

---- INPUT DATA FOR PRECEDING PROGRAM FOLLOW -------------

```
W(1GO)+C(2*2)H, W RUMPLING(BEEBY), RFS, FULL SYMM., 5 PH. SH.
   4   2   1           IDEG NL1 NL2
   2  30   6           KNBS KNT NPUN
  12                   NPSI
 3.1600 0.0000         ARA1
 0.0000 3.1600         ARA2
 0.5000 0.5000         SS1
 0.0000 0.0000         SS2
 0.0000 0.0000         SS3
 0.0000 0.0000         SS4
 1.5800 0.0000 0.0000  ASA
 3.1600-3.1600         ARB1
 3.1600 3.1600         ARB2
 0.0000 0.0000         SO1
```

```
0.0000 0.0000           SO2
0.0000 0.0000           SO3
1.4000 0.0000 0.0000    ASB
0.5000 0.4000 0.1000    FR ASE STEP
   5                    NSTEP
  17                    KNB(1)
0.0000 0.0000    1   1  BEAMS
1.0000 0.0000    7   7
1.0000 1.0000    7   7
2.0000 0.0000    7   7
2.0000 1.0000   10  10
2.0000 2.0000    7   7
3.0000 0.0000    7   7
3.0000 1.0000   10  10
3.0000 2.0000   10  10
4.0000 0.0000    7   7
4.0000 1.0000   10  10
3.0000 3.0000    7   7
4.0000 2.0000   10  10
5.0000 0.0000    7   7
5.0000 1.0000   10  10
4.0000 3.0000   10  10
5.0000 2.0000   10  10
  13                    KNB(2)
0.5000 0.5000    7   7
1.5000 0.5000   10  10
1.5000 1.5000    7   7
2.5000 0.5000   10  10
2.5000 1.5000   10  10
3.5000 0.5000   10  10
2.5000 2.5000    7   7
3.5000 1.5000   10  10
3.5000 2.5000   10  10
4.5000 0.5000   10  10
4.5000 1.5000   10  10
4.5000 2.5000   10  10
3.5000 3.5000    7   7
0.0020                  TST
   1   2   3   4  18  19  NPU
   0.00    0.00         THETA FI
   0.00   10.00         VO VV
0.0010                  EPS
   5                    LITER
   1   1   0            IT1 IT2 IT3
400.0000 183.8500    1.4000    1.4000    0.0000    THDB1 AM1 FPER1 FPAR1 DRO1
5402.0000    1.0080    2.0000    2.0000    0.0000    THDB2 AM2 FPER2 FPAR2 DRO2
   1.0000    1.0000    1.0000    1.0000    0.0000    THDB3 AM3 FPER3 FPAR3 DRO3
300.0000                TI TF DT
   4                    LMAX
   2                    NEL
 .3000
-.8877 -.3222-2.5871   .0032    .0001    .0000    .0000    .0000    .0000    .0000
1.3979 0.0211 0.0003 0.0000 0.0000 0.0000 0.0000
 .4000
-1.0151 -.4178-1.5835   .0082    .0003    .0000    .0000    .0000    .0000    .0000
1.3321 0.0320 0.0006 0.0000 0.0000 0.0000 0.0000
 .5000
-1.1255 -.5032-1.0993   .0169    .0008    .0000    .0000    .0000    .0000    .0000
1.2785 0.0440 0.0010 0.0000 0.0000 0.0000 0.0000
 .7500
-1.3563 -.6825 -.9244   .0612    .0042    .0002    .0000    .0000    .0000    .0000
1.1743 0.0772 0.0025 0.0001 0.0000 0.0000 0.0000
1.0000
-1.5484 -.8282 -.9594   .1479    .0127    .0009    .0001    .0000    .0000    .0000
1.0944 0.1125 0.0050 0.0001 0.0000 0.0000 0.0000
1.5000
-1.8682-1.0650-1.0764   .4547    .0565    .0060    .0006    .0000    .0000    .0000
```

```
 0.9739 0.1818 0.0125 0.0006 0.0000 0.0000 0.0000
 2.0000
-2.1338-1.2639-1.1761  .7973  .1489  .0205  .0026  .0003  .0000  .0000
 0.8847 0.2416 0.0233 0.0014 0.0001 0.0000 0.0000
 3.0000
-2.5582-1.6008-1.3333 1.1383  .4604  .0950  .0180  .0028  .0004  .0000
 0.7610 0.3215 0.0527 0.0049 0.0003 0.0000 0.0000
 4.0000
-2.8837-1.8755-1.4771 1.2583  .7767  .2267  .0591  .0129  .0023  .0003
 0.6829 0.3567 0.0877 0.0113 0.0009 0.0001 0.0000
 5.0000
-3.1446-2.1007-1.6131 1.3196 1.0178  .3728  .1271  .0359  .0083  .0016
 0.6334 0.3644 0.1227 0.0207 0.0022 0.0002 0.0000
 6.0000
-3.3643-2.2904-1.7352 1.3479 1.2153  .4993  .2083  .0735  .0212  .0050
 0.6021 0.3577 0.1534 0.0328 0.0043 0.0004 0.0000
 7.0000
-3.5565-2.4564-1.8434 1.3564 1.3807  .6050  .2868  .1215  .0425  .0122
 0.5819 0.3447 0.1778 0.0469 0.0074 0.0008 0.0001
 2                       NLAY
 0.0000 0.0000 0.0000    FPOS(1,J)
 0.1000 3.1600 0.0000    FPOS(2,J)
 0                       NVAR
 0.0000 0.0000 0.0000    VPOS(1,J)
 1                       NTAU
 1   1                   LPS
 2                       NINV
 0   0                   INV
 40.00 126.00   5.00     EI EF DE
-10.00                   EI EF DE
----------------------------------------------------------
---- FIRST PART OF PRINTER OUTPUT FOLLOWS ---------------
----(NEGLECTING MOST BLANK LINES AND PAGE --------------
---- EJECTS AND TRUNCATING LONG LINES)    --------------
----------------------------------------------------------
 W(100)+C(2*2)H, W RUMPLING(BEEBY), RFS, FULL SYMM., 5 PH. SH.
 IDEG =   4 NL1 =   2 NL2 =    1
 PARAMETERS FOR INTERIOR:
 SURF VECS        3.1600  0.0000
                  0.0000  3.1600
 SS     .5000   .5000
 SS     0.0000 0.0000
 SS     0.0000 0.0000
 SS     0.0000 0.0000
 ASA    1.5800 0.0000 0.0000
 PARAMETERS FOR SURFACE:
 SURF VECS        3.1600 -3.1600
                  3.1600  3.1600
 SO     0.0000 0.0000
 SO     0.0000 0.0000
 SO     0.0000 0.0000
 ASB    1.4000 0.0000 0.0000
 FR =   .5000 ASE =   .4000 STEP =   .1000 NSTEP =    5
         PQ1    PQ2        SYM

     1   0.000  0.000      1   1
     2   1.000  0.000      7   7
     3   1.000  1.000      7   7
     4   2.000  0.000      7   7
     5   2.000  1.000     10  10
     6   2.000  2.000      7   7
     7   3.000  0.000      7   7
     8   3.000  1.000     10  10
     9   3.000  2.000     10  10
    10   4.000  0.000      7   7
    11   4.000  1.000     10  10
    12   3.000  3.000      7   7
```

```
13      4.000   2.000       10 10
14      5.000   0.000        7  7
15      5.000   1.000       10 10
16      4.000   3.000       10 10
17      5.000   2.000       10 10

18       .500    .500        7  7
19      1.500    .500       10 10
20      1.500   1.500        7  7
21      2.500    .500       10 10
22      2.500   1.500       10 10
23      3.500    .500       10 10
24      2.500   2.500        7  7
25      3.500   1.500       10 10
26      3.500   2.500       10 10
27      4.500    .500       10 10
28      4.500   1.500       10 10
29      4.500   2.500       10 10
30      3.500   3.500        7  7
TST =   .0020
PUNCHED BEAMS         1   2   3   4 18 19
THETA  FI =      0.00     0.00
VO       0.00VV         10.00
EPS =    .0010 LITER =    5
IT1 =    1 IT2 =   1 IT3 =   0
THDB =  400.0000 AM =  183.8500 FPER =    1.4000 FPAR =    1.4000 DRO =
THDB = 5402.0000 AM =    1.0080 FPER =    2.0000 FPAR =    2.0000 DRO =
THDB =    1.0000 AM =    1.0000 FPER =    1.0000 FPAR =    1.0000 DRO =
LMAX =   4
        PHASE SHIFTS

E =    .3000  1ST ELEMENT      -.8877  -.3222 -2.5871    .0032    .0001
              2ND ELEMENT     1.3979   .0211   .0003   0.0000   0.0000

E =    .4000  1ST ELEMENT     -1.0151  -.4178 -1.5835    .0082    .0003
              2ND ELEMENT     1.3321   .0320   .0006   0.0000   0.0000

E =    .5000  1ST ELEMENT     -1.1255  -.5032 -1.0993    .0169    .0008
              2ND ELEMENT     1.2785   .0440   .0010   0.0000   0.0000

E =    .7500  1ST ELEMENT     -1.3563  -.6825  -.9244    .0612    .0042
              2ND ELEMENT     1.1743   .0772   .0025    .0001   0.0000

E =   1.0000  1ST ELEMENT     -1.5484  -.8282  -.9594    .1479    .0127
              2ND ELEMENT     1.0944   .1125   .0050    .0001   0.0000

E =   1.5000  1ST ELEMENT     -1.8682 -1.0650 -1.0764    .4547    .0565
              2ND ELEMENT      .9739   .1818   .0125    .0006   0.0000

E =   2.0000  1ST ELEMENT     -2.1338 -1.2639 -1.1761    .7973    .1489
              2ND ELEMENT      .8847   .2416   .0233    .0014    .0001

E =   3.0000  1ST ELEMENT     -2.5582 -1.6008 -1.3333   1.1383    .4604
              2ND ELEMENT      .7610   .3215   .0527    .0049    .0003

E =   4.0000  1ST ELEMENT     -2.8837 -1.8755 -1.4771   1.2583    .7767
              2ND ELEMENT      .6829   .3567   .0877    .0113    .0009

E =   5.0000  1ST ELEMENT     -3.1446 -2.1007 -1.6131   1.3196   1.0178
              2ND ELEMENT      .6334   .3644   .1227    .0207    .0022

E =   6.0000  1ST ELEMENT     -3.3643 -2.2904 -1.7352   1.3479   1.2153
              2ND ELEMENT      .6021   .3577   .1534    .0328    .0043

E =   7.0000  1ST ELEMENT     -3.5565 -2.4564 -1.8434   1.3564   1.3807
              2ND ELEMENT      .5819   .3447   .1778    .0469    .0074
COMPOSITE LAYER VECTOR :   0.0000 0.0000 0.0000
```

```
COMPOSITE LAYER VECTOR :    .1000 3.1600 0.0000
VARYING POSITION IN COMPOSITE LAYER :  0.0000 0.0000 0.0000
NTAU = 1
PHASE SHIFT ASSIGNMENT IN COMPOSITE LAYER   1  1
 2 SUBPLANES IN BEEBY INVERSION :   0  0
NO. OF ELEMENTS IN CAAA :   759
ENERGY = 1.8396 H  OR    40.00 EV
 10 BEAMS USED :   6    4
   0.000 0.000   1.000 0.000   1.000 1.000   2.000 0.000   2.000 1.000
  2.500  .500
VPIS =    -.1840 VPIO =    -.1840 DCUTS =   52.1344 DCUTO =   52.1344
TEMP = 300.0000
MSMF   OK
NO. OF LATT. PTS. FOR GH :   120
NO. OF LATT. PTS. FOR GH :   120
MTINV  OK
MSMF   OK
MSMF   OK
SURFACE GEOMETRY : 1.4800 0.0000 0.0000
SURFACE GEOMETRY : 1.4000 0.0000 0.0000
IMAX = 9      BNORM =    .3133E-03
ANORM =   .3736E-01

IMAX = 2      BNORM =    .2588E-01
ANORM =   .3741E-01

IMAX = 2      BNORM =    .2541E-03
ANORM =   .3741E-01

E=   1.8396 THETA=    0.0000 FI=   0.0000

   PQ1    PQ2      REF INT    SYM
 0.000  0.000   .76154E-02    1
 1.000  0.000   .16527E-02    7
 1.000  1.000   .19366E-02    7
  .500   .500   .14379E-04    7
 1.500   .500   .87869E-04   10
SURFACE GEOMETRY : 1.5000 0.0000 0.0000
IMAX = 9      BNORM =    .3051E-03
ANORM =   .3090E-01

IMAX = 2      BNORM =    .2370E-01
ANORM =   .3096E-01

IMAX = 2      BNORM =    .2511E-03
ANORM =   .3096E-01

E=   1.8396 THETA=    0.0000 FI=   0.0000

   PQ1    PQ2      REF INT    SYM
 0.000  0.000   .63069E-02    1
 1.000  0.000   .11411E-02    7
 1.000  1.000   .13471E-02    7
  .500   .500   .12059E-05    7
 1.500   .500   .12928E-03   10
SURFACE GEOMETRY : 1.6000 0.0000 0.0000
IMAX = 9      BNORM =    .2958E-03
ANORM =   .2611E-01

IMAX = 2      BNORM =    .2344E-01
ANORM =   .2617E-01

IMAX = 2      BNORM =    .2343E-03
ANORM =   .2617E-01

E=   1.8396 THETA=    0.0000 FI=   0.0000
```

```
   PQ1     PQ2      REF INT      SYM
  0.000   0.000    .49243E-02     1
  1.000   0.000    .68164E-03     7
  1.000   1.000    .83833E-03     7
   .500    .500    .23941E-04     7
  1.500    .500    .19487E-03    10
SURFACE GEOMETRY : 1.7000 0.0000 0.0000
IMAX =  9     BNORM =      .2855E-03
ANORM =     .2354E-01


IMAX =  2     BNORM =      .2482E-01
ANORM =     .2359E-01


IMAX =  2     BNORM =      .2015E-03
ANORM =     .2359E-01


E=    1.8396 THETA=     0.0000 FI=     0.0000

   PQ1     PQ2      REF INT      SYM
  0.000   0.000    .39253E-02     1
  1.000   0.000    .41449E-03     7
  1.000   1.000    .53103E-03     7
   .500    .500    .73124E-04     7
  1.500    .500    .25439E-03    10
SURFACE GEOMETRY : 1.8000 0.0000 0.0000
IMAX =  9     BNORM =      .2747E-03
ANORM =     .2284E-01


IMAX =  2     BNORM =      .2693E-01
ANORM =     .2289E-01


IMAX =  2     BNORM =      .1471E-03
ANORM =     .2289E-01


E=    1.8396 THETA=     0.0000 FI=     0.0000

   PQ1     PQ2      REF INT      SYM
  0.000   0.000    .35729E-02     1
  1.000   0.000    .39096E-03     7
  1.000   1.000    .45979E-03     7
   .500    .500    .13581E-03     7
  1.500    .500    .27426E-03    10
SURFACE GEOMETRY : 1.4300 0.0000 0.0000
SURFACE GEOMETRY : 1.4000 0.0000 0.0000
IMAX =  9     BNORM =      .3287E-03
ANORM =     .4076E-01


IMAX =  2     BNORM =      .3221E-01
ANORM =     .4083E-01


IMAX =  2     BNORM =      .3753E-03
ANORM =     .4083E-01


E=    1.8396 THETA=     0.0000 FI=     0.0000

   PQ1     PQ2      REF INT      SYM
  0.000   0.000    .83495E-02     1
  1.000   0.000    .18664E-02     7
  1.000   1.000    .21712E-02     7
   .500    .500    .18538E-04     7
  1.500    .500    .75328E-04    10
SURFACE GEOMETRY : 1.5000 0.0000 0.0000
IMAX =  9     BNORM =      .3190E-03
ANORM =     .3351E-01


IMAX =  2     BNORM =      .2956E-01
ANORM =     .3359E-01
```

```
IMAX =  2     BNORM =     .3523E-03
ANORM =    .3359E-01

E=   1.8396 THETA=   0.0000 FI=    0.0000

    PQ1    PQ2      REF INT      SYM
  0.000  0.000   .69840E-02       1
  1.000  0.000   .13362E-02       7
  1.000  1.000   .15216E-02       7
   .500   .500   .52262E-06       7
  1.500   .500   .11209E-03      10
```

10.9 Main Program for an Upright Diatomic Molecule on an fcc or bcc (100) Substrate, Using Layer Doubling

We consider the particular case of Ni(100) + c(2×2)CO, assuming that the molecule stands upright in a site of reasonably high symmetry. The C-O bond length may be expected to be about 1.13 Å, as in the free molecule, and so as not to risk convergence problems with RFS between the layers of C and O, we choose Layer Doubling instead. This requires a number of additional matrices for working space (S1, S2, S3, S4, RS, RN in the program). To reduce the total storage requirements of these matrices, the following symmetry is assumed and used: the structure shall have two orthogonal mirror planes and normal incidence is required. Several adsorption sites have this symmetry: the top, hollow and two bridge sites. The molecule cannot be inclined away from the surface normal without breaking this symmetry.

To investigate inclined molecular orientations, a lower symmetry must be assumed, which increases matrix dimensions greatly. In that case it may be more advantageous to treat the molecular layer as a composite layer, because then RFS could be used between the layers, requiring fewer matrices; also many more adsorption models (dissociated molecule, coplanar adatoms, etc.) could then easily be considered as well. Of course, one may in any case alternatively or simultaneously limit oneself to lower energies (e.g., $E \lesssim 120$ eV), which also reduces matrix dimensions (fewer phase shifts, fewer beams).

The main ideas of the main program presented for this surface structure are the same as those in Sect. 10.7. Although lower symmetry is assumed here than in Sect. 10.7, we still have to select a propagation track that coincides with the x-axis: cf. Fig. 10.7; this is because the propagation track must be contained in each of the symmetry elements, in this case the two orthogonal mirror planes, which intersect in the x-axis.

The program listed below puts the admolecules in top sites. To put them in hollow sites one merely has to tip the substrate slab upside down, as in Sect. 10.7. The main program contains instructions to do this in comment lines: one simply uses the reflection matrix for the bottom rather than the top of the substrate slab produced by Layer Doubling from the substrate layers. An alternative is to exchange the substrate layer registries SS1 and SS2 in the input (requiring a fresh run).

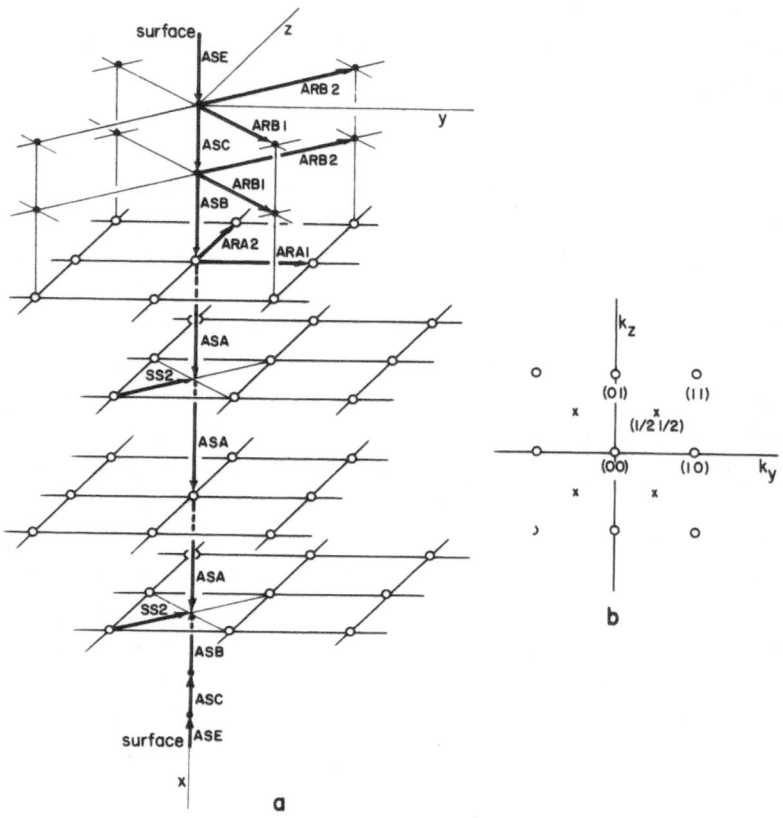

Fig. 10.7 (a) Geometry for a double c(2×2) overlayer on fcc or bcc(100). The diatomic molecule stands upright in top adsorption sites at top of figure, in hollow adsorption sites at bottom of figure (SS2 = (0,0)). The (xy) and (xz) planes are assumed to be mirror planes. (b) Reciprocal space (o = integral-order beams, x = fractional-order beams)

To obtain the bridge site, one must be careful to shift one of the two symmetry planes so that both still intersect the adatoms. To do this, one would leave SO1 = (0,0) for the registry of the two overlayers, and set SS1 = (1/2, 0) and SS2 = (0,1/2). Two 90°-rotated overlayer domains can exist with bridge sites: interchanging SS1 and SS2 (or tipping the substrate slab upside down) produces the other bridge site, rotated 90° from the first one; however, at normal incidence (assumed here) a simple permutation of the beam intensities makes a separate calculation for the second bridge-site domain unnecessary.

Here we have assumed that the carbon atoms are closest to the substrate, the oxygen atoms sticking into the vacuum. To put the molecule upside down, it is sufficient to interchange the use of the overlayer diffraction matrices ROM,TOM (for carbon) and ROV,TOV (for oxygen).

Note that the innermost overlayer is added onto the substrate first, by subroutine DBG, producing a reflection matrix for the system composed of the substrate and that overlayer. That system is then considered to be a substrate for the addition of the second outermost overlayer. An equivalent approach would be to combine the two overlayers first (using subroutine DBLG instead) and to then consider that double overlayer as a single overlayer to be put onto the substrate with subroutine ADREF1. This however requires more storage matrices, because it is desirable to save the individual overlayer matrices in order to allow more geometries to be considered (DBLG overwrites the matrices input to it, DBG does not).

Two loops over the two layer spacings are included in the program to allow a search for the best combination of spacings. (Subroutine READIN is not sufficient for input of all the geometrical data for this double-overlayer case and is therefore supplemented by some additional input lines in the main program; for more complicated surface structures subroutine READCL might be used, neglecting its primary purpose for composite layers).

```
C   MAIN + DATA FOR NI(100)+C(2*2)CO, LAY. DBLNG., (XY) AND (XZ) MIRR. PL
C   MAIN LEED PROGRAM FOR A DOUBLE C(2*2) OVERLAYER ON FCC OR BCC (100),
C   USING 2 ORTHOGONAL MIRROR PLANES, BY LAYER DOUBLING.
        DIMENSION ARA1(2),ARA2(2),RAR1(2),RAR2(2),SS1(2),SS2(2),SS3(2),
       1SS4(2),ASA(3),AS(3),ARB1(2),ARB2(2),RBR1(2),RBR2(2),SO1(2),
       2SO2(2),SO3(2),ASB(3),ASV(3),ASW(3),ASC(3)
C   2=NL, 120=KLM
        COMPLEX VL(2,2),FLMS(2,120),FLM(120)
C   2(1ST DIMENSION)=NL=NL1*NL2, 4(2ND DIMENSION)=IDEG
        DIMENSION V(2,2),JJS(2,4)
C   2=KNBS=NO. OF BEAM SETS READ IN, 21=KNT=NO. OF BEAMS READ IN
        DIMENSION KNB(2),NB(2),SPQF(2,21),SPQ(2,21),PQF(2,21),PQ(2,21)
        INTEGER KSYM(2,21),SYM(2,21)
C   20 MUST BE .GE. NPUN
        DIMENSION NPU(20),NPUC(20)
C   12=NPSI, 24=NEL*(LMAX+1)
        DIMENSION ES(12),PHSS(12,24)
C   7680=NLM FROM DATA STATEMENT, 225=NN, 15=N (FOR CELMG)
C   YLM IS ALSO USED, AS COMPLEX VECTOR OF SMALLER CORE SIZE, IN MSMF,ETC
        DIMENSION NLMS(7),CLM(7680),YLM(225),FAC2(225),FAC1(15)
C   15=NN1, 8=NN2, 8=NN3
        DIMENSION PPP(15,8,8)
C   64=LMMAX
        DIMENSION LX(64),LXI(64),LXM(64),LT(64)
C   8=LMAX+1=L1, 3 AND 10 FIXED FOR COMMON /MPT/ IF PRESENT
        COMPLEX AF(8),CAF(8),TSFO(3,10),TSF(3,10)
        COMPLEX ROV(21,21),TOV(21,21),ROM(21,21),TOM(21,21)
        COMPLEX RS(21,21),RN(21,21)
C   13=KNB(1)
        COMPLEX RA1(13,13),TA1(13,13),RA2(13,13),TA2(13,13)
C   8=KNB(2)
        COMPLEX RB1(8,8),TB1(8,8),RB2(8,8),TB2(8,8)
C   13=KNP
        COMPLEX PP(2,21),XS(21),S1(21,21),S2(21,21),
       1S3(13,13),S4(13,13)
        DIMENSION JNT(21)
C   36=LEV, 28=LOD
        COMPLEX AMULT(21),CYLM(21,64),XEV(36,36),XOD(28,28),
       1YLME(36),YLMO(28)
        DIMENSION IPLE(36),IPLO(28),AT(21)
        COMPLEX XI(21)
        COMMON  E3,AK21,AK31,VPI1,TV,EMACH
        COMMON /X4/E,VPI,AK2,AK3,ASA
        COMMON  /SL/ARA1, ARA2, ARB1, ARB2, RBR1, RBR2, NL1, NL2
        COMMON /MS/LMAX,EPS,LITER,SO1,SO2,SO3,SS1, SS2, SS3,SS4
        COMMON /ADS/ASL, FR, ASE, VPIS, VPIO, VO, VV
        COMMON /BT/IT,T,TO,DRPER,DRPAR,DRO
        COMMON /TEMP/IT1,IT2,IT3,TI,TF,DT,TO1,DRPER1,DRPAR1,DRO1,DRPER2,
       1DRPAR2,DRO2,DRPER3,DRPAR3,DRO3
C   NLMS IS DIMENSION OF CLM AS FUNCTION OF LMAX
        DATA NLMS(1),NLMS(2),NLMS(3),NLMS(4),NLMS(5),NLMS(6),NLMS(7)/
       10,76,284,809,1925,4032,7680/
    160 FORMAT(3F7.4)
    161 FORMAT(3F7.2)
    165 FORMAT(26H INPUT INTERLAYER VECTOR :,3F7.4)
    200 FORMAT(20I3)
    240 FORMAT(8HOIDEG = ,1I3,7H NL1 = ,1I3,7H NL2 = ,1I3)
    370 FORMAT(9H1ENERGY =,F7.4,7H H  OR ,1F7.2,3H EV)
    380 FORMAT(8H VPIS = ,F9.4,8H VPIO = ,F9.4,9H DCUTS = ,F9.4,9H DCUTO =
       1 ,F9.4)
    390 FORMAT(8HOTEMP = ,F9.4)
    410 FORMAT(10H MSMF   OK)
    420 FORMAT(10H SUBREF OK)
    430 FORMAT(20HOSURFACE GEOMETRY : ,3F7.4)
    450 FORMAT(20H CORRECT TERMINATION)
    460 FORMAT(20A4)
    465 FORMAT(1H1,20A4)
```

```
C
C
C     READ, WRITE AND PUNCH A DESCRIPTIVE TITLE
          READ (5,460)(CLM(I),I=1,20)
          WRITE(6,465)(CLM(I),I=1,20)
          WRITE(7,460)(CLM(I),I=1,20)
C     EMACH IS MACHINE ACCURACY (USED BY ZGE AND ZSU)
          EMACH=1.0E-6
C     IDEG: EACH LAYER HAS AN IDEG-FOLD SYMMETRY AXIS
C     NL1, NL2: SUPERLATTICE CHARACTERIZATION  NL1    NL2
C                              P(1*1)            1      1
C                              C(2*2)            2      1
C                              P(2*1)            2      1
C                              P(1*2)            1      2
C                              P(2*2)            2      2
          READ(5,200)IDEG,NL1,NL2
          WRITE(6,240)IDEG,NL1,NL2
          NL=NL1*NL2
C     KNBS= NO.OF BEAM SETS TO BE READ IN (.LE.NL)
C     KNT= TOTAL NO. OF BEAMS TO BE READ IN
C     NPUN= NO. OF BEAMS FOR WHICH INTENSITIES ARE TO BE PUNCHED OUT
          READ(5,200)KNBS,KNT,NPUN
C     NPSI= NO.OF ENERGIES AT WHICH PHASE SHIFTS WILL BE READ IN
          READ(5,200)NPSI
C     READ IN GEOMETRY, PHYSICAL PARAMETERS, CONVERGENCE CRITERIA
          CALL READIN(TVA,RAR1,RAR2,ASA,TVB,ASB,STEP,NSTEP,IDEG,NL,V,VL,
         1JJS,KNBS,KNB,KNT,SPQF,KSYM,SPQ,TST,TSTS,NPUN,NPU,THETA,FI,
         2LMMAX,NPSI,ES,PHSS,L1)
C     READ IN ADDITIONAL GEOMETRY INFORMATION FOR THE PRESENT TWO-
C     OVERLAYER SITUATION.
C     NVAR= NO. OF SPACINGS BETWEEN OVERLAYERS.
C      ASC= INITIAL INTEROVERLAYER VECTOR.
C     STEPC= STEP IN VARIATION OF SPACING ASC(1).
          READ(5,200)NVAR
          READ(5,160)(ASC(J),J=1,3)
          WRITE(6,165)(ASC(J),J=1,3)
          READ(5,160)STEPC
          DO 610 J=1,3
      610 ASC(J)=ASC(J)/0.529
          STEPC=STEPC/0.529
          TO=TO1
          KLM=(2*LMAX+1)*(2*LMAX+2)/2
          LEV=(LMAX+1)*(LMAX+2)/2
          LOD=LMMAX-LEV
      630 N=2*LMAX+1
          NN=N*N
          NLM=NLMS(LMAX)
C     CLM= CLEBSCH-GORDON COEFFICIENTS FOR MATRICES X AND TAU
          CALL CELMG(CLM,NLM,YLM,FAC2,NN,FAC1,N,LMAX)
C     LX,LXI,LT,LXM: PERMUTATIONS OF (L,M)-SEQUENCE
          CALL LXGENT(LX,LXI,LT,LXM,LMAX,LMMAX)
          IF (IT1+IT2+IT3) 640,650,640
      640 NN3=LMAX+1
          NN2=LMAX+1
          NN1=NN2+NN3-1
C     PPP= CLEBSCH-GORDON COEFFICIENTS FOR COMPUTATION OF TEMPERATURE-
C     DEPENDENT PHASE SHIFTS (SKIPPED IF NOT NEEDED)
          CALL  CPPP (PPP, NN1, NN2, NN3)
      650 CONTINUE
C
C
C     READ ENERGY RANGE AND STEP; IF EF AND DE ARE BLANK, PROGRAM COMPUTES
C     FOR ENERGY EI. PROGRAM ALWAYS RETURNS TO THIS LINE TO READ A NEW
C     ENERGY RANGE AND STEP. IF A NEGATIVE EI IS READ, RUN IS TERMINATED
      680 READ (5,161)  EI, EF, DE
          IF (EI)  1000, 690, 690
      690 EI=EI/27.18+VV
```

```
      EF=EF/27.18+VV
      DE=DE/27.18
      E = EI
      EC = E
      E3 = E
C
C
C   START LOOP OVER ENERGIES IN GIVEN ENERGY RANGE
  700 EEV=(E-VV)*27.18
      WRITE (6,370)  E,EEV
C   COMPUTE COMPONENTS OF INCIDENT WAVEVECTOR PARALLEL TO SURFACE
      AK=SQRT(2.0*(E-VV))*SIN(THETA)
      AK2=AK*COS(FI)
      AK3=AK*SIN(FI)
      AK21=AK2
      AK31=AK3
C   SELECT BEAMS APPROPRIATE FOR CURRENT ENERGY
      CALL BEAMS(KNBS,KNB,SPQ,SPQF,KSYM,KNT,NPU,NPUN,AK2,AK3,E,TST,
     1NB,PQ,PQF,SYM,NPUC,MPU,NT,NP)
C   SET OPTICAL POTENTIAL (IMAGINARY PART OF MUFFIN-TIN CONSTANT,
C   REPRESENTING DAMPING). VPIS FOR SUBSTRATE, VPIO FOR OVERLAYER.
C   HERE VPIS=VPIO PROPORTIONAL TO E**1/3
      VPIS=-(3.8/27.18)*EXP((ALOG((E)/(90.0/27.18+VV)))*0.333333)
      VPIO=VPIS
      DCUTO = SQRT(2.0 * E)
C   SET LIMITING RADII ON LATTICE SUMS, POSSIBLY DIFFERENT FOR SUBSTRATE
C   (DCUTS) AND OVERLAYER (DCUTO)
      DCUTS =  - 5.0 * DCUTO/(AMIN1(VPIS, - 0.05))
      DCUTO =  - 5.0 * DCUTO/(AMIN1(VPIO, - 0.05))
      WRITE (6,380)  VPIS, VPIO, DCUTS, DCUTO
C   FIX TEMPERATURE T. T MAY BE INCREASED STEPWISE IN A LOOP OVER THE
C   INPUT RANGE (TI,TF) WITH INPUT STEP DT
  770 T = TI
  780 WRITE (6,390)   T
C
C
C   START WITH OVERLAYER: E NOW BECOMES ENERGY IN OVERLAYER
      E = EC - VO
      E3 = E
C   PRODUCE ATOMIC T-MATRIX ELEMENTS FROM PHASE SHIFTS, CORRECTED FOR
C   THERMAL VIBRATION (CAF AND TSF) OR NOT (AF AND TSFO). AF AND CAF
C   ARE USED FOR BRAVAIS-LATTICE LAYERS. TSFO AND TSF ARE USED FOR
C   COMPOSITE LAYERS.
C   THIS CALL IS FOR OXYGEN OVERLAYER (IEL=3 IN INPUT SEQUENCE OF PHASE
C   SHIFTS)
      CALL TSCATF(3,L1,ES,PHSS,NPSI,IT3,E,O.,PPP,NN1,NN2,NN3,DRO3,
     1DRPER3,DRPAR3,TO,T,TSFO,TSF,AF,CAF)
C   CHOOSE PARAMETERS RELEVANT TO OXYGEN OVERLAYER
      DRPER=DRPER3
      DRPAR=DRPAR3
      DRO=DRO3
C   IT=(0),1 MEANS INCLUDE (NO) EXPLICIT DEBYE-WALLER FACTORS IN MSMF
      IT=0
      TV = TVB
      VPI = VPIO
      VPI1 = VPIO
C   PERFORM PLANAR LATTICE SUM IN FMAT. IF MUFFIN-TIN CONSTANTS AND
C   DAMPINGS IN SUBSTRATE AND OVERLAYER ARE EQUAL, DO LATTICE SUMS FOR
C   BOTH TYPES OF LAYERS NOW (NLS=NL IMPLIES: DO ALL SUBLATTICES).
C   OTHERWISE DO ONLY OVERLAYER LATTICE SUM (NLS=1 IMPLIES: DO ONLY THE
C   SUBLATTICE EQUAL TO THE OVERLAYER LATTICE)
      NLS = 1
      IF (ABS(VPIS-VPIO)+ABS(VO) .LE. 1.0E-4)  NLS = NL
      CALL  FMAT (FLMS, V, JJS, NL, NLS, DCUTO, IDEG, LMAX,KLM)
C   PRODUCE REFLECTION (ROV) AND TRANSMISSION (TOV) MATRICES FOR THE OXY.
C   OVERLAYER. ONLY ONE REGISTRY: ID=1. ONLY ONE SUBLATTICE: NLL=1.
C   OVERLAYER SYMMETRIES AND REGISTRIES APPLY: LAY=1. TEMPERATURE-
```

```
C   DEPENDENT PHASE SHIFTS USED: CAF
        CALL MSMF(ROV,TOV,ROV,TOV,ROV,TOV,ROV,TOV,NT,NT,NT,AMULT,CYLM,
       1PQ,SYM,NT,FLMS,FLM,V,NL,0,0,1,1,CAF,CAF,L1,LX,LXI,LMMAX,KLM,
       2XEV,XOD,LEV,LOD,YLM,YLME,YLMO,IPLE,IPLO,CLM,NLM,1)
        WRITE (6,410)
C
C   REPEAT ABOVE OPERATIONS FOR THE CARBON OVERLAYER (SAME LATTICE SUMS
C   ARE APPLICABLE).
C   ATOMIC T-MATRIX ELEMENTS FOR CARBON (IEL=2 IN INPUT SEQUENCE)
        CALL TSCATF(2,L1,ES,PHSS,NPSI,IT2,E,0.,PPP,NN1,NN2,NN3,DRO2,
       1DRPER2,DRPAR2,TO,T,TSFO,TSF,AF,CAF)
C   CHANGED PARAMETERS FOR CARBON OVERLAYER
        DRPER=DRPER2
        DRPAR=DRPAR2
        DRO=DRO2
C   DIFFRACTION MATRICES (ROM,TOM) FOR THE CARBON OVERLAYER
        CALL MSMF(ROM,TOM,ROM,TOM,ROM,TOM,ROM,TOM,NT,NT,NT,AMULT,CYLM,
       1PQ,SYM,NT,FLMS,FLM,V,NL,0,0,1,1,CAF,CAF,L1,LX,LXI,LMMAX,KLM,
       2XEV,XOD,LEV,LOD,YLM,YLME,YLMO,IPLE,IPLO,CLM,NLM,1)
        WRITE (6,410)
C
C
C   CONSIDER SUBSTRATE NEXT. CHOOSE CORRECT ENERGY
        E = EC
        E3 = E
C   PRODUCE ATOMIC T-MATRIX ELEMENTS FOR SUBSTRATE
        CALL TSCATF(1,L1,ES,PHSS,NPSI,IT1,E,0.,PPP,NN1,NN2,NN3,DRO1,
       1DRPER1,DRPAR1,TO,T,TSFO,TSF,AF,CAF)
        DRPER=DRPER1
        DRPAR=DRPAR1
        DRO=DRO1
        IT=0
        TV = TVA
        VPI = VPIS
        VPI1 = VPIS
        IF (ABS(VPIS-VPIO)+ABS(VO) .LE. 1.0E-4)  GO TO 930
C   PERFORM PLANAR LATTICE SUMS FOR SUBSTRATE IF NOT DONE BEFORE
        CALL  FMAT (FLMS,  V,  JJS,  NL,  NL,  DCUTS,  IDEG,  LMAX,KLM)
C   PRODUCE REFLECTION (RA1,RA2) AND TRANSMISSION (TA1,TA2) MATRICES
C   FOR SUBSTRATE LAYERS. TWO REGISTRIES: ID=2 (RA1,TA1 FOR REGISTRY
C   SS1, RA2,TA2 FOR REGISTRY SS2). NLL=NL SUBLATTICES. SUBSTRATE LAYER
C   SYMMETRIES AND REGISTRIES APPLY: LAY=2.
C   FIRST BEAM SET FIRST (NAA BEAMS)
  930 NAA = NB(1)
        CALL MSMF(RA1,TA1,RA2,TA2,RA1,TA1,RA1,TA1,NAA,NAA,NAA,AMULT,CYLM,
       1PQ,SYM,NT,FLMS,FLM,V,NL,   0,   0,2,NL,CAF,CAF,L1,LX,LXI,LMMAX,KLM,
       2XEV,XOD,LEV,LOD,YLM,YLME,YLMO,IPLE,IPLO,CLM,NLM,2)
        WRITE (6,410)
C   LAYER DOUBLING IS DONE FOR FIRST BEAM SET, PRODUCING REFLECTION AND
C   TRANSMISSION MATRICES FOR A THICK SUBSTRATE SLAB
        CALL SUBREF(RA1,TA1,RA2,TA2,NAA,S1,S2,S3,S4,PP,XS,JNT,NP,0,NT,PQ)
        WRITE(6,420)
C   SECOND BEAM SET NEXT (NBB BEAMS; NOTE OFFSETS NA=NAA, NS=0)
        NBB = NB(2)
        CALL MSMF(RB1,TB1,RB2,TB2,RB1,TB1,RB1,TB1,NBB,NBB,NBB,AMULT,CYLM,
       1PQ,SYM,NT,FLMS,FLM,V,NL,NAA,   0,2,NL,CAF,CAF,L1,LX,LXI,LMMAX,KLM,
       2XEV,XOD,LEV,LOD,YLM,YLME,YLMO,IPLE,IPLO,CLM,NLM,2)
        WRITE (6,410)
C   LAYER DOUBLING IS DONE FOR SECOND BEAM SET
        N=NAA
        CALL SUBREF(RB1,TB1,RB2,TB2,NBB,S1,S2,S3,S4,PP,XS,JNT,NP,N,NT,PQ)
        WRITE(6,420)
C
C
C   NOW ADD THE INNERMOST OVERLAYER BY LAYER DOUBLING.
C   COPY SUBSTRATE REFLECTION MATRICES ONTO A BLOCK-DIAGONAL MATRIX RS.
C   HERE THE TOP FACE OF THE SUBSTRATE SLAB IS CHOSEN, GIVING TOP
```

```
C   ADSORPTION SITES, SINCE SS1=(0,0) AND SO1=(0,0). TO OBTAIN THE HOLLOW
C   ADSORPTION SITE, LET MATCOP COPY RA2 AND RB2 ONTO RS: THAT SELECTS
C   THE BOTTOM FACE OF THE SUBSTRATE SLAB, WHERE THE REGISTRY IS SS2=
C   (.5,.5)
        CALL MATCOP(RA1,NAA,RB1,NBB,RB1,NBB,RB1,NBB,RS,NT,2)
C   ASW BECOMES THE 3-D VECTOR FROM THE INNER OVERLAYER TO THE SUBSTRATE.
C   THE SPACING ASW(1) WILL BE VARIED IN LOOP 969
        DO 935 J=1,3
  935 ASW(J)=ASB(J)
        DO 969 ISTEP=1,NSTEP
        DO 936 J=1,3
  936 AS(J)=ASW(J)*0.529
        WRITE(6,430)(AS(J),J=1,3)
C   RN BECOMES THE REFLECTION MATRIX FOR SUBSTRATE + 1ST (CARBON)
C   OVERLAYER
        CALL DBG(ROM,TOM,ROM,TOM,RS,RN,ASW,NT,0,PQ,NT,S1,S2,XS,PP,JNT,NT,
       1EMACH)
C
C   ASV BECOMES THE 3-D VECTOR FROM OUTER OVERLAYER TO INNER OVERLAYER
C   AND WILL BE VARIED IN LOOP 960
        DO 937 J=1,3
  937 ASV(J)=ASC(J)
        DO 960 I = 1, NVAR
        DO 940 J=1,3
  940 AS(J)=ASV(J)*0.529
        A=AS(1)
        WRITE (6,430) (AS(J),J=1,3)
C   THE OUTER (OXYGEN) OVERLAYER IS ADDED BY LAYER DOUBLING, PRODUCING
C   REFLECTED BEAM AMPLITUDES
C   SOME STORAGE IS SAVED BY USING S2 HERE, ALTHOUGH IT HAS
C   IMPROPER (BUT SUFFICIENT) DIMENSIONS
        CALL ADREF1(ROV,TOV,ROV,TOV,RN,S1,S2,XS,JNT,XI,PQ,NT,EMACH,ASV)
C   COMPUTE REFLECTED INTENSITIES (AT) FROM AMPLITUDES (XI), PRINT AND
C   PUNCH (NPNCH=1; PUNCH LAYER SPACING A ALSO FOR LATER IDENTIFICATION)
        CALL RINT(NT,XI,AT,PQ,PQF,SYM,VV,THETA,FI,MPU,NPUC,EEV,A,1)
  960 ASV(1)=ASV(1)+STEPC
  969 ASW(1)=ASW(1)+STEP
        IF (IT1+IT2+IT3) 980,990,980
  980 T = T + DT
        IF (DT.LE.0.0001) GO TO 990
        IF (T-TF)  780, 780, 990
  990 E = E + DE
        EC = E
        E3 = E
        IF (E-EF)  700, 700, 680
 1000 WRITE (6,450)
        STOP
        END
```

---- INPUT DATA FOR PRECEDING PROGRAM FOLLOW ------------

```
NI(100)+C(2*2)CO, LAY. DBLNG., 2 MIRROR PLANES
  4  2  1              IDEG NL1 NL2
  2 21 10              KNBS KNT NPUN
12                     NPSI
2.4900 0.0000          ARA1
0.0000 2.4900          ARA2
0.0000 0.0000          SS1
0.5000 0.5000          SS2
0.0000 0.0000          SS3
0.0000 0.0000          SS4
1.7650 0.0000 0.0000   ASA
2.4900-2.4900          ARB1
2.4900 2.4900          ARB2
0.0000 0.0000          SO1
0.0000 0.0000          SO2
0.0000 0.0000          SO3
```

```
1.7000  0.0000  0.0000      ASB
0.6290  0.7300  0.1000      FR ASE STEP
  3                         NSTEP
 13                         KNB(1)
0.0000  0.0000    1   1     BEAMS
1.0000  0.0000    2   2
0.0000  1.0000    2   2
1.0000  1.0000    9   9
2.0000  0.0000    2   2
0.0000  2.0000    2   2
2.0000  1.0000    9   9
1.0000  2.0000    9   9
2.0000  2.0000    9   9
3.0000  0.0000    2   2
0.0000  3.0000    2   2
3.0000  1.0000    9   9
1.0000  3.0000    9   9
  8                         KNB(2)
0.5000  0.5000    9   9
1.5000  0.5000    9   9
0.5000  1.5000    9   9
1.5000  1.5000    9   9
2.5000  0.5000    9   9
0.5000  2.5000    9   9
2.5000  1.5000    9   9
1.5000  2.5000    9   9
0.0020                      TST
  1   2   3   4   5   6  14 15 16 17     NPU
   0.00     0.00            THETA FI
   0.00    11.20            VO VV
0.0010                      EPS
  5                         LITER
  1   1   1                 IT1 IT2 IT3
440.0000      58.7100      1.4000      1.4000      0.0000      THDB1 AM1 FPER1 FPAR1 DRO1
973.0000      12.0112      2.0000      2.0000      0.0000      THDB2 AM2 FPER2 FPAR2 DRO2
843.0000      15.9994      2.0000      2.0000      0.0000      THDB3 AM3 FPER3 FPAR3 DRO3
300.0000                                                       TI TF DT
  7                         LMAX
  3                         NEL
 .3000
2.8398 -.0137 -.1876   .0011   .0000   .0000   .0000   .0000
2.5056 -.0739 -.0027 -.0000 -.0000 -.0000 -.0000 -.0000 -.0000 -.0000
-.5573 -.0514 -.0014 -.0000 -.0000 -.0000 -.0000 -.0000 -.0000 -.0000
 .4000
2.7639 -.0216 -.3164   .0024   .0001   .0000   .0000   .0000
2.4088 -.1066 -.0054 -.0001 -.0000 -.0000 -.0000 -.0000 -.0000 -.0000
-.6427 -.0752 -.0028 -.0000 -.0000 -.0000 -.0000 -.0000 -.0000 -.0000
 .5000
2.6930 -.0322 -.2630   .0048   .0002   .0000   .0000   .0000
2.3239 -.1403 -.0090 -.0003 -.0000 -.0000 -.0000 -.0000 -.0000 -.0000
-.7177 -.1000 -.0047 -.0001 -.0000 -.0000 -.0000 -.0000 -.0000 -.0000
 .7500
2.5276 -.0668 -.2322   .0160   .0009   .0000   .0000   .0000
2.1455 -.2254 -.0221 -.0010 -.0000 -.0000 -.0000 -.0000 -.0000 -.0000
-.8762 -.1644 -.0118 -.0004 -.0000 -.0000 -.0000 -.0000 -.0000 -.0000
1.0000
2.3803 -.1132 -.2323   .0386   .0028   .0002   .0000   .0000
1.9976 -.3088 -.0406 -.0025 -.0001 -.0000 -.0000 -.0000 -.0000 -.0000
-1.0084 -.2295 -.0224 -.0010 -.0000 -.0000 -.0000 -.0000 -.0000 -.0000
1.5000
2.1240 -.2247 -.2118   .1202   .0125   .0012   .0001   .0000
1.7566 -.4660 -.0900 -.0088 -.0005 -.0000 -.0000 -.0000 -.0000 -.0000
-1.2267 -.3562 -.0523 -.0036 -.0001 -.0000 -.0000 -.0000 -.0000 -.0000
2.0000
1.9101 -.3448 -.1758   .2494   .0335   .0042   .0004   .0000
1.5621 -.6089 -.1506 -.0202 -.0015 -.0001 -.0000 -.0000 -.0000 -.0000
-1.4065 -.4756 -.0914 -.0087 -.0005 -.0000 -.0000 -.0000 -.0000 -.0000
```

154

```
3.0000
1.5727 -.5697 -.1288  .5535  .1138  .0216  .0034  .0004
1.2601 -.8551 -.2863 -.0597 -.0073 -.0006 -.0000 -.0000 -.0000 -.0000
-1.6964 -.6914 -.1869 -.0282 -.0024 -.0001 -.0000 -.0000 -.0000 -.0000
4.0000
1.3160 -.7574 -.1272  .7884  .2257  .0583  .0125  .0022
1.0377-1.0555 -.4240 -.1175 -.0201 -.0022 -.0002 -.0000 -.0000 -.0000
-1.9261 -.8792 -.2929 -.0601 -.0071 -.0005 -.0000 -.0000 -.0000 -.0000
5.0000
1.1099 -.9122 -.1445  .9491  .3361  .1107  .0305  .0069
 .8746-1.2162 -.5532 -.1870 -.0415 -.0059 -.0006 -.0000 -.0000 -.0000
-2.1137-1.0427 -.4007 -.1028 -.0159 -.0016 -.0001 -.0000 -.0000 -.0000
6.0000
 .9366-1.0434 -.1666 1.0690  .4287  .1691  .0571  .0159
 .7593-1.3416 -.6690 -.2621 -.0712 -.0126 -.0015 -.0001 -.0000 -.0000
-2.2683-1.1849 -.5055 -.1535 -.0295 -.0036 -.0003 -.0000 -.0000 -.0000
7.0000
 .7862-1.1581 -.1902 1.1630  .5045  .2251  .0894  .0297
 .6828-1.4352 -.7690 -.3380 -.1079 -.0231 -.0034 -.0004 -.0000 -.0000
-2.3954-1.3077 -.6049 -.2096 -.0480 -.0070 -.0007 -.0001 -.0000 -.0000
    4              NVAR
0.9000 0.0000 0.0000   ASC
0.1000                 STEPC
  40.00 201.00    5.00 EI EF DE
 -10.00                EI EF DE
```

----- FIRST PART OF PRINTER OUTPUT FOLLOWS ----------------
-----(NEGLECTING MOST BLANK LINES AND PAGE --------------
----- EJECTS AND TRUNCATING LONG LINES)   -------------

```
NI(100)+C(2*2)CO, LAY. DBLNG., 2 MIRROR PLANES
IDEG =  4 NLI =  2 NL2 =  1
PARAMETERS FOR INTERIOR:
SURF VECS        2.4900  0.0000
                 0.0000  2.4900
SS     0.0000 0.0000
SS      .5000  .5000
SS     0.0000 0.0000
SS     0.0000 0.0000
ASA    1.7650 0.0000 0.0000
PARAMETERS FOR SURFACE:
SURF VECS        2.4900 -2.4900
                 2.4900  2.4900
SO     0.0000 0.0000
SO     0.0000 0.0000
SO     0.0000 0.0000
ASB    1.7000 0.0000 0.0000
FR =   .6290 ASE =   .7300 STEP =   .1000 NSTEP =   3
         PQ1    PQ2        SYM

   1    0.000  0.000     1  1
   2    1.000  0.000     2  2
   3    0.000  1.000     2  2
   4    1.000  1.000     9  9
   5    2.000  0.000     2  2
   6    0.000  2.000     2  2
   7    2.000  1.000     9  9
   8    1.000  2.000     9  9
   9    2.000  2.000     9  9
  10    3.000  0.000     2  2
  11    0.000  3.000     2  2
  12    3.000  1.000     9  9
  13    1.000  3.000     9  9

  14     .500   .500     9  9
  15    1.500   .500     9  9
  16     .500  1.500     9  9
```

```
  17      1.500    1.500       9   9
  18      2.500     .500       9   9
  19       .500    2.500       9   9
  20      2.500    1.500       9   9
  21      1.500    2.500       9   9
TST =    .0020
PUNCHED BEAMS      1   2   3   4   5   6  14  15  16  17
THETA  FI =      0.00    0.00
VO       0.00VV          11.20
EPS =   .0010  LITER =    5
IT1 =   1 IT2 =    1 IT3 =    1
THDB =  440.0000 AM =   58.7100 FPER =    1.4000 FPAR =    1.4000 DRO =
THDB =  973.0000 AM =   12.0112 FPER =    2.0000 FPAR =    2.0000 DRO =
THDB =  843.0000 AM =   15.9994 FPER =    2.0000 FPAR =    2.0000 DRO =
LMAX =    7
        PHASE SHIFTS
```

```
E =    .3000   1ST ELEMENT     2.8398   -.0137   -.1876    .0011   0.0000   0.0000
               2ND ELEMENT     2.5056   -.0739   -.0027   0.0000   0.0000   0.0000
               3RD ELEMENT     -.5573   -.0514   -.0014   0.0000   0.0000   0.0000

E =    .4000   1ST ELEMENT     2.7639   -.0216   -.3164    .0024    .0001   0.0000
               2ND ELEMENT     2.4088   -.1066   -.0054   -.0001   0.0000   0.0000
               3RD ELEMENT     -.6427   -.0752   -.0028   0.0000   0.0000   0.0000

E =    .5000   1ST ELEMENT     2.6930   -.0322   -.2630    .0048    .0002   0.0000
               2ND ELEMENT     2.3239   -.1403   -.0090   -.0003   0.0000   0.0000
               3RD ELEMENT     -.7177   -.1000   -.0047   -.0001   0.0000   0.0000

E =    .7500   1ST ELEMENT     2.5276   -.0668   -.2322    .0160    .0009   0.0000
               2ND ELEMENT     2.1455   -.2254   -.0221   -.0010   0.0000   0.0000
               3RD ELEMENT     -.8762   -.1644   -.0118   -.0004   0.0000   0.0000

E =   1.0000   1ST ELEMENT     2.3803   -.1132   -.2323    .0386    .0028    .0002
               2ND ELEMENT     1.9976   -.3088   -.0406   -.0025   -.0001   0.0000
               3RD ELEMENT    -1.0084   -.2295   -.0224   -.0010   0.0000   0.0000

E =   1.5000   1ST ELEMENT     2.1240   -.2247   -.2118    .1202    .0125    .0012
               2ND ELEMENT     1.7566   -.4660   -.0900   -.0088   -.0005   0.0000
               3RD ELEMENT    -1.2267   -.3562   -.0523   -.0036   -.0001   0.0000

E =   2.0000   1ST ELEMENT     1.9101   -.3448   -.1758    .2494    .0335    .0042
               2ND ELEMENT     1.5621   -.6089   -.1506   -.0202   -.0015   -.0001
               3RD ELEMENT    -1.4065   -.4756   -.0914   -.0087   -.0005   0.0000

E =   3.0000   1ST ELEMENT     1.5727   -.5697   -.1288    .5535    .1138    .0216
               2ND ELEMENT     1.2601   -.8551   -.2863   -.0597   -.0073   -.0006
               3RD ELEMENT    -1.6964   -.6914   -.1869   -.0282   -.0024   -.0001

E =   4.0000   1ST ELEMENT     1.3160   -.7574   -.1272    .7884    .2257    .0583
               2ND ELEMENT     1.0377  -1.0555   -.4240   -.1175   -.0201   -.0022
               3RD ELEMENT    -1.9261   -.8792   -.2929   -.0601   -.0071   -.0005

E =   5.0000   1ST ELEMENT     1.1099   -.9122   -.1445    .9491    .3361    .1107
               2ND ELEMENT      .8746  -1.2162   -.5532   -.1870   -.0415   -.0059
               3RD ELEMENT    -2.1137  -1.0427   -.4007   -.1028   -.0159   -.0016

E =   6.0000   1ST ELEMENT      .9366  -1.0434   -.1666   1.0690    .4287    .1691
               2ND ELEMENT      .7593  -1.3416   -.6690   -.2621   -.0712   -.0126
               3RD ELEMENT    -2.2683  -1.1849   -.5055   -.1535   -.0295   -.0036

E =   7.0000   1ST ELEMENT      .7862  -1.1581   -.1902   1.1630    .5045    .2251
               2ND ELEMENT      .6828  -1.4352   -.7690   -.3380   -.1079   -.0231
               3RD ELEMENT    -2.3954  -1.3077   -.6049   -.2096   -.0480   -.0070
INPUT INTERLAYER VECTOR :    .9000 0.0000 0.0000
ENERGY = 1.8837 H   OR     40.00 EV
   9 BEAMS USED :    6    3
```

```
      0.000 0.000    1.000 0.000    0.000 1.000    1.000 1.000    2.000 0.000

VPIS =     -.1114 VPIO =     -.1114 DCUTS =    87.1160 DCUTO =    87.1160
TEMP = 300.0000
MSMF    OK
MSMF    OK
MSMF    OK
X =       .1831E+01
X =       .1866E+01
X =       .1876E+01
X =       .1875E+01
   4  ITER IN SUBREF
SUBREF OK
MSMF    OK
X =       .4142E+00
X =       .3272E+00
X =       .3422E+00
X =       .3408E+00
X =       .3408E+00
NO CONV IN SUBREF AFTER  5  ITER, X =       .3408E+00
SUBREF OK
SURFACE GEOMETRY :  1.7000 0.0000 0.0000
SURFACE GEOMETRY :   .9000 0.0000 0.0000
E=    1.8837 THETA=    0.0000 FI=    0.0000

   PQ1     PQ2      REF INT     SYM
 0.000   0.000   .54008E-02     1
 1.000   0.000   .62319E-03     2
 0.000   1.000   .62320E-03     2
  .500    .500   .17939E-03     9
SURFACE GEOMETRY :  1.0000 0.0000 0.0000
E=    1.8837 THETA=    0.0000 FI=    0.0000

   PQ1     PQ2      REF INT     SYM
 0.000   0.000   .41683E-02     1
 1.000   0.000   .87098E-03     2
 0.000   1.000   .87098E-03     2
  .500    .500   .17697E-03     9
SURFACE GEOMETRY :  1.1000 0.0000 0.0000
E=    1.8837 THETA=    0.0000 FI=    0.0000

   PQ1     PQ2      REF INT     SYM
 0.000   0.000   .40521E-02     1
 1.000   0.000   .99021E-03     2
 0.000   1.000   .99021E-03     2
  .500    .500   .20259E-03     9
SURFACE GEOMETRY :  1.2000 0.0000 0.0000
E=    1.8837 THETA=    0.0000 FI=    0.0000

   PQ1     PQ2      REF INT     SYM
 0.000   0.000   .48322E-02     1
 1.000   0.000   .95366E-03     2
 0.000   1.000   .95366E-03     2
  .500    .500   .23705E-03     9
SURFACE GEOMETRY :  1.8000 0.0000 0.0000
SURFACE GEOMETRY :   .9000 0.0000 0.0000
E=    1.8837 THETA=    0.0000 FI=    0.0000

   PQ1     PQ2      REF INT     SYM
 0.000   0.000   .39647E-02     1
 1.000   0.000   .82491E-03     2
 0.000   1.000   .82491E-03     2
  .500    .500   .17219E-03     9
```

11. Subroutine Listings

This chapter contains a complete listing of all subroutines (or functions) required by the main programs of the preceding chapter. A fair degree of generality is ensured through, e.g., variable dimensions and interchangeability of subroutines, so that many other main programs can make use of this particular set of subroutines.

It is recommended for the user's convenience that all supplied subroutines be compiled onto a single permanent library file, from which each run can select the subset of subroutines needed by it. In this compilation, the highest possible optimization is advisable for speed of computation (but a test against low-optimization compilation is necessary for safety). A non-standard Fortran feature of some subroutines (namely GHD and GHMAT) is the use of the function ACOS (arc cos): to some compilers this function is known as ARCOS, and the corresponding substitution should be made. The unit numbers for input, print and punch are set to 5, 6 and 7, respectively.

Because of the very large number of situations that this set of subroutines can handle, no exhaustive test of their performance has been possible (although each has been used successfully in a number of different programs). The authors welcome feedback from users on any problem that might occur. We can then also communicate to the users any new developments concerning these programs.

To guide the user through the supplied subroutines (listed in alphabetical order in this chapter), we give below a brief description of each subroutine's purpose; more detailed explanations are included in the listings as comments, while many features are further discussed in other chapters. The relationship between subroutines (i.e., which subroutine calls which subroutine) can be obtained from Fig. 10.1.

ADREF1   adds an overlayer onto a substrate, by Layer Doubling, producing a vector of reflected amplitudes;

BEAMS   selects the beams needed at a given energy;

BLM   produces a Clebsch-Gordon coefficient for CELMG;

CA   produces a Clebsch-Gordon coefficient for CAAA;

CAAA	computes a set of Clebsch-Gordon-type coefficients for composite layers;
CELMG	computes a set of Clebsch-Gordon-type coefficients for Bravais-lattice and composite layers;
CPPP	computes a set of Clebsch-Gordon-type coefficients for PSTEMP;
CXMTXT	solves a set of linear equations;
DBG	combines two layers by Layer Doubling, producing one reflection matrix only;
DBLG	combines two layers by Layer Doubling, producing all reflection and transmission matrices;
DEBWAL	computes Debye-Waller factors;
FACT	produces factorials;
FMAT	evaluates lattice sums for Bravais-lattice and composite layers;
GHD	evaluates lattice sums for composite layers;
GHMAT	produces interlayer propagators in angular-momentum space;
GHSC	produces interlayer propagators in angular-momentum space for GHMAT;
LXGENT	produces permutations of the (ℓ,m)-sequence;
MATCOP	copies variably-dimensioned matrices onto one another;
MFOLD	produces individual diffraction matrix elements for Bravais-lattice layers;
MFOLT	produces individual diffraction matrix elements for composite layers;
MPERTI	generates diffraction matrices by RSP and partial Matrix Inversion for composite layers;
MSMF	generates diffraction matrices for Bravais-lattice layers;
MTINV	generates diffraction matrices by Matrix Inversion for composite layers;
PRPGAT	propagates the wavefield by one layer in the RFS scheme;
PSTEMP	produces temperature-dependent phase shifts;
READCL	reads in input data for composite layers;
READIN	reads in most input data;
RFS02	performs the RFS computation (2-layer bulk periodicity, one overlayer);
RFS02V	performs the RFS computation (2-layer bulk periodicity, one overlayer, variable top substrate spacing);

RFS03   performs the RFS computation (3-layer bulk periodicity, one overlayer);

RFS20V   performs the RFS computation (2-layer bulk periodicity, two overlayers);

RINT   produces beam intensities from beam amplitudes;

SB   computes spherical Bessel functions;

SH   computes spherical harmonics;

SLIND   analyzes the relationship between sublattices under rotation;

SPHRM   computes spherical harmonics;

SRTLAY   sorts subplanes in a composite layer;

SUBREF   performs Layer Doubling in the bulk;

TAUMAT   produces single-subplane multiple-scattering matrices for composite layers;

TAUY   produces quantities needed by MPERTI and MTINV;

THINV   performs Beeby-type inversion;

THMAT   produces the matrix to be inverted in THINV;

TLRT   performs one Layer Doubling step in the general case;

TLRTA   performs one Layer Doubling step in the symmetrical case;

TPERTI   directs the RSP iteration;

TPSTPI   performs one step in the RSP iteration;

TRANS   shifts the reference point in a Bravais-lattice layer;

TRANSP   shifts the reference points in a composite layer;

TSCATF   generates atomic scattering amplitudes;

UNFOLD   finds the beams related by symmetry to a given beam;

XM   produces the intra-layer multiple-scattering matrix for Bravais-lattice layers;

XMT   produces an intra-layer multiple-scattering matrix for TAUMAT;

ZGE   prepares the solution of a set of linear equations;

ZSU   concludes the solution of a set of linear equations;

To help understand the internal organization of the more complex subroutines (MSMF, MTINV, MPERTI, MFOLD and MFOLT), Figs. 11.1, 11.2 and 11.3 show simplified flowcharts of their structure.

160

Fig. 11.1 Flowchart of subroutine MSMF (indicating subroutines called, and subroutine nestings in square brackets)

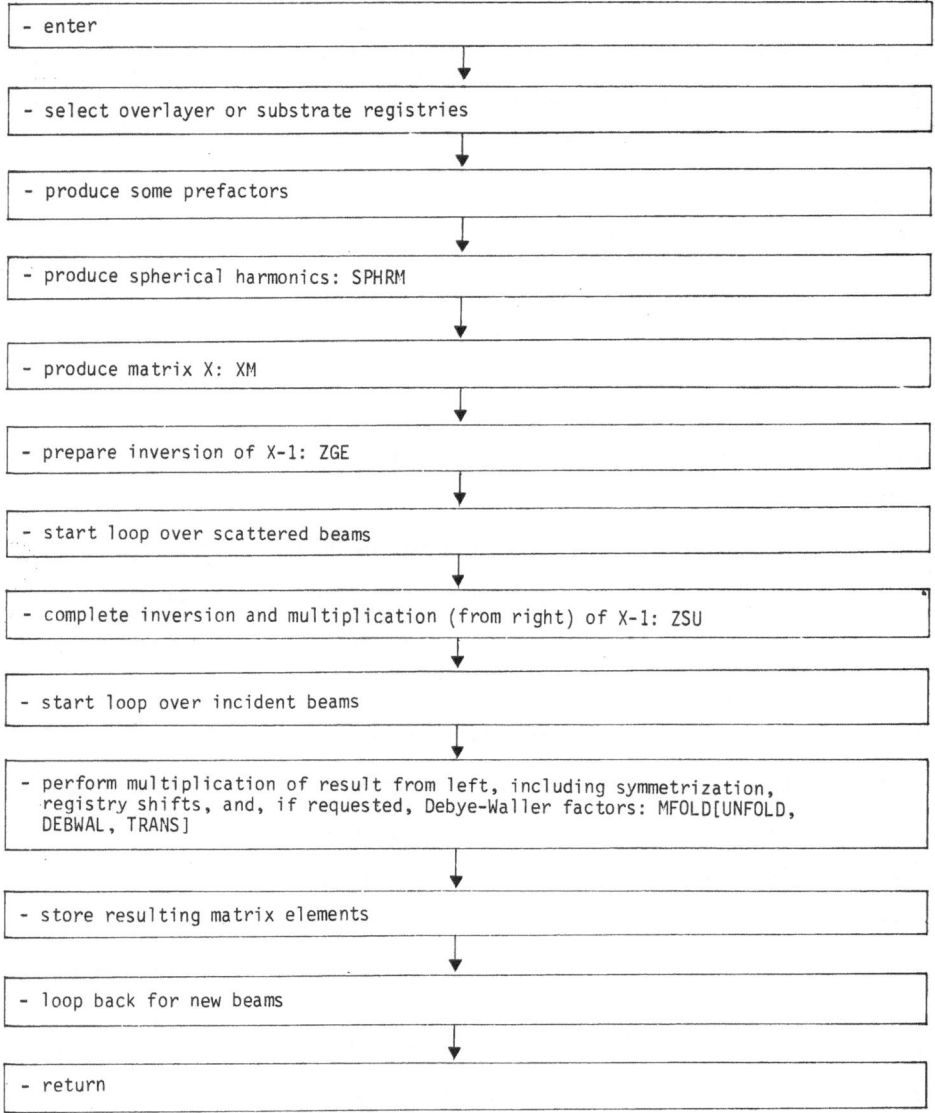

- enter

- select overlayer or substrate registries

- produce some prefactors

- produce spherical harmonics: SPHRM

- produce matrix X: XM

- prepare inversion of X-1: ZGE

- start loop over scattered beams

- complete inversion and multiplication (from right) of X-1: ZSU

- start loop over incident beams

- perform multiplication of result from left, including symmetrization, registry shifts, and, if requested, Debye-Waller factors: MFOLD[UNFOLD, DEBWAL, TRANS]

- store resulting matrix elements

- loop back for new beams

- return

Fig. 11.2 Flowchart of subroutines MTINV and MPERTI (indicating subroutines called, and subroutine nestings in square brackets)

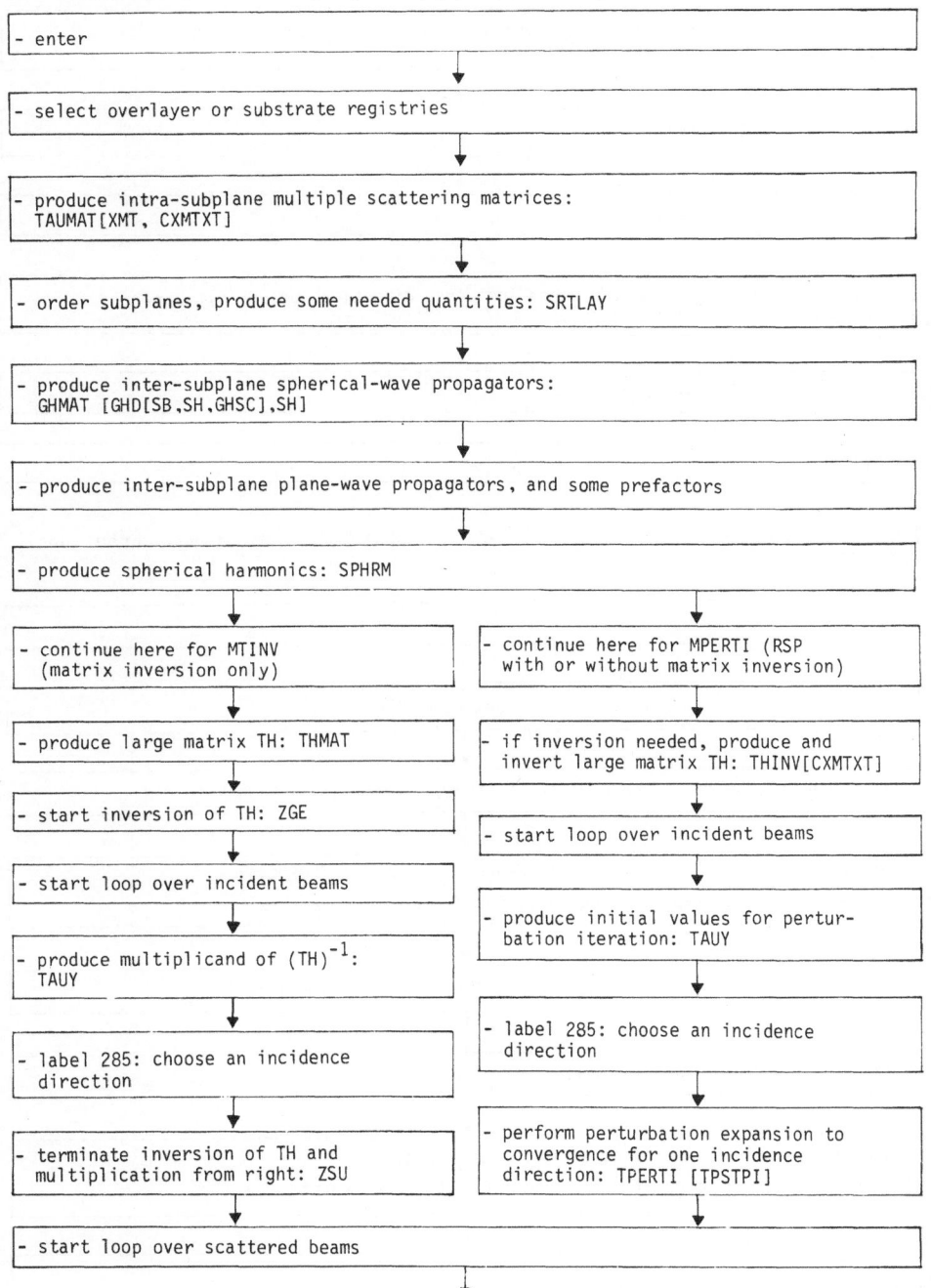

162

Fig. 11.2 (continued)

```
- perform multiplication of result from left, incl. symmetrization and
  registry shifts: MFOLT[UNFOLD,TRANSP]
```
↓
```
- store resulting matrix elements
```
↓
```
- loop back for new scattered beam
```
↓
```
- return to label 285 with the other incidence direction, unless symmetry
  in ±x has been notified in input
```
↓
```
- loop back for new incident beam
```
↓
```
- return
```

Fig. 11.3 Flowchart of subroutines MFOLD and MFOLT (indicating subroutines called)

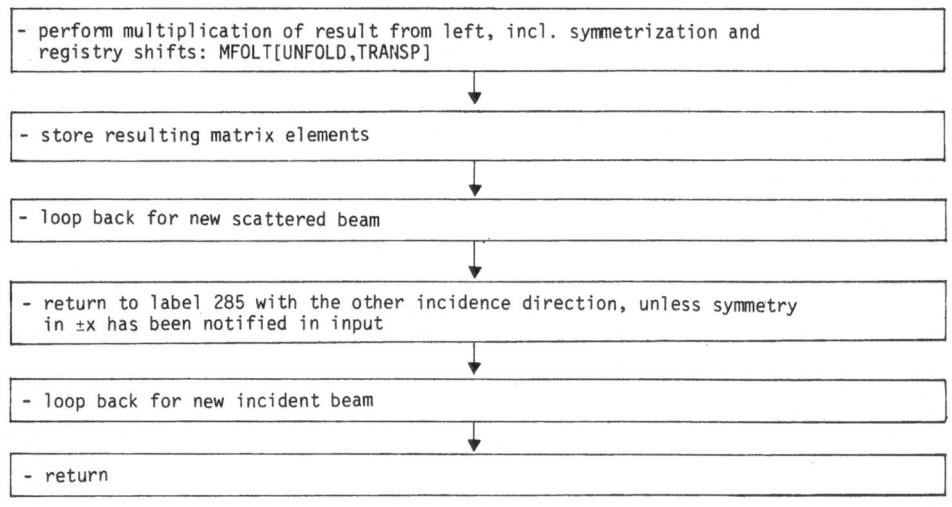

```
- enter
```
↓
```
- find symmetrical counterparts of current beam: UNFOLD
```
↓
```
- in MFOLD only: produce Debye-Waller factors, if requested: DEBWAL
```
↓
```
- perform registry shifts: TRANS (if MFOLD) or TRANSP (if MFOLT);
- repeat preceding step, if necessary, for different registries
```
↓
```
- produce quantities Z (and EGS, if MFOLT) relevant to symmetrical beams
```
↓
```
- in MFOLT only: sum over subplanes
```
↓
```
- sum over quantities relevant to symmetrical beams;
- repeat preceding step, if necessary, for different registries
```
↓
```
- return
```

```
C   SUBROUTINE ADREF1 COMPUTES A REFLECTION VECTOR FOR A SUBSTRATE WITH
C   AN OVERLAYER, GIVEN REFLECTION AND TRANSMISSION MATRICES FOR
C   SUBSTRATE AND OVERLAYER. THE OVERLAYER MAY BE ASYMMETRICAL IN +-X.
C   LAYER DOUBLING IS USED. (TO PRODUCE A COMPLETE REFLECTION MATRIX
C   RATHER THAN VECTOR, USE SUBROUTINE DBG.)
C   RA,TA= REFLECTION AND TRANSMISSION MATRICES OF OVERLAYER FOR
C     INCIDENCE TOWARDS +X.
C   RAM,TAM= SAME AS RA,TA FOR INCIDENCE TOWARDS -X.
C   RB= REFLECTION MATRIX OF SUBSTRATE.
C   ST,WK,U,INT= WORKING SPACES.
C   XI= OUTPUT REFLECTION AMPLITUDES OF EACH BEAM (FOR INCIDENCE IN
C     (00) BEAM).
C   PQ= LIST OF BEAMS G.
C   N= NO. OF BEAMS USED.
C   EMACH= MACHINE ACCURACY.
C   S= INTERLAYER VECTOR FROM OVERLAYER TO SUBSTRATE.
C   IN COMMON BLOCKS:
C   E= CURRENT ENERGY.
C   VPI: NOT USED.
C   AK2,AK3= PARALLEL COMPONENTS OF PRIMARY INCIDENT K-VECTOR.
C   AS: NOT USED.
C   ASL: NOT USED.
C   FR= FRACTION OF S(1) ALLOTTED TO SUBSTRATE.
C   ASE= SPACING BETWEEN SURFACE AND OVERLAYER.
C   VPIS,VPIO= DAMPING IN SUBSTRATE AND OVERLAYER, RESP.
C   VO= MUFFIN-TIN CONSTANT OF OVERLAYER RELATIVE TO THAT OF SUBSTRATE.
C   VV: NOT USED.
      SUBROUTINE  ADREF1 (RA, TA, RAM, TAM, RB, ST, WK, U, INT, XI,
     1PQ, N, EMACH, S)
      INTEGER  INT
      COMPLEX  RA, TA, RB, WK, U, CZ, RU, IU, XX, YY, RAM, TAM, ST,XI
      COMPLEX CSQRT,CEXP
      DIMENSION  RA(N,N), TA(N,N), RB(N,N), WK(3,N), U(N), RAM(N,N)
      DIMENSION  TAM(N,N), ST(N,N), PQ(2,N), XI(N), AS(3)
      DIMENSION  INT(N),S(3)
      COMMON /X4/E, VPI, AK2, AK3,  AS
      COMMON /ADS/ASL, FR, ASE, VPIS, VPIO, VO, VV
      CZ = CMPLX(0.0,0.0)
      RU = CMPLX(1.0,0.0)
      IU = CMPLX(0.0,1.0)
      AK = 2.0 * E
C   COMPUTE INTERLAYER PROPAGATORS:
C   WK(1,I) FOR PROPAGATION BETWEEN SURFACE PLANE AND OVERLAYER
C   NUCLEAR PLANE.
C   WK(2,I),WK(3,I) FOR PROPAGATION BETWEEN OVERLAYER AND SUBSTRATE.
      DO 470 I = 1, N
      BK2 = AK2 + PQ(1,I)
      BK3 = AK3 + PQ(2,I)
C   XX IS K(PERP) IN SUBSTRATE
      XX = CMPLX(AK - BK2 * BK2 - BK3 * BK3, - 2.0 * VPIS + 0.000001)
      XX = CSQRT(XX)
      YY = CMPLX(AK - 2.0 * VO - BK2 * BK2 - BK3 * BK3, - 2.0 * VPIO +
     10.000001)
C   YY IS K(PERP) IN OVERLAYER
      YY = CSQRT(YY)
      X=BK2*S(2)+BK3*S(3)
      WK(1,I)=CEXP(IU*(YY*ASE))
      WK(2,I)=CEXP(IU*(YY*(1.0-FR)*S(1)+XX*FR*S(1)+X))
  470 WK(3,I)=CEXP(IU*(YY*(1.0-FR)*S(1)+XX*FR*S(1)-X))
C
C
C   PRODUCE MATRIX TO BE INVERTED
      DO 490 I = 1, N
      DO 480 J = 1, N
      ST(I,J) = CZ
      DO 480 K = 1, N
  480 ST(I,J) = ST(I,J) - RAM(I,K) * WK(3,K) * RB(K,J) * WK(2,J)
```

```
  490 ST(I,I) = ST(I,I) + RU
C   GAUSSIAN ELIMINATION STEP OF MATRIX INVERSION
        CALL   ZGE (ST, INT, N, N, EMACH)
        DO 500 I = 1, N
  500 U(I) = TA(I,1) * WK(1,1)
C   SUBSTITUTION STEP OF MATRIX INVERSION
        CALL   ZSU (ST, INT, U, N, N, EMACH)
        DO 510 I = 1, N
        XI(I) = CZ
        DO 510 K = 1, N
  510 XI(I) = XI(I) + RB(I,K) * WK(2,K) * U(K)
        DO 520 I = 1, N
        U(I) = WK(1,I) * RA(I,1) * WK(1,1)
        DO 520 K = 1, N
  520 U(I) = U(I) + WK(1,I) * TAM(I,K) * WK(3,K) * XI(K)
        DO 530 I = 1, N
  530 XI(I) = U(I)
        RETURN
        END
C-------------------------------------------------------------------------
C   SUBROUTINE BEAMS SELECTS THOSE BEAMS FROM THE INPUT LIST THAT ARE
C   NEEDED AT THE CURRENT ENERGY, BASED ON THE PARAMETER TST WHICH LIMITS
C   THE DECAY OF PLANE WAVES FROM ONE LAYER TO THE NEXT (THE INTERLAYER
C   SPACING MUST ALREADY BE INCORPORATED IN TST,WHICH IS DONE BY
C   SUBROUTINE READIN)
C   KNBS=NO. OF INPUT BEAM SETS
C   KNB= NO. OF BEAMS IN EACH INPUT BEAM SET
C   SPQ,SPQF= LIST OF INPUT BEAMS (RECIPROCAL LATTICE VECTORS G)
C   KSYM= CORRESPONDING SYMMETRY PROPERTIES
C   KNT= TOTAL NO. OF INPUT BEAMS
C   NPU: INDICATES FOR WHICH BEAMS INTENSITIES ARE TO BE PUNCHED OUT
C   NPUN= NO. OF BEAMS FOR WHICH INTENSITIES ARE TO BE PUNCHED OUT
C   AK2,AK3= COMPONENTS OF INCIDENT WAVEVECTOR PARALLEL TO SURFACE
C   E= CURRENT ENERGY
C   TST= CRITERION FOR BEAM SELECTION
C   NB,PQ,PQF,SYM,NPUC,MPU,NT: AS KNB,SPQ,SPQF,KSYM,NPU,NPUN,KNT AFTER
C     BEAM SELECTION
C   NP= LARGEST NO. OF BEAMS SELECTED IN ANY ONE BEAM SET
        SUBROUTINE BEAMS(KNBS,KNB,SPQ,SPQF,KSYM,KNT,NPU,NPUN,AK2,AK3,
       1E,TST,NB,PQ,PQF,SYM,NPUC,MPU,NT,NP)
        DIMENSION KNB(KNBS),SPQ(2,KNT),SPQF(2,KNT),KSYM(2,KNT),NPU(NPUN),
       1NB(KNBS),PQ(2,KNT),PQF(2,KNT),NPUC(NPUN)
        INTEGER SYM(2,KNT)
  705 FORMAT(1H0,1I3,13H BEAMS USED :,8I4)
  706 FORMAT(1H ,10(9(2X,2F6.3),/))
        KNBJ=0
        NT=0
        NP=0
        MPU=0
        DO 704 J=1,KNBS
        N=KNB(J)
        NB(J)=0
        DO 703 K=1,N
        KK=K+KNBJ
        MPU1=MPU+1
        IF (KK.NE.NPU(MPU1)) GO TO 700
        MPU=MPU+1
        NPUC(MPU)=0
  700 IF ((2.0*E-(AK2+SPQ(1,KK))**2-(AK3+SPQ(2,KK))**2)
       1-(-TST)) 703,701,701
  701 NB(J)=NB(J)+1
        NT=NT+1
        IF (KK.NE.NPU(MPU1)) GO TO 707
        NPUC(MPU)=NT
  707 DO 702 I=1,2
        SYM(I,NT)=KSYM(I,KK)
        PQ(I,NT)=SPQ(I,KK)
```

```
  702 PQF(I,NT)=SPQF(I,KK)
  703 CONTINUE
      KNBJ=KNBJ+KNB(J)
  704 NP=MAXO(NP,NB(J))
      WRITE(6,705)NT,(NB(J),J=1,KNBS)
      WRITE(6,706)((PQF(I,K),I=1,2),K=1,NT)
      RETURN
      END
C--------------------------------------------------------------------------
COMMENT FUNCTION BLM PROVIDES THE INTEGRAL OF THE PRODUCT
C          OF THREE SPHERICAL HARMONICS, EACH OF WHICH CAN BE
C          EXPRESSED AS A PREFACTOR TIMES A LEGENDRE FUNCTION.
C          THE THREE PREFACTORS ARE LUMPED TOGETHER AS FACTOR
C          'C'; AND THE INTEGRAL OF THE THREE LEGENDRE FUNCTIONS
C          FOLLOWS GAUNT'S SUMMATION SCHEME SET OUT BY SLATER:
C          ATOMIC STRUCTURE, VOL1, 309,310
C   AUTHOR: PENDRY
      FUNCTION  BLM (L1, M1, L2, M2, L3, M3, LMAX)
      DOUBLE PRECISION FAC(35)
      COMMON /F/ FAC
   40 FORMAT(28H INVALID ARGUMENTS FOR BLM: ,6(I3,1H,))
      PI = 3.14159265
      IF (M1+M2+M3)   420, 350, 420
  350 IF (L1-LMAX-LMAX)   360, 360, 530
  360 IF (L2-LMAX)   370, 370, 530
  370 IF (L3-LMAX)   380, 380, 530
  380 IF (L1-IABS(M1))   530, 390, 390
  390 IF (L2-IABS(M2))   530, 400, 400
  400 IF (L3-IABS(M3))   530, 410, 410
  410 IF (MOD(L1+L2+L3,2))   420, 430, 420
  420 BLM = 0.0
      RETURN
  430 NL1 = L1
      NL2 = L2
      NL3 = L3
      NM1 = IABS(M1)
      NM2 = IABS(M2)
      NM3 = IABS(M3)
      IC = (NM1 + NM2 + NM3)/2
      IF (MAXO(NM1,NM2,NM3)-NM1)   470, 470, 440
  440 IF (MAXO(NM2,NM3)-NM2)   450, 450, 460
  450 NL1 = L2
      NL2 = L1
      NM1 = NM2
      NM2 = IABS(M1)
      GO TO 470
  460 NL1 = L3
      NL3 = L1
      NM1 = NM3
      NM3 = IABS(M1)
  470 IF (NL2-NL3)   480, 490, 490
  480 NTEMP = NL2
      NL2 = NL3
      NL3 = NTEMP
      NTEMP = NM2
      NM2 = NM3
      NM3 = NTEMP
  490 IF (NL3-IABS(NL2-NL1))   500, 510, 510
  500 BLM = 0.0
      RETURN
C
C          CALCULATION OF FACTOR 'A'
C
  510 IS = (NL1 + NL2 + NL3)/2
      IA1 = IS - NL2 - NM3
      IA2 = NL2 + NM2
      IA3 = NL2 - NM2
```

```
      IA4 = NL3 + NM3
      IA5 = NL1 + NL2 - NL3
      IA6 = IS - NL1
      IA7 = IS - NL2
      IA8 = IS - NL3
      IA9 = NL1 + NL2 + NL3 + 1
      A=((-1.0)**IA1)/FAC(IA3+1)*FAC(IA2+1)/FAC(IA6+1)*FAC(IA4+1)
      A=A/FAC(IA7+1)*FAC(IA5+1)/FAC(IA8+1)*FAC(IS+1)/FAC(IA9+1)
C
C        CALCULATION OF SUM 'B'
C
      IB1 = NL1 + NM1
      IB2 = NL2 + NL3 - NM1
      IB3 = NL1 - NM1
      IB4 = NL2 - NL3 + NM1
      IB5 = NL3 - NM3
      IT1 = MAX0(0, - IB4) + 1
      IT2 = MIN0(IB2,IB3,IB5) + 1
      B = 0.
      SIGN = ( - 1.0)**(IT1)
      IB1 = IB1 + IT1 - 2
      IB2 = IB2 - IT1 + 2
      IB3 = IB3 - IT1 + 2
      IB4 = IB4 + IT1 - 2
      IB5 = IB5 - IT1 + 2
      DO 520 IT = IT1, IT2
      SIGN = - SIGN
      IB1 = IB1 + 1
      IB2 = IB2 - 1
      IB3 = IB3 - 1
      IB4 = IB4 + 1
      IB5 = IB5 - 1
      BN=SIGN/FAC(IT)*FAC(IB1+1)/FAC(IB3+1)*FAC(IB2+1)
      BN=BN/FAC(IB4+1)/FAC(IB5+1)
  520 B=B+BN
C
C        CALCULATION OF FACTOR 'C'
C
      IC1 = NL1 - NM1
      IC2 = NL1 + NM1
      IC3 = NL2 - NM2
      IC4 = NL2 + NM2
      IC5 = NL3 - NM3
      IC6 = NL3 + NM3
      CN = FLOAT((2 * NL1 + 1) * (2 * NL2 + 1) * (2 * NL3 + 1))/PI
      C=CN/FAC(IC2+1)*FAC(IC1+1)/FAC(IC4+1)*FAC(IC3+1)
      C=C/FAC(IC6+1)*FAC(IC5+1)
      C = (SQRT(C))/2.
C
C
      BLM = (( - 1.0)**IC) * A * B * C
      RETURN
  530 WRITE(6,40)L1,M1,L2,M2,L3,M3
      RETURN
      END
C----------------------------------------------------------------------
C FUNCTION CA PERFORMS SAME PURPOSE FOR CAAA AS BLM DOES FOR CELMG
      FUNCTION  CA (L1, MA1, L2, M2, L3, MA3)
      M1 =  - MA1
      M3 =  - MA3
      IF ((IABS(M1) .GT. L1) .OR. (IABS(M2) .GT. L2) .OR. (IABS(M3) .GT.
     1 L3))  GO TO 1570
      IF ((M1+M2) .NE. (-M3))  GO TO 1570
      IF ( .NOT. ((L3 .GE. IABS(L1-L2)) .AND. (L3 .LE. (L1+L2))))  GO
     1TO 1570
      IF (MOD(L1+L2+L3,2) .NE. 0)  GO TO 1570
      IF ((L1 .EQ. 0) .OR. (L2 .EQ. 0) .OR. (L3 .EQ. 0))  GO TO 1580
```

```
      DELT = FACT(L1 + L2 - L3) * FACT(L2 + L3 - L1) * FACT(L3 + L1 -
     1L2)/FACT(L1 + L2 + L3 + 1) * (( - 1)**(L1 - L2 - M3 + ((L1 + L2 +
     2L3)/2)))
      PREF = FACT(L1 + M1) * FACT(L1 - M1) * FACT(L2 + M2) * FACT(L2 -
     1M2) * FACT(L3 - M3) * FACT(L3 + M3)
      PREFA = SQRT(PREF) * (10.0**(L1 + L2 + L3))
      PREFB = FACT((L1 + L2 + L3)/2)/(FACT((L1 + L2 - L3)/2) * FACT((L2
     1+ L3 - L1)/2) * FACT((L3 + L1 - L2)/2)) * (0.1)
      PREFT = DELT * PREFA * PREFB
      IGT = MINO(L2 + M2,L1 - M1,L1 + L2 - L3)
      ILST = MAXO(L1 + M2 - L3, - M1 + L2 - L3,0)
      NUM = IGT - ILST + 1
      SUM = 0.0
      DO 1560 IT = 1, NUM
      ITA = ILST + IT - 1
      SUM = SUM + (( - 1.0)**ITA)/(FACT(ITA) * FACT(L3 - L2 + ITA + M1)
     1* FACT(L3 - L1 + ITA - M2) * FACT(L1 + L2 - L3 - ITA) * FACT(L1 -
     2ITA - M1) * FACT(L2 - ITA + M2)) * (10.0**( - L1 - L2 - L3))
 1560 CONTINUE
      YINT = ((2.0 * FLOAT(L1) + 1.0) * (2.0 * FLOAT(L2) + 1.0) * (2.0 *
     1 FLOAT(L3) + 1.0)/(12.56637))**(0.5) * PREFT * SUM
      GO TO 1590
 1570 YINT = 0.0
      GO TO 1590
 1580 YINT = ((2.0 * FLOAT(L1) + 1.0) * (2.0 * FLOAT(L2) + 1.0) * (2.0 *
     1 FLOAT(L3) + 1.0)/(12.56637))**(0.5) * ( - 1.0)**((IABS(M1) +
     2IABS(M2) + IABS(M3))/2)/(FLOAT(IABS(L1) + IABS(L2) + IABS(L3)) +
     31.0)
 1590 CA = (( - 1.0)**(MA1 + MA3)) * YINT
      RETURN
      END
C-----------------------------------------------------------------------
C  SUBROUTINE CAAA COMPUTES CLEBSCH-GORDON COEFFICIENTS FOR USE BY
C  SUBROUTINE GHD IN THE SAME ORDER AS GHD USES THEM
      SUBROUTINE  CAAA (CAA,NCAA,LMMAX)
      DIMENSION  CAA(NCAA)
      II = 1
      DO 60 I = 1, LMMAX
      L1 = INT(SQRT(FLOAT(I - 1)))
      M1 = I - L1 - L1 * L1 - 1
      DO 50 J = 1, LMMAX
      L2 = INT(SQRT(FLOAT(J - 1)))
      M2 = J - L2 - L2 * L2 - 1
      M3 = M2 - M1
      IL = IABS(L1 - L2)
      IM = IABS(M3)
      LMIN = MAXO(IL, IM + MOD(IL + IM,2))
      LMAX = L1 + L2
      LMIN = LMIN + 1
      LMAX = LMAX + 1
      IF (I-J) 20,10,20
  10  IF (M1) 30,30,50
  20  IF (L1.LT.L2.OR.(L1.EQ.L2.AND.(IABS(M1).LT.IABS(M2))))
     1GO TO 50
  30  DO 40 ILA = LMIN, LMAX, 2
      LA = ILA - 1
      CAA(II) = CA(L1,M1,L2,M2,LA,M3)
  40  II = II + 1
  50  CONTINUE
  60  CONTINUE
      II=II-1
      WRITE(6,70)II
  70  FORMAT(27H NO. OF ELEMENTS IN CAAA : ,I5)
      RETURN
      END
C-----------------------------------------------------------------------
COMMENT ROUTINE TO TABULATE THE CLEBSCH-GORDON TYPE COEFFICIENTS
```

```
C         CLM AND ELM, FOR USE WITH SUBROUTINES XM AND XMT.
C         THE NON-ZERO VALUES ARE TABULATED FIRST FOR (L2+M2) AND
C         (L3+M3) ODD; AND THEN FOR (L2+M2) AND (L3+M3) EVEN -
C         THE SAME SCHEME AS THAT BY WHICH THEY ARE RETRIEVED IN
C         SUBROUTINES XM AND XMT.
C    AUTHOR: PENDRY
C    DIMENSION 35 IS SET FOR LMAX.LE.8
      SUBROUTINE  CELMG (CLM,        NLM, YLM, FAC2, NN, FAC1, N, LMAX)
      DIMENSION  CLM(NLM),                YLM(NN), FAC2(NN), FAC1(N)
      DOUBLE PRECISION FAC(35)
      COMMON /F/ FAC
      PI = 3.14159265
      L2MAX = LMAX + LMAX
      NF=4*LMAX+1
      FAC(1) = 1.0
      DO 340 I=1,NF
  340 FAC(I + 1) = FLOAT(I) * FAC(I)
C
COMMENT THE ARRAY YLM IS FIRST LOADED WITH SPHERICAL
C        HARMONICS, ARGUMENTS THETA=PI/2.0, FI=0.0
      LM = 0
      CL = 0.0
      A = 1.0
      B = 1.0
      ASG = 1.0
      LL = L2MAX + 1
COMMENT MULTIPLICATIVE FACTORS REQUIRED
      DO 240 L = 1, LL
      FAC1(L) = ASG * SQRT((2.0 * CL + 1.0) * A/(4.0 * PI * B * B))
      CM =  - CL
      LN = L + L - 1
      DO 230 M = 1, LN
      LO = LM + M
      FAC2(LO) = SQRT((CL + 1.0 + CM) * (CL + 1.0 - CM)/((2.0 * CL + 3.
     10) * (2.0 * CL + 1.0)))
  230 CM = CM + 1.0
      CL = CL + 1.0
      A = A * 2.0 * CL * (2.0 * CL - 1.0)/4.0
      B = B * CL
      ASG =  - ASG
  240 LM = LM + LN
COMMENT FIRST ALL THE YLM FOR M=+-L AND M=+-(L-1) ARE
C        ARE CALCULATED BY EXPLICIT FORMULAE
      LM = 1
      CL = 1.0
      ASG =  - 1.0
      YLM(1) = FAC1(1)
      DO 250 L = 1, L2MAX
      LN = LM + L + L + 1
      YLM(LN) = FAC1(L + 1)
      YLM(LM + 1) = ASG * FAC1(L + 1)
      YLM(LN - 1) = 0.0
      YLM(LM + 2) = 0.0
      CL = CL + 1.0
      ASG =  - ASG
  250 LM = LN
COMMENT USING YLM AND YL(M-1) IN A RECURRENCE RELATION
C        YL(M+1) IS CALCULATED
      LM = 1
      LL = L2MAX - 1
      DO 270 L = 1, LL
      LN = L + L - 1
      LM2 = LM + LN + 4
      LM3 = LM - LN
      DO 260 M = 1, LN
      LO = LM2 + M
      LP = LM3 + M
```

```
      LQ = LM + M + 1
      YLM(LO) =   - (FAC2(LP) * YLM(LP))/FAC2(LQ)
  260 CONTINUE
  270 LM = LM + L + L + 1
C
      K = 1
      II = 0
  280 LL = LMAX + II
      DO 310 IL2 = 1, LL
      L2 = IL2 - II
      M2 =   - L2 + 1 - II
      DO 310 I2 = 1, IL2
      DO 300 IL3 = 1, LL
      L3 = IL3 - II
      M3 =   - L3 + 1 - II
      DO 300 I3 = 1, IL3
      LA1 = MAXO(IABS(L2 - L3),IABS(M2 - M3))
      LB1 = L2 + L3
      LA11 = LA1 + 1
      LB11 = LB1 + 1
      M1 = M3 - M2
      DO 290 L11 = LA11, LB11, 2
      L1 = LA11 + LB11 - L11 - 1
      L = (L3 - L2 - L1)/2 + M3
      M = L1 * (L1 + 1) - M1 + 1
      ALM = (( - 1.0)**L) * 4.0 * PI * BLM(L1,M1,L2,M2,L3,  - M3,LMAX)
      CLM(K) = YLM(M) * ALM
  290 K = K + 1
  300 M3 = M3 + 2
  310 M2 = M2 + 2
      IF (II)   320, 320, 330
  320 II = 1
      GO TO 280
  330    CONTINUE
      RETURN
      END
C-----------------------------------------------------------------------
COMMENT CPPP TABULATES THE FUNCTION PPP(I1,I2,I3), EACH ELEMENT
C        CONTAINING THE INTEGRAL OF THE PRODUCT OF THREE LEGENDRE
C        FUNCTIONS P(I1),P(I2),P(I3). THE INTEGRALS ARE CALCULATED
C        FOLLOWING GAUNT'S SUMMATION SCHEME SET OUT BY SLATER:
C        ATOMIC STRUCTURE, VOL1, 309,310
C   PPP IS USED BY SUBROUTINE PSTEMP IN COMPUTING TEMPERATURE-DEPENDENT
C   PHASE SHIFTS
C   AUTHOR: PENDRY
C   DIMENSION 46 REQUIRES N1+N2+N3-1.LE.46
      SUBROUTINE  CPPP (PPP, N1, N2, N3)
      DIMENSION  PPP(N1,N2,N3), F(46)
      DOUBLE PRECISION F
      F(1) = 1.0
      DO 370 I = 1, 45
  370 F(I + 1) = F(I) * FLOAT(I)
      DO 460 I1 = 1, N1
      DO 460 I2 = 1, N2
      DO 460 I3 = 1, N3
      IM1 = I1
      IM2 = I2
      IM3 = I3
      IF (I1-I2)   380, 390, 390
  380 IM1 = I2
      IM2 = I1
  390 IF (IM2-I3)   400, 420, 420
  400 IM3 = IM2
      IM2 = I3
      IF (IM1-IM2)   410, 420, 420
  410 J = IM1
      IM1 = IM2
```

```
          IM2 = J
    420 A = 0.0
          IS = I1 + I2 + I3 - 3
          IF (MOD(IS,2)-1)  430, 460, 430
    430 IF (IABS(IM2-IM1)-IM3+1)  440, 440, 460
    440 SUM = 0.0
          IS = IS/2
          SIGN = 1.0
          DO 450 IT = 1, IM3
          SIGN =  - SIGN
          IA1 = IM1 + IT - 1
          IA2 = IM1 - IT + 1
          IA3 = IM3 - IT + 1
          IA4 = IM2 + IM3 - IT
          IA5 = IM2 - IM3 + IT
    450 SUM = SUM - SIGN * F(IA1) * F(IA4)/(F(IA2) * F(IA3) * F(IA5) * F(
         1IT))
          IA1 = 2 + IS - IM1
          IA2 = 2 + IS - IM2
          IA3 = 2 + IS - IM3
          IA4 = 3 + 2 * (IS - IM3)
          A =  - ( -1.0)**(IS - IM2) * F(IA4) * F(IS + 1) * F(IM3) * SUM/(
         1F(IA1) * F(IA2) * F(IA3) * F(2 * IS + 2))
    460 PPP(I1,I2,I3) = A
          RETURN
          END
C-----------------------------------------------------------------------
C    SUBROUTINE CXMTXT SOLVES A SET OF LINEAR EQUATIONS.
C       A = INPUT MATRIX.  A  IS DESTROYED DURING THE COMPUTATION AND
C             IS REPLACED BY THE UPPER TRIANGULAR MATRIX RESULTING FROM THE
C             GAUSSIAN ELIMINATION PROCESS (WITH PARTIAL PIVOTING).
C       NC = FIRST DIMENSION OF A IN CALLING ROUTINE. NC.GE.NR.
C       NR = ORDER OF A
C       NSYS = NO. OF SYSTEMS TO BE SOLVED.  IF INVERSE OPTION IS CHOSEN
C             NSYS MUST BE AT LEAST AS LARGE AS NR .
C             COEFFICIENT MATRIX MUST BE STORED IN A(I,J), I=1,NR, J=1,NR
C             CONSTANT VECTORS MUST BE STORED IN A(I,NR+1), I=1,NR#
C             A(I,NR+2), I=1,NR#  .... A(I,NR+NSYS), I=1,NR.
C             RESULT OVERWRITES CONSTANT VECTORS
C       NTOT = NR + NSYS
C       MARK = SINGULARITY INDICATOR (MARK=1 FOR SINGULAR A)
C       DET = DET(A)
C       INOPT = -1 FOR SYSTEM SOLN. AND DET
C                0 FOR DET ONLY
C               +1 FOR INVERSE AND DET
C       DIM. OF X ARRAY MUST BE AT LEAST AS LARGE AS FIRST DIM. OF A ARRAY
          SUBROUTINE  CXMTXT (A, NC,NR, NSYS, NTOT, MARK, DET, INOPT)
          REAL   AMAX, QZ
          COMPLEX  A, X
          COMPLEX  AGG, DET, CONST, SIGN, TEMP
          DIMENSION A(NC, NTOT), X(128)
    300 FORMAT(5X,6HDETC = ,2E20.5)
    310 FORMAT(///, 1X, 16HSINGULAR MATRIX , //)
C
C       PRESET PARAMETERS
          SIGN = (1.0E + 00,0.0E + 00)
          MARK = 0
          IFLAG = INOPT
          N = NR
          NPL = N + 1
          NMI = N - 1
          NN = N + N
          NPLSY = N + NSYS
          IF (IFLAG)  950, 950, 920
C
C       INVERSE OPTION - PRESET AUGMENTED PART TO I
    920 DO 930 I = 1, N
```

```
            DO 930 J = NPL, NN
     930 A(I,J) = (0.0E + 00,0.0E + 00)
            DO 940 I = 1, N
            J = I + N
     940 A(I,J) = (1.0E + 00,0.0E + 00)
            NPLSY = NN
   C
   C     TRIANGULARIZE A
     950 DO 1020 I = 1, NMI
            IPL = I + 1
   C     DETERMINE PIVOT ELEMENT
            MAX = I
            AMAX = CABS(A(I,I))
            DO 970 K = IPL, N
            QZ = CABS(A(K,I))
            IF (AMAX-QZ)  960, 970, 970
     960 MAX = K
            AMAX = CABS(A(K,I))
     970 CONTINUE
            IF (MAX-I)  980, 1000, 980
   C     PIVOTING NECESSARY - INTERCHANGE ROWS
     980 DO 990 L = I, NPLSY
            TEMP = A(I,L)
            A(I,L) = A(MAX,L)
     990 A(MAX,L) = TEMP
            SIGN =  - SIGN
   C     ELIMINATE A(I+1,I)---A(N,I)
    1000 DO 1020 J = IPL, N
            TEMP = A(J,I)
            QZ = CABS(TEMP)
            IF(QZ .LT. 1.0E-10) GO TO 1020
            CONST =  - TEMP/A(I,I)
            DO 1010 L = I, NPLSY
    1010 A(J,L) = A(J,L) + A(I,L) * CONST
    1020 CONTINUE
   C
   C     COMPUTE VALUE OF DETERMINANT
            TEMP = (1.0E + 00,0.0E + 00)
            DO 1030 I = 1, N
            AGG = A(I,I)
            QZ = CABS(AGG)
            IF(QZ .GT. 1.0E-10) GO TO 1030
   C     MATRIX SINGULAR
            WRITE(6,310)
            MARK = 1
            GO TO 1040
    1030 TEMP = TEMP * AGG
            DET = SIGN * TEMP
   C
   C     WRITE(6,300) DET
   C
   C     EXIT IF DET ONLY OPTION
    1040 IF (IFLAG)  1050, 1160, 1050
   C     CHECK FOR INVERSE OPTION OR SYSTEMS OPTION
    1050 IF (IFLAG-1)  1070, 1060, 1070
   C     INVERSE OPTION - ABORT IF A IS SINGULAR
    1060 IF (MARK-1)  1070, 1160, 1070
   C
   C     BACK SUBSTITUTE TO OBTAIN INVERSE OR SYSTEM SOLUTION(S)
    1070 DO 1150 I = NPL, NPLSY
            K = N
    1080 X(K) = A(K,I)
            IF (K-N)  1090, 1110, 1090
    1090 DO 1100 J = KPL, N
    1100 X(K) = X(K) - A(K,J) * X(J)
    1110 X(K) = X(K)/A(K,K)
            IF (K-1)  1120, 1130, 1120
    1120 KPL = K
```

```
      K = K - 1
      GO TO 1080
C     PUT SOLN. VECT. INTO APPROPRIATE COLUMN OF A
 1130 DO 1140 L = 1, N
 1140 A(L,I) = X(L)
 1150 CONTINUE
C
 1160 RETURN
      END
C---------------------------------------------------------------------
C     SUBROUTINE DBG COMPUTES THE REFLECTION MATRIX FOR A PAIR OF LAYERS
C     FOR INCIDENCE FROM ONE SIDE ONLY. LAYER DOUBLING IS USED. NO
C     OVERWRITING OF INPUT TAKES PLACE.
C     NOTE: SUBROUTINE DBLG PRODUCES ADDITIONNALLY THE REFLECTION FROM
C     THE OTHER SIDE OF THE LAYERS AND THE TRANSMISSIONS IN BOTH DIRECTIONS
C     RA1,TA1,RA2,TA2= REFLECTION AND TRANSMISSION MATRICES FOR FIRST
C       LAYER (ON THE SIDE OF THE INCIDENT BEAMS). 1 AND 2 REFER TO
C       INCIDENCE ON FIRST LAYER FROM SIDE OPPOSITE SECOND LAYER AND SIDE
C       TOWARDS SECOND LAYER, RESP.
C     RB1= REFLECTION MATRIX OF SECOND LAYER (E.G. SUBSTRATE).
C     RAB= OUTPUT REFLECTION MATRIX OF COMBINED LAYER.
C     ASD= INTERLAYER VECTOR FROM FIRST LAYER TO SECOND LAYER.
C     N=NO. OF BEAMS USED.
C     NA= OFFSET IN LIST PQ OF PARTICULAR SUBSET OF BEAMS USED.
C     PQ= LIST OF BEAMS G.
C     NT= TOTAL NO. OF BEAMS IN MAIN PROGRAM AT CURRENT ENERGY.
C     S1,S2,XS,PP,IPL= WORKING SPACES.
C     NP=DIMENSION FOR WORKING SPACES, NP.GE.N.
C     EMACH= MACHINE ACCURACY.
C     IN COMMON BLOCKS:
C     E= CURRENT ENERGY.
C     VPI= IMAGINARY PART OF ENERGY.
C     AK2,AK3= PARALLEL COMPONENTS OF PRIMARY INCIDENT K-VECTOR.
C     AS: NOT USED.
      SUBROUTINE DBG(RA1,TA1,RA2,TA2,RB1,RAB,ASD,N,
     1NA, PQ, NT, S1, S2, XS, PP, IPL, NP, EMACH)
      COMPLEX RAB,RA1,TA1,RB1
      COMPLEX  S1, S2, PP, XS, XX, CZ, IU
      COMPLEX RA2(N,N),TA2(N,N)
      DIMENSION RA1(N,N),TA1(N,N),RB1(N,N),RAB(N,N),S1(NP,NP),S2(NP,NP)
      DIMENSION PP(2,NP), XS(NP), PQ(2,NT), IPL(NP)
      DIMENSION ASD(3), AS(3)
      COMMON /X4/E, VPI, AK2, AK3,  AS
      CZ = CMPLX(0.0,0.0)
      IU = CMPLX(0.0,1.0)
      AK = 2.0 * E
C COMPUTE INTERLAYER PROPAGATORS PP
      DO 70 I = 1, N
      BK2 = AK2 + PQ(1,I + NA)
      BK3 = AK3 + PQ(2,I + NA)
      XX = CMPLX(AK - BK2 * BK2 - BK3 * BK3, - 2.0 * VPI + 0.000001)
C XX IS K(PERP)
      XX = CSQRT(XX)
      X = BK2 * ASD(2) + BK3 * ASD(3)
      PP(1,I) = CEXP(IU * (XX * ASD(1) + X))
   70 PP(2,I) = CEXP( - IU * ( - XX * ASD(1) + X))
C
C
C PRODUCE MATRIX TO BE INVERTED
      DO 90 J = 1, N
      DO 80 K = 1, N
      S1(J,K) = CZ
      DO 80 L = 1, N
   80 S1(J,K) = S1(J,K) - RA2(J,L) * PP(2,L) * RB1(L,K) * PP(1,K)
   90 S1(J,J) = S1(J,J) + 1.0
C GAUSSIAN ELIMINATION STEP OF MATRIX INVERSION
      CALL  ZGE (S1, IPL, NP, N, EMACH)
```

```
      DO 110 K = 1, N
      DO 100 J = 1, N
100 XS(J) = TA1(J,K)
C  SUBSTITUTION STEP OF MATRIX INVERSION
      CALL  ZSU (S1, IPL, XS, NP, N, EMACH)
      DO 110 J = 1, N
110 S2(J,K) = XS(J)
      DO 130 K = 1, N
      DO 120 J = 1, N
      S1(J,K) = CZ
      XS(J) = CZ
      DO 120 L = 1, N
120 XS(J) = XS(J) + RB1(J,L) * PP(1,L) * S2(L,K)
      DO 130 J = 1, N
      S2(J,K) = RA1(J,K)
      DO 130 L = 1, N
130 S2(J,K) = S2(J,K) + TA2(J,L) * PP(2,L) * XS(L)
      DO 140 I=1,N
      DO 140 K=1,N
140 RAB(I,K)=S2(I,K)
      RETURN
      END
C----------------------------------------------------------------------
C  SUBROUTINE DBLG COMPUTES (USING LAYER DOUBLING) THE REFLECTION AND
C  TRANSMISSION MATRICES FOR A PAIR OF LAYERS FOR INCIDENCE FROM BOTH
C  SIDES. EACH LAYER IS ASSUMED TO HAVE EQUAL DIFFRACTION MATRIX
C  ELEMENTS FOR INCIDENCE FROM EITHER SIDE OF IT. OPTION II=0 PRODUCES
C  SAME OUTPUT AS SUBROUTINE DBG, BUT WITH OVERWRITING OF INPUT.
C  RA1,TA1= REFLECTION AND TRANSMISSION MATRICES OF FIRST LAYER
C    (OVERWRITTEN BY REFLECTION AND TRANSMISSION MATRICES FOR INCIDENCE
C    FROM SIDE OF FIRST LAYER).
C  RB1,TB1=SAME AS RA1,TA1 FOR SECOND LAYER (OVERWRITTEN BY REFLECTION
C    AND TRANSMISSION MATRICES FOR INCIDENCE FROM SIDE OF SECOND LAYER).
C  ASD= INTERLAYER VECTOR FROM FIRST LAYER TO SECOND LAYER.
C  N=NO. OF BEAMS USED.
C  NA= OFFSET IN LIST PQ OF PARTICULAR SUBSET OF BEAMS USED.
C  PQ= LIST OF BEAMS G.
C  NT= TOTAL NO. OF BEAMS IN MAIN PROGRAM AT CURRENT ENERGY.
C  S1,S2,S3,S4,XS,PP,IPL= WORKING SPACES.
C  NP=DIMENSION FOR WORKING SPACES, NP.GE.N.
C  EMACH= MACHINE ACCURACY.
C  II=0: PRODUCE ONLY REFLECTION MATRIX FOR INCIDENCE ON SIDE OF
C    FIRST LAYER (I.E. ONLY RA1).
C  II.NE.0: PRODUCE ALL REFLECTION AND TRANSMISSION MATRICES.
C  IN COMMON BLOCKS:
C  E= CURRENT ENERGY.
C  VPI= IMAGINARY PART OF ENERGY.
C  AK2,AK3= PARALLEL COMPONENTS OF PRIMARY INCIDENT K-VECTOR.
C  AS: NOT USED.
      SUBROUTINE DBLG(RA1,TA1,RB1,TB1,ASD,N,
     1NA,PQ,NT,S1,S2,S3,S4,XS,PP,IPL,NP,EMACH,II)
      COMPLEX RA1,TA1,RB1,TB1
      COMPLEX  S1,S2,S3,S4,PP,XS,XX,CZ,IU
      DIMENSION RA1(N,N),TA1(N,N),RB1(N,N),TB1(N,N),S1(NP,NP),S2(NP,NP)
      DIMENSION  S3(NP,NP),S4(NP,NP),PP(2,NP),XS(NP),PQ(2,NT),IPL(NP)
      DIMENSION  ASD(3), AS(3)
      COMMON  /X4/E, VPI, AK2, AK3,   AS
      CZ = CMPLX(0.0,0.0)
      IU = CMPLX(0.0,1.0)
      AK = 2.0 * E
C  COMPUTE INTERLAYER PROPAGATORS PP
      DO 70 I = 1, N
      BK2 = AK2 + PQ(1,I + NA)
      BK3 = AK3 + PQ(2,I + NA)
      XX = CMPLX(AK - BK2 * BK2 - BK3 * BK3, - 2.0 * VPI + 0.000001)
C  XX IS K(PERP)
      XX = CSQRT(XX)
```

```
      X = BK2 * ASD(2) + BK3 * ASD(3)
      PP(1,I) = CEXP(IU * (XX * ASD(1) + X))
   70 PP(2,I) = CEXP( - IU * ( - XX * ASD(1) + X))
C
C
C  S1 AT LABEL 90 AND S3 AT LABEL 150 ARE MATRICES TO BE INVERTED
      DO 90 J = 1, N
      DO 80 K = 1, N
      S1(J,K) = CZ
      DO 80 L = 1, N
   80 S1(J,K) = S1(J,K) - RA1(J,L) * PP(2,L) * RB1(L,K) * PP(1,K)
   90 S1(J,J) = S1(J,J) + 1.0
C  GAUSSIAN ELIMINATION STEP OF MATRIX INVERSION
      CALL  ZGE (S1, IPL, NP, N, EMACH)
      DO 110 K = 1, N
      DO 100 J = 1, N
  100 XS(J) = TA1(J,K)
C  SUBSTITUTION STEP OF MATRIX INVERSION
      CALL  ZSU (S1, IPL, XS, NP, N, EMACH)
      DO 110 J = 1, N
  110 S2(J,K) = XS(J)
      DO 130 K = 1, N
      DO 120 J = 1, N
      XS(J) = CZ
      DO 120 L = 1, N
  120 XS(J) = XS(J) + RB1(J,L) * PP(1,L) * S2(L,K)
      DO 130 J = 1, N
      S4(J,K) = RA1(J,K)
      DO 130 L = 1, N
  130 S4(J,K) = S4(J,K) + TA1(J,L) * PP(2,L) * XS(L)
C  BRANCH TO 135 IF MORE THAN RA1 IS DESIRED
      IF (II .NE. 0)  GO TO 135
C
      DO 132 J = 1, N
      DO 132 K = 1, N
  132 RA1(J,K) = S4(J,K)
      RETURN
C
  135 DO 137 K=1,N
      DO 137 J=1,N
      S1(J,K)=CZ
      DO 137 L=1,N
  137 S1(J,K)=S1(J,K)+TB1(J,L)*PP(1,L)*S2(L,K)
      DO 150 J = 1, N
      DO 140 K = 1, N
      S3(J,K) = CZ
      DO 140 L = 1, N
  140 S3(J,K) = S3(J,K) - RB1(J,L) * PP(1,L) * RA1(L,K) * PP(2,K)
  150 S3(J,J) = S3(J,J) + 1.0
C
C  GAUSSIAN ELIMINATION STEP OF MATRIX INVERSION
      CALL  ZGE (S3, IPL, NP, N, EMACH)
      DO 190 K = 1, N
      DO 180 J = 1, N
  180 XS(J) = TB1(J,K)
C  SUBSTITUTION STEP OF MATRIX INVERSION
      CALL  ZSU (S3, IPL, XS, NP, N, EMACH)
      DO 190 J = 1, N
  190 S2(J,K) = XS(J)
      DO 210 K = 1, N
      DO 200 J = 1, N
      S3(J,K) = CZ
      XS(J) = CZ
      DO 200 L = 1, N
      XS(J) = XS(J) + RA1(J,L) * PP(2,L) * S2(L,K)
  200 S3(J,K) = S3(J,K) + TA1(J,L) * PP(2,L) * S2(L,K)
      DO 210 J = 1, N
```

```
      S2(J,K) = RB1(J,K)
      DO 210 L = 1, N
  210 S2(J,K) = S2(J,K) + TB1(J,L) * PP(1,L) * XS(L)
C
      DO 220 J = 1, N
      DO 220 K = 1, N
      RA1(J,K)=S4(J,K)
      TA1(J,K) = S1(J,K)
      RB1(J,K) = S2(J,K)
  220 TB1(J,K) = S3(J,K)
      RETURN
      END
C-------------------------------------------------------------------
C  SUBROUTINE DEBWAL COMPUTES DEBYE-WALLER FACTORS FOR
C  DIFFRACTION FROM BEAM G(PRIME) INTO BEAM G (AND ITS SYMMETRICAL
C  COUNTERPARTS) FOR, IN GENERAL, ANISOTROPIC ATOMIC VIBRATION,
C  INCLUDING ZERO-TEMPERATURE VIBRATION.
C   NG= NO. OF BEAMS CORRESPONDING TO G.
C   G= SET OF SYMMETRICAL BEAMS G.
C   GP= INCIDENT BEAM G(PRIME).
C   E= ENERGY.
C   VPI= OPTICAL POTENTIAL
C   AK2,AK3= PARALLEL COMPONENTS OF PRIMARY INCIDENT K-VECTOR.
C   T= ACTUAL TEMPERATURE.
C   TO= REFERENCE TEMPERATURE FOR VIBRATION AMPLITUDES.
C   DRX= RMS VIBRATION AMPLITUDE PERPENDICULAR TO SURFACE.
C   DRY= RMS VIBRATION AMPLITUDE PARALLEL TO SURFACE.
C   DO4= FOURTH POWER OF RMS ZERO-TEMPERATURE VIBRATION AMPLITUDE
C   (ISOTROPIC).
C   EDW= OUTPUT DEBYE-WALLER FACTOR (EDW(1,I) FOR REFLECTION, EDW(2,I)
C   FOR TRANSMISSION).
      SUBROUTINE  DEBWAL (NG, G, GP, E, VPI, AK2, AK3, T, TO, DRX, DRY,
     1DO4, EDW)
      COMPLEX CSQRT
      COMPLEX EDW
      DIMENSION  GP(2), G(2,12), EDW(2,NG)
      A1 = GP(1) + AK2
      A2 = GP(2) + AK3
      CC = REAL(CSQRT(CMPLX(2.0 * E - A1 * A1 - A2 * A2, - 2.0 * VPI)))
      A1 = G(1,1) + AK2
      A2 = G(2,1) + AK3
      DD = REAL(CSQRT(CMPLX(2.0 * E - A1 * A1 - A2 * A2, - 2.0 * VPI)))
C  C IS PERPENDICULAR COMPONENT OF SCATTERING VECTOR FOR REFLECTION,
C  D IS SAME FOR TRANSMISSION
      C = CC + DD
      D = CC - DD
      DO 330 I = 1, NG
      A1 = GP(1) - G(1,I)
      A2 = GP(2) - G(2,I)
C  D2 IS MEAN-SQUARE VIBRATION AMPLITUDE PARALLEL TO SURFACE AT ACTUAL
C  TEMPERATURE
      D2 = DRY * DRY * T/TO
C  ZERO-TEMPERATURE VIBRATION IS NOW MIXED IN
      D2 = SQRT(D2 * D2 + DO4)
      A1 = (A1 * A1 + A2 * A2) * D2
C  D2 IS SAME AS ABOVE, BUT FOR PERPENDICULAR COMPONENTS
      D2 = DRX * DRX * T/TO
      D2 = SQRT(D2 * D2 + DO4)
      EDW(1,I) = CMPLX(EXP( - 0.166667 * (A1 + C * C * D2)),0.0)
  330 EDW(2,I) = CMPLX(EXP( - 0.166667 * (A1 + D * D * D2)),0.0)
      RETURN
      END
C-------------------------------------------------------------------
C  FUNCTION FACT COMPUTES FACTORIAL(L)/10**L, USING AN ASYMPTOTIC
C  EXPANSION FOR L.GT.4. USED IN CA.
      FUNCTION  FACT (L)
      DOUBLE PRECISION  DFACT, X
```

```
      IF (L .GT. 4)  GO TO 1600
      IF (L .EQ. 0)  FACT = 1.0
      IF (L .EQ. 1)  FACT = 0.1
      IF (L .EQ. 2)  FACT = 0.02
      IF (L .EQ. 3)  FACT = 6.0 * 0.001
      IF (L .EQ. 4)  FACT = 24.0 * 0.0001
      RETURN
 1600 X = L + 1
      DFACT = DEXP( - X) * (10.0D0**((X - 0.5D0) * DLOG10(X) - (X - 1.
     1JD0))) * (DSQRT(6.283185307179586D0)) * (1.0 + (1.0/(12.0 * X)) +
     2(1.0/(288.0D0 * (X**2))) - (139.0D0/(51840.0D0 * (X**3))) - (571.
     3D0/(2488320.0D0 * (X**4))))
      FACT = SNGL(DFACT)
      RETURN
      END
C------------------------------------------------------------------
COMMENT FMAT CALCULATES THE VALUES OF THE SUM FLMS(JS,LM),
C        OVER LATTICE POINTS OF EACH SUBLATTICE JS, WHERE
C        LM=(0,0),(1,-1),(1,1),.....
C        NOTE: FOR ODD (L+M), FLMS IS ZERO
C   AUTHOR: PENDRY
C   DIMENSIONS 96 AND 33 REQUIRE LMAX.LE.8.
C    FLMS= OUTPUT LATTICE SUMS.
C    V,JJS= INPUT FROM SUBROUTINE SLIND.
C    NL= NO. OF SUBLATTICES
C    NLS= ACTUAL NO. OF SUBLATTICE SUMS DESIRED (E.G. 1 OR NL).
C    DCUT= CUTOFF DISTANCE FOR LATTICE SUM.
C    IDEG= DEGREE OF SYMMETRY OF LATTICE (IDEG-FOLD ROTATION AXIS).
C     NOTE: DO NOT USE IDEG=1. IDEG=3 PREFERABLE OVER IDEG=6.
C    LMAX= LARGEST VALUE OF L.
C    KLM=(2*LMAX+1)*(2*LMAX+2)/2.
C    IN COMMON BLOCKS:
C    E= CURRENT ENERGY.
C    AK: PARALLEL COMPONENTS OF PRIMARY INCIDENT K-VECTOR.
C    VPI= IMAGINARY PART OF ENERGY.
C    BR1,BR2,RAR1,RAR2,NL1,NL2: NOT USED.
C    AR1,AR2= BASIS VECTORS OF SUPERLATTICE.
      SUBROUTINE  FMAT (FLMS, V, JJS, NL, NLS, DCUT, IDEG, LMAX,KLM)
      COMPLEX  FLMS, SCC, SA, RTAB, CZERO, CI, KAPPA, SC, SD, SE, Z,
     1ACS, ACC, RF
      COMPLEX CSQRT,CEXP,CCOS,CSIN
      DIMENSION  FLMS(NL,KLM), V(NL,2), JJS(NL,IDEG), BR1(2)
      DIMENSION  BR2(2), SCC(6,33), SA(96), ANC(6), ANS(6), RTAB(4)
      DIMENSION  AK(2), AR1(2), AR2(2), RAR1(2), RAR2(2), R(2)
      COMMON  E, AK, VPI
      COMMON  /SL/BR1, BR2, AR1, AR2, RAR1, RAR2, NL1, NL2
      PI = 3.14159265
      CZERO = CMPLX(0.0,0.0)
      CI = CMPLX(0.0,1.0)
      KAPPA = CMPLX(2.0 * E, - 2.0 * VPI + 0.000001)
      KAPPA = CSQRT(KAPPA)
      AG=SQRT(AK(1)*AK(1)+AK(2)*AK(2))
C
COMMENT ANC,ANS AND SA ARE PREPARED TO BE USED IN THE SUM
C        OVER SYMMETRICALLY RELATED SECTORS OF THE LATTICE
      L2MAX = LMAX + LMAX
      LIM = L2MAX * IDEG
      LIML = L2MAX + 1
      ANG = 2.0 * PI/FLOAT(IDEG)
      D = 1.0
      DO 10 J = 1, IDEG
      ANC(J) = COS(D * ANG)
      ANS(J) = SIN(D * ANG)
   10 D = D + 1.0
      D = 1.0
      DO 20 J = 1, LIM
      SA(J) = CEXP( - CI * D * ANG)
```

```
   20 D = D + 1.0
      DO 30 J = 1, NL
      DO 30 K = 1, KLM
   30 FLMS(J,K) = CZERO
COMMENT THE LATTICE SUM STARTS.THE SUM IS DIVIDED INTO ONE
C        OVER A SINGLE SECTOR,THE OTHER (IDEG-1) SECTORS
C        ARE RELATED BY SYMMETRY EXCEPT FOR FACTORS
C        INVOLVING THE DIRECTION OF R
C   THE RANGE OF SUMMATION IS LIMITED BY DCUT
      D = SQRT(AR1(1) * AR1(1) + AR1(2) * AR1(2))
      LI1 = INT(DCUT/D) + 1
      D = SQRT(AR2(1) * AR2(1) + AR2(2) * AR2(2))
      LI2 = INT(DCUT/D) + 1
C   ONE SUBLATTICE AT A TIME IS TREATED IN THE FIRST SECTOR
      DO 160 JS = 1, NLS
      LI11 = LI1
      LI22 = LI2
      ASST = 0.0
      ADD = 1.0
      ANT = - 1.0
      ADR1 = V(JS,1) * COS(V(JS,2))
      ADR2 = V(JS,1) * SIN(V(JS,2))
   40 AST = - 1.0
      IF ((ADR1*AR2(1)+ADR2*AR2(2))-1.0E-6)  50, 50, 60
   50 AST = AST * ASST
   60 AN1 = ANT
      DO 130 I1 = 1, LI11
      AN1 = AN1 + ADD
      AN2 = AST
      DO 130 I2 = 1, LI22
      AN2 = AN2 + 1.0
COMMENT R=THE CURRENT LATTICE VECTOR IN THE SUM
C        AR=MOD(R)
C        RTAB(1)=-EXP(I*FI(R))
      R(1) = AN1 * AR1(1) + AN2 * AR2(1) + ADR1
      R(2) = AN1 * AR1(2) + AN2 * AR2(2) + ADR2
      AR = SQRT(R(1) * R(1) + R(2) * R(2))
      RTAB(1) = - CMPLX(R(1)/AR,R(2)/AR)
      ABC = 1.0
      ABB = 0.0
      IF (AG-1.0E-4)  80, 80, 70
   70 ABC = (AK(1) * R(1) + AK(2) * R(2))/(AG * AR)
      ABB = ( - AK(2) * R(1) + AK(1) * R(2))/(AG * AR)
   80 SC = CI * AG * AR
COMMENT SCC CONTAINS FACTORS IN THE SUMMATION DEPENDENT ON
C        THE DIRECTION OF R. CONTRIBUTIONS FROM SYMMETRICALLY
C        RELATED SECTORS CAN BE GENERATED SIMPLY AND ARE
C        ACCUMULATED FOR EACH SECTOR, INDEXED BY THE SUBSCRIPT J.
C        THE SUBSCRIPT M IS ORDERED:M=(-L2MAX),(-L2MAX+1)....
C        (+L2MAX)
      DO 90 J = 1, IDEG
      AD = ABC * ANC(J) - ABB * ANS(J)
      SD = CEXP(SC * AD)
      SCC(J,LIML) = SD
      MJ = 0
      SE = RTAB(1)
      DO 90 M = 1, L2MAX
      MJ = MJ + J
      MP = LIML + M
      MM = LIML - M
      SCC(J,MP) = SD * SA(MJ)/SE
      SCC(J,MM) = SD * SE/SA(MJ)
   90 SE = SE * RTAB(1)
      Z = AR * KAPPA
      ACS = CSIN(Z)
      ACC = CCOS(Z)
COMMENT RTAB(3)=SPHERICAL HANKEL FUNCTION OF THE FIRST KIND,L=0
```

```
C         RTAB(4)=SPHERICAL HANKEL FUNCTION OF THE FIRST KIND,L=1
       RTAB(3) = (ACS - CI * ACC)/Z
       RTAB(4) = ((ACS/Z - ACC) - CI * (ACC/Z + ACS))/Z
       AL = 0.0
COMMENT THE SUMMATION OVER FACTORS INDEPENDENT OF THE
C         DIRECTION OF R IS ACCUMULATED IN FLM, FOR EACH
C         SUBLATTICE INDEXED BY SUBSCRIPT JSP.  THE SECOND
C         SUBSCRIPT ORDERS L AND M AS: (0,0),(1,-1),(1,1),(2,-2),
C         (2,0),(2,2)...
       JF = 1
       DO 120 JL = 1, LIML
       RF = RTAB(3) * CI
       JM = L2MAX + 2 - JL
       DO 110 KM = 1, JL
C CONSIDER THE CORRESPONDING LATTICE POINTS IN THE OTHER SECTORS AND
C GIVE THEIR CONTRIBUTION TO THE APPROPRIATE SUBLATTICE
       DO 100 J = 1, IDEG
       JSP = JJS(JS,J)
  100 FLMS(JSP,JF) = FLMS(JSP,JF) + SCC(J,JM) * RF
       JF = JF + 1
       JM = JM + 2
  110 CONTINUE
COMMENT SPHERICAL HANKEL FUNCTIONS FOR HIGHER L ARE
C         GENERATED BY RECURRENCE RELATIONS
       ACS = (2.0 * AL + 3.0) * RTAB(4)/Z - RTAB(3)
       RTAB(3) = RTAB(4)
       RTAB(4) = ACS
       AL = AL + 1.0
  120 CONTINUE
  130 CONTINUE
COMMENT SPECIAL TREATMENT IS REQUIRED IF IDEG=2
C TWO SECTORS REMAIN TO BE SUMMED OVER
       IF (IDEG-2)  140, 140, 160
  140 IF (ASST)  150, 150, 160
  150 ASST = 1.0
       ADD = - 1.0
       ANT = 0.0
       IF ((ADR1*AR1(1)+ADR2*AR1(2)) .LE. 1.0E-4)  LI11 = LI1 - 1
       IF ((ADR1*AR2(1)+ADR2*AR2(2)) .LE. 1.0E-4)  LI22 = LI2 + 1
       GO TO 40
  160 CONTINUE
       RETURN
       END
C----------------------------------------------------------------------
C SUBROUTINE GHD COMPUTES DIRECT LATTICE SUMS FOR (LM)-SPACE
C PROPAGATORS GH BETWEEN TWO SUBPLANES (HAVING BRAVAIS LATTICES) OF
C A COMPOSITE LAYER. THE SUBPLANES MAY BE COPLANAR.
C FOR QUANTITIES NOT EXPLAINED BELOW SEE GHMAT, MPERTI OR MTINV
C IZ= SERIAL NO. OF CURRENT INTERPLANAR VECTOR DRL.
C IS=1 FOR PROPAGATION FROM FIRST TO SECOND SUBPLANE.
C IS=2 FOR PROPAGATION FROM SECOND TO FIRST SUBPLANE.
C GH= OUTPUT INTERPLANAR PROPAGATOR.
C S= WORKING SPACE (LATTICE SUM).
C LMS= (2*LMAX+1)**2.
C Y= WORKING SPACE (SPHERICAL HARMONICS).
C L2M= 2*LMAX+1.
C KO= COMPLEX MAGNITUDE OF WAVEVECTOR.
C DCUT= CUTOFF RADIUS FOR LATTICE SUM.
C CAA= CLEBSCH-GORDON COEFFICIENTS FROM SUBROUTINE CAAA.
C NCAA= NO. OF C.-G. COEFFICIENTS IN CAA.
C LXM= PERMUTATION OF (LM) SEQUENCE FROM SUBROUTINE LXGENT.
C NOTE: DIMENSION 21 SET FOR LMAX.LE.10.
       SUBROUTINE GHD(IZ,IS,GH,LMG,LMMAX,S,LMS,Y,L2M,DRL,NLAY2,
      1KO,DCUT,CAA,NCAA,LXM)
       COMPLEX GH(LMG,LMMAX),Y(L2M,L2M),H(21),S(LMS)
       COMPLEX RU,CI,CZ,KO,FF,Z,Z1,Z2,Z3,ST
       DIMENSION DRL(NLAY2,3),ARA1(2),ARA2(2),ARB1(2),ARB2(2)
```

```
         DIMENSION RBR1(2),RBR2(2),CAA(NCAA)
         DIMENSION V(3),LXM(LMMAX)
         COMMON E,AK2,AK3,VPI
         COMMON /SL/ ARA1,ARA2,ARB1,ARB2,RBR1,RBR2,NL1,NL2
         RU=(1.0,0.0)
         CZ=(0.0,0.0)
         CI=(0.0,1.0)
         DCUT2=DCUT*DCUT
         DO 5 I=1,LMS
5        S(I)=CZ
         DO 10 I=1,3
10       V(I)=DRL(IZ,I)
C   TURN INTERPLANAR VECTOR AROUND IF IS=2
         IF (IS-2) 40,20,20
20       DO 30 I=1,3
30       V(I)=-V(I)
C
C   START OF TWO 1-DIMENSIONAL LATTICE LOOPS FOR SUMMATION OVER 1 QUADRAN
40       NUMR=0
         JJ1=0
50       JJ1=JJ1+1
         JJ2=0
60       JJ2=JJ2+1
         NOR=0
         J1=JJ1-1
         J2=JJ2-1
C   START OF LOOP OVER QUADRANTS
         DO 140 KK=1,4
         GO TO (70,80,90,100),KK
70       NR1=J1
         NR2=J2
         GO TO 110
80       IF (J1.EQ.0.AND.J2.EQ.0) GO TO 150
         IF (J2.EQ.0) GO TO 140
         NR1=J1
         NR2=-J2
         GO TO 110
90       IF (J1.EQ.0) GO TO 140
         NR1=-J1
         NR2=J2
         GO TO 110
100      IF (J1.EQ.0.OR.J2.EQ.0) GO TO 140
         NR1=-J1
         NR2=-J2
C
110      PX=NR1*ARB1(1)+NR2*ARB2(1)
         PY=NR1*ARB1(2)+NR2*ARB2(2)
         X1=(PX+V(2))**2+(PY+V(3))**2
C   CUTOFF OF LATTICE SUMMATION AT RADIUS DCUT
         IF (X1.GT.DCUT2) GO TO 140
         NOR=1
         NUMR=NUMR+1
         Z1=CEXP(-CI*(PX*AK2+PY*AK3))
         X2=SQRT(X1+V(1)**2)
         X1=SQRT(X1)
         Z2=K0*X2
         Z=CMPLX(V(1)/X2,0.0)
         FY=0.0
         IF (ABS(X1).LT.1.E-6) GO TO 118
         CFY=(PX+V(2))/X1
         IF (ABS(ABS(CFY)-1.).LE.1.E-6) GO TO 115
         FY= ACOS(CFY)
         GO TO 117
115      IF (CFY.LT.0.0) FY=3.14159265
117      IF ((PY+V(3)).LT.0.0) FY=-FY
C   COMPUTE REQUIRED BESSEL FUNCTIONS AND SPHERICAL HARMONICS
118      CALL SB(Z2,H,L2M)
```

```
      CALL SH(L2M,Z,FY,Y)
C
      ST=RU
      DO 130 L=1,L2M
      ST=ST*CI
      Z3=ST*H(L)*Z1
      L1=L*L-L
      DO 120 M=1,L
      M1=M-1
      IPM=L1+M
      IMM=L1-M+2
      S(IPM)=S(IPM)+Z3*Y(L,M)
      IF (M.EQ.1) GO TO 120
      S(IMM)=S(IMM)+Z3*Y(M-1,L)*(-1)**MOD(M1,2)
C  S NOW CONTAINS THE LATTICE SUM
120   CONTINUE
130   CONTINUE
140   CONTINUE
150   IF (NOR.EQ.1) GO TO 60
      IF (JJ2-1) 50,160,50
160   CONTINUE
C
C  PRINT NUMBER OF LATTICE POINTS USED IN SUMMATION
      WRITE(6,170)NUMR
170   FORMAT(28H NO. OF LATT. PTS. FOR GH : ,I5)
      FF=-8.0*3.14159265*K0
C  USE SUBROUTINE GHSC TO MULTIPLY LATTICE SUM INTO CLEBSCH-GORDON
C  COEFFICIENTS
      CALL GHSC(IZ,IS,GH,LMG,LMMAX,S,LMS,CAA,NCAA,FF,LXM,NLAY2)
      RETURN
      END
C-------------------------------------------------------------------------
C  SUBROUTINE GHMAT COMPUTES (LM)-SPACE INTERPLANAR PROPAGATORS GH FOR
C  THE COMPOSITE LAYER TREATED BY SUBROUTINES MPERTI OR MTINV. DEPENDING
C  ON THE INTERPLANAR SPACING, EITHER A RECIPROCAL-SPACE SUMMATION OR
C  A DIRECT-SPACE SUMMATION IS PERFORMED (THE LATTER IN SUBROUTINE GHD).
C  USING INFORMATION PROVIDED BY SUBROUTINE SRTLAY, GHMAT REUSES (DOES
C  NOT RECOMPUTE) EXISTING VALUES OF GH PRODUCED PREVIOUSLY, COMPUTES
C  NEW VALUES NOT PREVIOUSLY PRODUCED AND COPIES THESE NEW VALUES INTO
C  GH'S THAT ARE IDENTICAL (INSTEAD OF COMPUTING THE LATTER SEPARATELY).
C  ORDERING OF ELEMENTS: FOR EACH GH THE (LM) INDEX IS ORDERED THUS:
C  (00),(2-2),(20),(22),(4-4),(4-2),...,(1-1),(11),(3-3),(31),...
C  ,(10),(3-2),(30),(32),(5-4),...,(2-1),(21),(4-3),(4-1),(41),...
C  THE GH(I,J)(I,J=SUBPLANE INDICES) ARE STACKED ON TOP OF EACH OTHER
C  IN COLUMNAR MATRIX GH, SO THAT GH(I,J) IS FOUND IN THE K-TH
C  POSITION FROM THE TOP, WHERE K=MGH(I,J).
C   GH= MATRIX CONTAINING ALL INTERPLANAR PROPAGATORS FOR THE COMPOSITE
C    LAYER CONSIDERED BY MPERTI OR MTINV. THE INDIVIDUAL MATRICES ARE
C    STACKED VERTICALLY IN GH.
C   LMG= NLAY2*LMMAX.
C   LMMAX= (LMAX+1)**2.
C   MGH= MATRIX CONTAINING KEY TO POSITION OF INDIVIDUAL GH'S IN THE
C    MATRIX GH: MGH(I,J) IS SEQUENCE NUMBER OF GH(I,J) IN COLUMNAR
C    MATRIX GH.
C   NLAY= NO. OF SUBPLANES CONSIDERED.
C   NUGH= LIST OF THOSE GH'S THAT MUST BE COMPUTED.
C   NGEQ= LIST OF THOSE GH'S THAT CAN BE COPIED FROM FRESHLY COMPUTED
C    GH'S.
C   NGOL= LIST OF THOSE GH'S THAT CAN BE COPIED FROM PREVIOUSLY
C    PRODUCED GH'S.
C   NLAY2= NLAY*(NLAY-1).
C   TST= INPUT QUANTITY FOR DETERMINING NO. OF POINTS REQUIRED IN
C    RECIPROCAL LATTICE SUMMATION.
C   TEST,Y,S= WORKING SPACE.
C   L2M= 2*LMAX+1.
C   LM= LMAX+1.
C   LMS= (2*LMAX+1)**2.
```

```
C     DRL= SET OF INTERPLANAR VECTORS.
C     TV= AREA OF UNIT CELL OF EACH SUBPLANE.
C     LXM= PERMUTATION OF (LM) SEQUENCE.
C     LEV= (LMAX+1)*(LMAX+2)/2.
C     DCUT= CUTOFF RADIUS FOR LATTICE SUMMATION.
C     CAA= CLEBSCH-GORDON COEFFICIENTS.
C     NCAA= NO. OF C.-G. COEFFICIENTS.
C     IN COMMON BLOCKS:
C     E= CURRENT ENERGY.
C     AK2,AK3= PARALLEL COMPONENTS OF PRIMARY INCIDENT K-VECTOR.
C     VPI= IMAGINARY PART OF ENERGY.
C     ARA1,ARA2= BASIS VECTORS OF SUBSTRATE LAYER LATTICE.
C     ARB1,ARB2= BASIS VECTORS OF SUPERLATTICE.
C     RBR1,RBR2= RECIPROCAL LATTICE OF SUPERLATTICE.
C     NL1,NL2= SUPERLATTICE CHARACTERIZATION (SEE MAIN PROGRAM).
      SUBROUTINE GHMAT(GH,LMG,LMMAX,MGH,NLAY,NUGH,NGEQ,NGOL,NLAY2,
     1TST,TEST,Y1,L2M,Y,LM,S,LMS,DRL,TV,LXM,LEV,DCUT,CAA,NCAA)
      DIMENSION MGH(NLAY,NLAY),NUGH(NLAY2),NGEQ(NLAY2),NGOL(NLAY2)
      DIMENSION RBR1(2),RBR2(2),DRL(NLAY2,3),TEST(NLAY2),LXM(LMMAX)
      DIMENSION ARA1(2),ARA2(2),ARB1(2),ARB2(2),CAA(NCAA)
      COMPLEX GH(LMG,LMMAX),Y(LM,LM),Y1(L2M,L2M),S(LMS)
      COMPLEX CI,CZ,KPRG,KO,Z,T1,T1P,T2,T2P,T3,BS,CS,CFAC
      COMMON E,AK2,AK3,VPI
      COMMON /SL/ ARA1,ARA2,ARB1,ARB2,RBR1,RBR2,NL1,NL2
      CZ=(0.0,0.0)
      CI=(0.0,1.0)
      KO=CSQRT(CMPLX(2.0*E,-2.0*VPI+0.000001))
      CFAC=-16.0*(3.14159265)**2*CI/TV
      NLYLM=NLAY2*LMMAX
C
C     COPY OLD VALUES OF GH ONTO GH WHERE APPROPRIATE
      DO 6 I=1,NLAY2
      IF (NGOL(I).EQ.0) GO TO 6
      IS=(I-1)*LMMAX
      IT=(NGOL(I)-1)*LMMAX
      IU=IS+NLYLM
      IV=IT+NLYLM
      DO 2 J=1,LMMAX
      DO 2 K=1,LMMAX
      GH(IS+J,K)=GH(IT+J,K)
2     GH(IU+J,K)=GH(IV+J,K)
6     CONTINUE
C     INITIALIZE GH IN PREPARATION FOR NEW VALUES
      DO 8 I=1,NLAY2
      IF (NUGH(I).EQ.0) GO TO 8
      IS=(I-1)*LMMAX
      IT=IS+NLYLM
      DO 7 J=1,LMMAX
      DO 7 K=1,LMMAX
      GH(IS+J,K)=CZ
7     GH(IT+J,K)=CZ
8     CONTINUE
C
C     TSTS= ESTIMATED NO. OF POINTS IN DIRECT LATTICE SUM
      TSTS=DCUT*DCUT*3.14159265/TV
      DO 10 IZ=1,NLAY2
      IF (NUGH(IZ).EQ.0) GO TO 10
      IF (ABS(DRL(IZ,1)).LE.0.001) GO TO 85
C     AKP2= ESTIMATED NO. OF POINTS IN RECIPROCAL LATTICE SUM
      AKP2=(2.0*E+(ALOG(TST)/DRL(IZ,1))**2)*TV/(4.*3.1415926)
C     SKIP DIRECT LATTICE SUM, IF RECIPROCAL LATTICE SUM FASTER (BUT NUMBER
C     OF REC. LATT. POINTS IS TO BE RESTRICTED DUE TO A CONVERGENCE
C     PROBLEM)
      IF ((TSTS.GE.2.0*AKP2).AND.(AKP2.LT.80.0)) GO TO 10
C     PRODUCE GH(I,J) AND GH(J,I) WITH DIRECT LATTICE SUM FOR TWO
C     PROPAGATION DIRECTIONS
85    DO 9 K=1,2
```

```
9       CALL GHD(IZ,K,GH,LMG,LMMAX,S,LMS,Y1,L2M,DRL,NLAY2,KO,DCUT,
       1CAA,NCAA,LXM)
C     GH(I,J) AND GH(J,I) NOW NO LONGER NEED TO BE COMPUTED
       NUGH(IZ)=0
10      CONTINUE
C
       TSTS=0.0
       DO 1 I=1,NLAY2
       IF ((NUGH(I).EQ.0).OR.(ABS(DRL(I,1)).LE.0.001)) GO TO 1
       DRL(I,1)=DRL(I,1)
C     TEST(I) WILL SERVE AS CUTOFF IN RECIPROCAL LATTICE SUM
       TEST(I)=(ALOG(TST)/DRL(I,1))**2
       TSTS=AMAX1(TEST(I),TSTS)
1      CONTINUE
       IF (TSTS.LE.0.00001) GO TO 1460
C
C     START OF TWO 1-DIMENSIONAL SUMMATION LOOPS IN ONE QUADRANT OF
C     RECIPROCAL SPACE
       NUMG=0
       JJ1=0
1171   JJ1=JJ1+1
       JJ2=0
1172   JJ2=JJ2+1
       NOG=0
       J1=JJ1-1
       J2=JJ2-1
C     START OF LOOP OVER QUADRANTS
       DO 1370 KK=1,4
       GO TO (1180,1190,1200,1210),KK
1180   NG1=J1
       NG2=J2
       GO TO 1220
1190   IF (J1.EQ.0.AND.J2.EQ.0) GO TO 1380
       IF (J2.EQ.0) GO TO 1370
       NG1=J1
       NG2=-J2
       GO TO 1220
1200   IF (J1.EQ.0) GO TO 1370
       NG1=-J1
       NG2=J2
       GO TO 1220
1210   IF (J1.EQ.0.OR.J2.EQ.0) GO TO 1370
       NG1=-J1
       NG2=-J2
C     CURRENT RECIPROCAL LATTICE POINT
1220   GX=NG1*RBR1(1)+NG2*RBR2(1)
       GY=NG1*RBR1(2)+NG2*RBR2(2)
       GKX=GX+AK2
       GKY=GY+AK3
       GK2=GKX*GKX+GKY*GKY
       AKP2=2.0*E-GK2
C     TEST FOR CUTOFF
       IF (AKP2.LT.(-TSTS)) GO TO 1370
       NUMG=NUMG+1
       NOG=1
       KPRG=CSQRT(CMPLX(AKP2,-2.0*VPI+0.000001))
       Z=KPRG/KO
       FY=0.0
       IF (GK2.LE.1.0E-8) GO TO 1240
       CFY=GKX/SQRT(GK2)
       IF (ABS(ABS(CFY)-1.).LE.1.E-6) GO TO 1230
       FY= ACOS(CFY)
       GO TO 1235
1230   IF (CFY.LT.0.0) FY=3.14159265
1235   IF (GKY.LT.0.0) FY=-FY
C     FIND APPROPRIATE SPHERICAL HARMONICS
1240   CALL SH(LM,Z,FY,Y)
```

```
C   START OF LOOP OVER INTERPLANAR VECTORS
        DO 1360 IZ=1,NLAY2
C   SKIP IF NEW GH(I,J) NOT NEEDED
        IF (NUGH(IZ).EQ.0) GO TO 1360
C   SKIP IF THIS CONTRIBUTION OUTSIDE CUTOFF
        IF (AKP2.LT.(-TEST(IZ))) GO TO 1360
        T2=GKX*DRL(IZ,2)+GKY*DRL(IZ,3)
        T2P=-T2
        T3=KPRG*ABS(DRL(IZ,1))
        T1=(CEXP(CI*(T2+T3))/KPRG)*CFAC
        T1P=(CEXP(CI*(T2P+T3))/KPRG)*CFAC
C
C   PRODUCE A LIMITED SET OF MATRIX ELEMENTS
        DO 1350 I=1,LMMAX
        IP=LXM(I)
        NII=IP+(IZ-1)*LMMAX
        NNI=NII+NLYLM
        L1=INT(SQRT(FLOAT(I-1)))
        M1=I-L1-L1*L1-1
        IF (M1) 1250,1250,1260
1250    BS=(-1)**(MOD(M1,2))*Y(L1+1,-M1+1)
        GO TO 1270
1260    BS=Y(M1,L1+1)
1270    DO 1340 J=1,LMMAX
        JP=LXM(J)
        L2=INT(SQRT(FLOAT(J-1)))
        M2=J-L2-L2*L2-1
        IF (I-J) 1290,1280,1290
1280    IF (M1) 1300,1300,1340
1290    IF (L1.LT.L2.OR.(L1.EQ.L2.AND.(IABS(M1).LT.IABS(M2))))
       1GO TO 1340
1300    IF (M2) 1310,1320,1320
1310    CS=(-1)**(MOD(M2,2))*Y(-M2,L2+1)
        GO TO 1330
1320    CS=Y(L2+1,M2+1)
C   GH(I,J) AND GH(J,I) ARE PRODUCED TOGETHER
1330    GH(NNI,JP)=GH(NNI,JP)+T1P*BS*CS
        GH(NII,JP)=GH(NII,JP)+(-1)**(MOD(L1+M1+L2+M2,2))*T1*BS*CS
1340    CONTINUE
1350    CONTINUE
1360    CONTINUE
1370    CONTINUE
1380    IF (NOG.EQ.1) GO TO 1172
        IF (JJ2-1) 1171,1400,1171
1400    CONTINUE
C
        DO 1450 IZ=1,NLAY2
        IF (NUGH(IZ).EQ.0) GO TO 1450
C   FILL IN MISSING MATRIX ELEMENTS FROM ELEMENTS JUST PRODUCED, USING
C   SYMMETRY RELATIONS
        DO 1440 I=1,LMMAX
        IP=LXM(I)
        NII=IP+(IZ-1)*LMMAX
        NNI=NII+NLYLM
        L1=INT(SQRT(FLOAT(I-1)))
        M1=I-L1-L1*L1-1
        IM1=I-2*M1
        IMP=LXM(IM1)
        DO 1430 J=1,LMMAX
        JP=LXM(J)
        L2=INT(SQRT(FLOAT(J-1)))
        M2=J-L2-L2*L2-1
        JM2=J-2*M2
        JMP=LXM(JM2)
        NJM=JMP+(IZ-1)*LMMAX
        NNM=NJM+NLYLM
        SIGN=(-1)**MOD(M1+M2,2)
```

```
          IF (I.NE.J) GO TO 1410
          IF (M1.LE.0) GO TO 1430
          GO TO 1420
1410   IF (L1.GE.L2.AND.(L1.NE.L2.OR.(IABS(M1).GE.IABS(M2))))
       1GO TO 1430
1420   GH(NNI,JP)=SIGN*GH(NNM,IMP)
          GH(NII,JP)=SIGN*GH(NJM,IMP)
1430   CONTINUE
1440   CONTINUE
1450   CONTINUE
C   PRINT NUMBER OF RECIPROCAL LATTICE POINTS (BEAMS) USED IN SUMMATION
          WRITE(6,3)NUMG
3         FORMAT(23H NO. OF BEAMS FOR GH : ,I5)
C
C   COPY NEW MATRICES GH(I,J) ONTO OTHERS THAT ARE IDENTICAL
1460   DO 5 IZ=1,NLAY2
          IF (NGEQ(IZ).EQ.0) GO TO 5
          II=(IZ-1)*LMMAX
          IIN=II+NLYLM
          IN=(NGEQ(IZ)-1)*LMMAX
          INN=IN+NLYLM
          DO 4 I=1,LMMAX
          DO 4 J=1,LMMAX
          GH(II+I,J)=GH(IN+I,J)
4         GH(IIN+I,J)=GH(INN+I,J)
5         CONTINUE
          RETURN
          END
C-------------------------------------------------------------------
C   SUBROUTINE GHSC COMPUTES THE (LM)-SPACE INTERPLANAR PROPAGATORS GH
C   FROM DIRECT LATTICE SUMS PRODUCED IN SUBROUTINE GHD AND CLEBSCH-
C   GORDON COEFFICIENTS FROM SUBROUTINE CAAA.
C   FOR QUANTITIES NOT EXPLAINED BELOW SEE SUBROUTINES GHD AND GHMAT.
C   IZ= SERIAL NO. OF CURRENT INTERPLANAR VECTOR DRL.
C   IS=1 FOR PROPAGATION FROM FIRST TO SECOND SUBPLANE.
C   IS=2 FOR PROPAGATION FROM SECOND TO FIRST SUBPLANE.
C   FF= PREFACTOR OF GH.
C   LXM= PERMUTATION OF (LM) SEQUENCE
          SUBROUTINE GHSC(IZ,IS,GH,LMG,LMMAX,S,LMS,CAA,NCAA,FF,LXM,NLAY2)
          COMPLEX GH(LMG,LMMAX),S(LMS),FF
          DIMENSION CAA(NCAA),LXM(LMMAX)
          II=1
          NLYLM=LMMAX*NLAY2
C   PRODUCE A LIMITED SET OF MATRIX ELEMENTS
          DO 1350 I=1,LMMAX
          IP=LXM(I)
          NII=IP+(IZ-1)*LMMAX
          IF (IS.EQ.2) NII=NII+NLYLM
          L1=INT(SQRT(FLOAT(I-1)))
          M1=I-L1-L1*L1-1
          DO 1340 J=1,LMMAX
          JP=LXM(J)
          L2=INT(SQRT(FLOAT(J-1)))
          M2=J-L2-L2*L2-1
          M3=M2-M1
          IL=IABS(L1-L2)
          IM=IABS(M3)
          LMIN=MAX0(IL,IM+MOD(IL+IM,2))
          LMAX=L1+L2
          LMIN=LMIN+1
          LMAX=LMAX+1
          IF (I-J) 1290,1280,1290
1280   IF (M1) 1300,1300,1340
1290   IF (L1.LT.L2.OR.(L1.EQ.L2.AND.(IABS(M1).LT.IABS(M2))))
       1GO TO 1340
1300   DO 203 ILA=LMIN,LMAX,2
          LA=ILA-1
```

```
         CC=CAA(II)
         II=II+1
203      GH(NII,JP)=GH(NII,JP)+CC*S(LA*LA+LA+M3+1)
         GH(NII,JP)=FF*GH(NII,JP)
1340  CONTINUE
1350  CONTINUE
C
C   FILL IN MISSING MATRIX ELEMENTS FROM ELEMENTS JUST PRODUCED
         DO 1440 I=1,LMMAX
         IP=LXM(I)
         NII=IP+(IZ-1)*LMMAX
         IF (IS.EQ.2) NII=NII+NLYLM
         L1=INT(SQRT(FLOAT(I-1)))
         M1=I-L1-L1*L1-1
         IM1=I-2*M1
         IMP=LXM(IM1)
         DO 1430 J=1,LMMAX
         JP=LXM(J)
         L2=INT(SQRT(FLOAT(J-1)))
         M2=J-L2-L2*L2-1
         JM2=J-2*M2
         JMP=LXM(JM2)
         NJM=JMP+(IZ-1)*LMMAX
         IF (IS.EQ.2) NJM=NJM+NLYLM
         SIGN=(-1)**MOD(M1+M2,2)
         IF (I.NE.J) GO TO 1410
         IF (M1.LE.0) GO TO 1430
         GO TO 1420
1410  IF (L1.GE.L2.AND.(L1.NE.L2.OR.(IABS(M1).GE.IABS(M2))))
        1GO TO 1430
1420  GH(NII,JP)=SIGN*GH(NJM,IMP)
1430  CONTINUE
1440  CONTINUE
         RETURN
         END
C-------------------------------------------------------------------
C  SUBROUTINE LXGENT GENERATES THE NEEDED RELATIONSHIPS BETWEEN
C  DIFFERENT ORDERING SEQUENCES OF THE (L,M) PAIRS (L.LE.LMAX, ABS(M)
C  .LE.L). THREE ORDERINGS ARE CONSIDERED, THE 'NATURAL' ONE (N), THE
C  'COPLANAR' ONE (C) AND THE 'SYMMETRIZED' ONE (S). THEY ARE TABULATED
C  BELOW FOR THE CASE OF LMAX=4.
C
C     I=  0 0 0 0 0 0 0 0 0 1 1 1 1 1 1 1 1 1 1 2 2 2 2 2 2
C         1 2 3 4 5 6 7 8 9 0 1 2 3 4 5 6 7 8 9 0 1 2 3 4 5
C
C  N: L=  0 1 1 1 2 2 2 2 2 3 3 3 3 3 3 3 4 4 4 4 4 4 4 4 4
C     M=  0-1 0 1 2-2-1 0 1 2-3-2-1 0 1 2 3-4-3-2-1 0 1 2 3 4
C
C  C: L=  0 1 1 2 2 2 3 3 3 3 4 4 4 4 1 2 2 3 3 3 4 4 4 4
C     M=  0-1 1-2 0 2-3-1 1 3-4-2 0 2 4 0-1 1-2 0 2-3-1 1 3
C              L+M=EVEN          *       L+M=ODD
C
C  S: L=  0 2 2 2 4 4 4 4 4 1 1 3 3 3 3 1 3 3 3 2 2 4 4 4 4
C     M=  0-2 0 2-4-2 0 2 4-1 1-3-1 1 3 0-2 0 2-1 1-3-1 1 3
C           L=EVEN       * L=ODD  * L=ODD * L=EVEN
C           M=EVEN       * M=ODD  * M=EVEN* M=ODD
C
C  TO DESCRIBE THE RELATIONSHIPS, A PARTICULAR PAIR (L,M) IN A PARTICU-
C  LAR SEQUENCE SHALL BE REPRESENTED HERE BY N(I), C(I) OR S(I) (I=1,
C  LMMAX): E.G. N(10)=(3,-3). THE RELATIONSHIPS LX,LXI,LT,LXM GENERATED
C  BY LXGENT ARE NOW DEFINED BY:
C  LX:  N(LX(I))= S(I).
C  LXI: C(I)= S(LXI(I)) IF I.LE.LEV,
C              S(LXI(I)+LEV) IF I.GT.LEV.
C  LT:  CSM(N(LT(I)))= S(I) WHERE CSM MEANS: CHANGE THE SIGN OF M.
C  LXM: N(I)= S(LXM(I)).
C
```

```
C    LX,LXI,LT,LXM= OUTPUT PERMUTATIONS OF (L,M) SEQUENCE.
C    LMAX= LARGEST VALUE OF L.
C    LMMAX= (LMAX+1)**2.
      SUBROUTINE LXGENT(LX,LXI,LT,LXM,LMAX,LMMAX)
      DIMENSION LX(LMMAX),LXM(LMMAX),LXI(LMMAX),LT(LMMAX)
      LEV=(LMAX+1)*(LMAX+2)/2
      LEE=(LMAX/2+1)**2
      LEO=((LMAX+1)/2+1)*((LMAX+1)/2)+LEE
      LOE=((LMAX-1)/2+1)**2+LEO
      LL=0
      L1=0
      LT1=0
      L=-1
1     L=L+1
      M=-L
2     LL=LL+1
      IF (MOD(L+M,2)-1) 3,6,3
3     LT1=LT1+1
      LT(LL)=LT1
      IF (MOD(L,2)-1) 4,5,4
4     L1=L1+1
      LX(L1)=LL
      GO TO 9
5     LEE=LEE+1
      LX(LEE)=LL
      GO TO 9
6     LEV=LEV+1
      LT(LL)=LEV
      IF (MOD(L,2)) 7,8,7
7     LEO=LEO+1
      LX(LEO)=LL
      GO TO 9
8     LOE=LOE+1
      LX(LOE)=LL
9     M=M+1
      IF (L-M) 10,2,2
10    IF (L-LMAX) 1,11,11
11    DO 12 L=1,LMMAX
      L1=LX(L)
      LT1=LT(L1)
12    LXI(LT1)=L
      L1=LEE+1
      DO 13 L=L1,LMMAX
13    LXI(L)=LXI(L)-LEE
      L1=LMAX+1
      JLM=1
      DO 16 L=1,L1
      LL=L+L-1
      DO 16 M=1,LL
      JLP=JLM+LL+1-2*M
      DO 14 LT1=1,LMMAX
      IF (JLM-LX(LT1)) 14,15,14
14    CONTINUE
15    LT(LT1)=JLP
      LXM(JLM)=LT1
16    JLM=JLM+1
      RETURN
      END
C-----------------------------------------------------------------------
C    SUBROUTINE MATCOP COPIES UP TO NL=4 SQUARE MATRICES OF INDEPENDENT
C    DIMENSIONS ONTO ANOTHER SQUARE MATRIX OF DIMENSION EQUAL TO THE SUM
C    OF THE INPUT ONES, PRODUCING A BLOCK-DIAGONALIZED MATRIX WHEN NL.GT.1
C    THIS SUBROUTINE IS USEFUL FOR DEALING WITH VARIABLY-DIMENSIONED
C    MATRICES.
      SUBROUTINE MATCOP(R1,N1,R2,N2,R3,N3,R4,N4,RS,NS,NL)
      COMPLEX R1(N1,N1),R2(N2,N2),R3(N3,N3),R4(N4,N4),RS(NS,NS)
      COMPLEX CZ
```

```
      CZ=(0.0,0.0)
      DO 5 I=1,NS
      DO 5 J=1,NS
5     RS(I,J)=CZ
      DO 10 I=1,N1
      DO 10 J=1,N1
10    RS(I,J)=R1(I,J)
      IF (NL.EQ.1) RETURN
      NSH=N1
      DO 20 I=1,N2
      DO 20 J=1,N2
20    RS(I+NSH,J+NSH)=R2(I,J)
      IF (NL.EQ.2) RETURN
      NSH=NSH+N2
      DO 30 I=1,N3
      DO 30 J=1,N3
30    RS(I+NSH,J+NSH)=R3(I,J)
      IF (NL.EQ.3) RETURN
      NSH=NSH+N3
      DO 40 I=1,N4
      DO 40 J=1,N4
40    RS(I+NSH,J+NSH)=R4(I,J)
      RETURN
      END
C-----------------------------------------------------------------------
C  SUBROUTINE MFOLD PRODUCES, FOR SUBROUTINE MSMF, THE INDIVIDUAL
C  REFLECTION AND TRANSMISSION MATRIX ELEMENTS FOR A BRAVAIS-LATTICE
C  LAYER. A DEBYE-WALLER FACTOR IS INCLUDED EXPLICITLY, IF REQUESTED
C  (IT.NE.0). AN ORIGIN SHIFT TO A SYMMETRY AXIS OR PLANE IS INCLUDED
C  FOR UP TO FOUR DIFFERENT SHIFTS (YIELDING FOUR DIFFERENT RESULTS).
C  SYMMETRY-INDUCED FOLDING OF THE MATRIX ELEMENTS IS ALSO CARRIED OUT
C  HERE.
C  FOR QUANTITIES NOT EXPLAINED BELOW SEE SUBROUTINE MSMF.
C  JG= INDEX OF CURRENT SCATTERED BEAM.
C  NA= OFFSET OF PRESENT BEAM SET IN LIST PQ.
C  YLM= RESULT OF (1-X)**(-1)*Y*T, FROM MSMF.
C  CYLM= SET OF SPHERICAL HARMONICS, FROM MSMF.
C  N: NOT USED.
C  ID= NO. OF ORIGIN (REGISTRY) SHIFTS TO BE USED.
C  RA,TA= OUTPUT MATRIX ELEMENTS.
C  IN COMMON BLOCKS:
C  GP= CURRENT INCIDENT BEAM.
C  LM= LMAX+1.
C  LAY= INDEX IN SYM(LAY,JGA) FOR READING OF SYMMETRY CODE NUMBERS.
C  S1,S2,S3,S4= SPACE FOR UP TO FOUR ORIGIN SHIFTS.
C  IT=0: NO EXPLICIT DEBYE-WALLER FACTOR TO BE USED (COMPLEX PHASE
C    SHIFTS INSTEAD).
C  IT=1: USE EXPLICIT DEBYE-WALLER FACTORS.
C  TEMP,TO,DRX,DRY,DRO= THERMAL VIBRATION DATA, SEE MAIN PROGRAM.
      SUBROUTINE  MFOLD (JG,NA,E,VPI,AK2,AK3,YLM,CYLM,N,LMMAX,PQ,SYM,NT,
     1ID,RA,TA)
      INTEGER  SYM
      COMPLEX  CYLM, YLM, CZ, RU, CI, RA, TA, RB, TB, ST, SM, SL, CM,
     1CY, CYM, CTR, CTT, R, T, Z
      COMPLEX  RB1, RB2, RB3, RB4, TB1, TB2, TB3, TB4
      COMPLEX EP6,EDW,FDW,GDW,HDW
      DIMENSION RA(ID),TA(ID),GDW(2,12),HDW(2,12),S3(2),S4(2)
      DIMENSION  G(2,12), GP(2), PQ(2,NT), EDW(2,12), FDW(2,12), S1(2)
      DIMENSION  S2(2),SYM(2,NT),CYLM(NT,LMMAX),YLM(LMMAX),R(12),T(12)
      DIMENSION  Z(12), I2(5)
      COMMON /MFB/ GP,LM,LAY,S1,S2,S3,S4
      COMMON /BT/IT, TEMP, TO, DRX, DRY, DRO
      DATA  I2(1), I2(2), I2(3), I2(4), I2(5)/5, 6, 2, 8, 4/
      CZ = (0.0,0.0)
      RU = (1.0,0.0)
      CI = (0.0,1.0)
      EP6=CMPLX(0.5,0.8660254)
```

```
      ASQ = 1.41421356
      ACU=1.7320508
      IGC = 0
      JGA = JG + NA
C  FIND THE SET OF BEAMS SYMMETRICAL TO THE SCATTERED BEAM
      CALL  UNFOLD (JGA, G, NG, LAY, PQ, SYM, NT)
      IF ((SYM(LAY,JGA) .NE. 10) .OR. (PQ(2,JGA) .GE. 0.0))  GO TO 130
      IGC = 1
      GP(2) =  - GP(2)
      AK3 =  - AK3
C  SELECT EXPLICIT DEBYE-WALLER FACTOR OR NOT
  130 IF (IT)  160, 140, 160
  140 DO 150 I = 1, 2
      DO 150 K = 1, 12
  150 EDW(I,K) = RU
      GO TO 170
C  COMPUTE DEBYE-WALLER FACTORS
  160 CALL  DEBWAL (NG, G, GP, E, VPI, AK2, AK3, TEMP, TO, DRX, DRY,
     1DRO, EDW)
  170 IF (ID-2) 176,175,172
  172 IF (ID-4) 174,173,173
C  APPLY ORIGIN (REGISTRY) SHIFTS (AS FACTORS TO DEBYE-WALLER FACTORS)
  173 CALL TRANS(HDW,EDW,NG,G,GP,S4)
  174 CALL TRANS(GDW,EDW,NG,G,GP,S3)
  175 CALL  TRANS (FDW, EDW, NG, G, GP, S2)
  176 CALL  TRANS (EDW, EDW, NG, G, GP, S1)
      DO 180 I = 1,12
      R(I) = CZ
  180 T(I) = CZ
C  START SUMMATION OVER (L,M)
      ISK = 0
      LAB = SYM(LAY,JG + NA)
      IF ((LAB .EQ. 7) .OR. (LAB .EQ. 10))  ISK = 1
      ST = RU
      SL = RU
      CM = RU
      JLM = 1
      L1 = 1
C  FOR 3- AND 6-FOLD SYMMETRIES GO TO 330, FOR OTHER SYMMETRIES (OR
C  NONE) GO TO 185
      IF (LAB-11) 185,330,330
  185 Z(3)=CZ
      Z(7)=CZ
      DO 240 L = 1, LM
      SM = SL
      LL = L + L - 1
      L1 = L1 + LL
      DO 230 M = 1, LL
      MM = L1 - M
      CY = CYLM(JG,MM)
      MM = L1 - LL - 1 + M
      CYM = CYLM(JG,MM)
      CTT = YLM(JLM) * ST
      CTR = CTT * SL
      CTT = CTT * SM
      R(1) = R(1) + CTR * CY
      T(1) = T(1) + CTT * CY
      IF (LAB .EQ. 1)  GO TO 220
      Z(2) = CYM * CM
      Z(4) = CYM
      Z(5) = CY * SM
      Z(6) = Z(2) * SM
      Z(8) = CYM * SM
      IF (ISK)  200, 200, 190
  190 Z(7) = CY * CM
      Z(3) = Z(7) * SM
C  CONSIDER UP TO 8 SYMMETRICAL BEAMS
```

```
200 DO 210 I = 2, 8
    R(I) = R(I) + CTR * Z(I)
210 T(I) = T(I) + CTT * Z(I)
220 JLM = JLM + 1
    CM = CM * CI
230 SM =  - SM
    CM = CONJG(CM)
    SL =  - SL
240 ST = ST * CI
C   CONSIDER ID DIFFERENT ORIGIN SHIFTS
    DO 300 II = 1, ID
    RB1 = R(1) * EDW(1,1)
    TB1 = T(1) * EDW(2,1)
    RB = RB1
    TB = TB1
    FAC = 1.0
    IF (LAB .EQ. 1)  GO TO 280
    IF (LAB .GT. 6)  GO TO 250
    JLAB = I2(LAB - 1)
    RB = RB1 + R(JLAB) * EDW(1,2)
    TB = TB1 + T(JLAB) * EDW(2,2)
    FAC = 1.0/ASQ
    GO TO 280
250 IF (LAB .EQ. 10)  GO TO 260
    RB3 = R(5) * EDW(1,3)
    TB3 = T(5) * EDW(2,3)
    JLAB = 3
    IF (LAB .EQ. 8)  JLAB = 2
    IF (LAB .EQ. 9)  JLAB = 4
    RB2 = R(JLAB) * EDW(1,2)
    TB2 = T(JLAB) * EDW(2,2)
    RB4 = R(JLAB + 4) * EDW(1,4)
    TB4 = T(JLAB + 4) * EDW(2,4)
    FAC = 0.5
    GO TO 270
260 RB1 = R(1) * EDW(1,1) + R(8) * EDW(1,8)
    TB1 = T(1) * EDW(2,1) + T(8) * EDW(2,8)
    RB2 = R(3) * EDW(1,3) + R(6) * EDW(1,6)
    TB2 = T(3) * EDW(2,3) + T(6) * EDW(2,6)
    RB3 = R(4) * EDW(1,4) + R(5) * EDW(1,5)
    TB3 = T(4) * EDW(2,4) + T(5) * EDW(2,5)
    RB4 = R(2) * EDW(1,2) + R(7) * EDW(1,7)
    TB4 = T(2) * EDW(2,2) + T(7) * EDW(2,7)
    FAC = 0.5/ASQ
270 RB = RB1 + RB2 + RB3 + RB4
    TB = TB1 + TB2 + TB3 + TB4
C   INCLUDE SYMMETRY NORMALIZATION FACTOR IN FINAL VALUE OF MATRIX
C   ELEMENT
280 RA(II) = FAC * RB
    TA(II) = FAC * TB
C   SHUFFLE FACTORS FROM HDW TO GDW TO FDW TO EDW SO THAT THEY APPEAR
C   IN EDW AT THE RIGHT TIME FOR INCLUSION OF THE RIGHT ORIGIN SHIFT
    DO 290 I = 1, 8
    DO 290 K = 1, 2
    EDW(K,I) = FDW(K,I)
    FDW(K,I)=GDW(K,I)
290 GDW(K,I)=HDW(K,I)
300 CONTINUE
    IF (IGC)  320, 320, 310
310 GP(2) =  - GP(2)
    AK3 =  - AK3
320 RETURN
C   DO SAME AS ABOVE FOR 3- AND 6-FOLD SYMMETRIES
330 IF (LAB.LE.13) EP6=EP6*EP6
    DO 440 L=1,LM
    SM=SL
    LL=L+L-1
```

```
      L1=L1+LL
      DO 430 M=1,LL
      MM=L1-M
      CY=CYLM(JG,MM)
      MM=L1-LL-1+M
      CYM=CYLM(JG,MM)
      CTT=YLM(JLM)*ST
      CTR=CTT*SL
      CTT=CTT*SM
      R(1)=R(1)+CTR*CY
      T(1)=T(1)+CTT*CY
      IF (LAB-11) 340,340,350
  340 Z(2)=CY*CM
      Z(3)=Z(2)*CM
      GO TO 410
  350 IF (LAB-13)360,365,370
  360 Z(3)=CY*CM
      Z(5)=Z(3)*CM
      Z(6)=CYM*SM
      Z(2)=Z(6)*CM
      Z(4)=Z(2)*CM
      GO TO 410
  365 Z(3)=CY*CM
      Z(5)=Z(3)*CM
      Z(2)=CYM
      Z(4)=CYM*CM
      Z(6)=Z(4)*CM
      GO TO 410
  370 IF (LAB-15) 375,390,390
  375 Z(2)=CY*CM
      DO 380 I=3,6
  380 Z(I)=Z(I-1)*CM
      GO TO 410
  390 Z(3)=CY*CM
      Z(12)=CYM*SM
      Z(2)=Z(12)*CM
      DO 400 I=5,11,2
      Z(I)=Z(I-2)*CM
  400 Z(I-1)=Z(I-3)*CM
  410 DO 420 I=2,NG
      R(I)=R(I)+CTR*Z(I)
  420 T(I)=T(I)+CTT*Z(I)
      JLM=JLM+1
      CM=CM*EP6
  430 SM=-SM
      CM=CONJG(CM)
      SL=-SL
  440 ST=ST*CI
      FAC=1.0/ACU
      IF ((LAB.GE.12).AND.(LAB.LE.14)) FAC=FAC/ASQ
      IF (LAB.EQ.15) FAC=FAC*0.5
      DO 470 II=1,ID
      RB=CZ
      TB=CZ
      DO 450 I=1,NG
      RB=RB+R(I)*EDW(1,I)
  450 TB=TB+T(I)*EDW(2,I)
      RA(II)=RB*FAC
      TA(II)=TB*FAC
      DO 460 I=1,12
      DO 460 K=1,2
      EDW(K,I)=FDW(K,I)
      FDW(K,I)=GDW(K,I)
  460 GDW(K,I)=HDW(K,I)
  470 CONTINUE
      RETURN
      END
```

```
C--------------------------------------------------------------------
C    SUBROUTINE MFOLT HAS THE SAME FUNCTION AS MFOLD, BUT IS DESIGNED FOR
C    THE CASE OF COMPOSITE LAYERS. MFOLT USES INPUT FROM MPERTI OR MTINV.
C    ONLY FEATURES DIFFERENT FROM THOSE OF MFOLD ARE LISTED BELOW.
C    TS= INPUT FROM MPERTI OR MTINV.
C    LMN= NLAY*LMMAX.
C    RG= PLANE-WAVE PROPAGATORS BETWEEN SUBPLANES.
C    NLAY= NO. OF SUBPLANES.
C    JGP= INDEX OF INCIDENT BEAM.
C    EGS= WORKING SPACE.
C    LXM= PERMUTATION OF (LM) SEQUENCE.
C    INC=+-1: INDICATES INCIDENCE IN DIRECTION OF +-X.
C    POSS= ATOMIC POSITIONS IN UNIT CELL.
      SUBROUTINE   MFOLT (JG,NA,E,VPI,AK2,AK3,CYLM,N,
     1LMMAX,PQ,SYM,NT,TS,LMN,RG,NLAY,JGP,EGS,LXM,ID,RA,TA,INC,POSS)
      INTEGER   SYM
      COMPLEX   CYLM, CZ, RU, CI, RA, TA, RB, TB, ST, SM, SL, CM,
     1CY, CYM, CTR, CTT, R, T, Z,CR,CT,CS
      COMPLEX EP6,EDW,FDW,GDW,HDW
      COMPLEX TS(LMN),RG(3,NLAY,N),EGS(2,12,NLAY)
      DIMENSION RA(ID),TA(ID),GDW(2,12),HDW(2,12),S3(2),S4(2)
      DIMENSION  G(2,12), GP(2), PQ(2,NT), EDW(2,12), FDW(2,12), S1(2)
      DIMENSION  S2(2),SYM(2,NT),CYLM(NT,LMMAX),R(12),T(12)
      DIMENSION  Z(12), LXM(LMMAX),POSS(NLAY,3)
      COMMON /MFB/ GP,LM,LAY,S1,S2,S3,S4
      CZ = (0.0,0.0)
      RU = (1.0,0.0)
      CI = (0.0,1.0)
      EP6=CMPLX(0.5,-0.8660254)
      ASQ = 1.41421356
      ACU=1.7320508
      IGC = 0
      JGA = JG + NA
      CALL   UNFOLD (JGA, G, NG, LAY, PQ, SYM, NT)
      IF ((SYM(LAY,JGA) .NE. 10) .OR. (PQ(2,JGA) .GE. 0.0)) GO TO 140
      IGC = 1
      GP(2) =  - GP(2)
      AK3 =  - AK3
  140 DO 150 I = 1, 2
      DO 150 K = 1, 12
  150 EDW(I,K) = RU
  170 IF (ID-2) 176,175,172
  172 IF (ID-4) 174,173,173
  173 CALL TRANSP(HDW,EDW,NG,G,GP,S4,POSS,NLAY,AK2,AK3,INC)
  174 CALL TRANSP(GDW,EDW,NG,G,GP,S3,POSS,NLAY,AK2,AK3,INC)
  175 CALL TRANSP(FDW,EDW,NG,G,GP,S2,POSS,NLAY,AK2,AK3,INC)
  176 CALL TRANSP(EDW,EDW,NG,G,GP,S1,POSS,NLAY,AK2,AK3,INC)
C   GENERATE FACTORS RELATING POSITIONS OF SUBPLANES, IN CONNECTION WITH
C   SYMMETRIZATION.
      DO 177 J=1,NLAY
      DO 177 I=1,NG
      EGS(1,I,J)=CEXP(CI*((G(1,I)-G(1,1))*(POSS(J,2)-POSS(1,2))+
     1(G(2,I)-G(2,1))*(POSS(J,3)-POSS(1,3))))
  177 EGS(2,I,J)=CEXP(-CI*((G(1,I)-G(1,1))*(POSS(J,2)-POSS(NLAY,2))+
     1(G(2,I)-G(2,1))*(POSS(J,3)-POSS(NLAY,3))))
      DO 180 I=1,NG
      R(I) = CZ
  180 T(I) = CZ
      LAB = SYM(LAY,JG + NA)
      ST = RU
      SL = RU
      CM = RU
      JLM = 1
      L1 = 1
      IF (LAB-11) 184,330,330
  184 DO 240 L = 1, LM
      SM = SL
```

```
            LL = L + L - 1
            L1 = L1 + LL
            DO 230 M = 1, LL
            MM = L1 - M
            CYM = CYLM(JG,MM)
            MM = L1 - LL - 1 + M
            CY = CYLM(JG,MM)
            IF (INC) 128,127,127
        127 CTR=SL*SM
            CTT=RU
            GO TO 129
        128 CTR=RU
            CTT=SL*SM
        129 CONTINUE
            Z(1)=CY
            IF (LAB .EQ. 1)  GO TO 200
            IF (LAB-5) 185,189,190
        185 IF (LAB-3)  186,187,188
        186 Z(2)=CY*SM
            GO TO 200
        187 Z(2)=CYM*CM
            GO TO 200
        188 Z(2)=CYM*CM*SM
            GO TO 200
        189 Z(2)=CYM*SM
            GO TO 200
        190 IF (LAB-9) 191,195,196
        191 IF (LAB-7) 192,193,194
        192 Z(2)=CYM
            GO TO 200
        193 Z(3)=CY*SM
            Z(2)=CY*CM
            Z(4)=Z(2)*SM
            GO TO 200
        194 Z(4)=CYM*CM
            Z(3)=CY*SM
            Z(2)=Z(4)*SM
            GO TO 200
        195 Z(2)=CYM
            Z(3)=CY*SM
            Z(4)=CYM*SM
            GO TO 200
        196 Z(6)=CYM*CM
            Z(4)=CYM
            Z(5)=CY*SM
            Z(2)=Z(6)*SM
            Z(3)=CY*CM
            Z(7)=Z(3)*SM
            Z(8)=CYM*SM
        200 KLM=LXM(JLM)
            DO 210 I=1,NG
            CR=CZ
            CT=CZ
C     SUM OVER ATOMS IN UNIT CELL
            DO 199 J=1,NLAY
            JJ=(J-1)*LMMAX
            IF (INC.EQ.-1) GO TO 198
            CS=TS(JJ+KLM)*RG(1,J,JGP)/RG(1,1,JGP)
            CR=CR+RG(2,J,JG)/RG(2,1,JG)*CS*EGS(1,I,J)
            CT=CT+RG(1,NLAY,JG)/RG(1,J,JG)*CS*EGS(2,I,J)
            GO TO 199
        198 CS=TS(JJ+KLM)*RG(2,NLAY,JGP)/RG(2,J,JGP)
            CR=CR+RG(1,NLAY,JG)/RG(1,J,JG)*CS*EGS(2,I,J)
            CT=CT+RG(2,J,JG)/RG(2,1,JG)*CS*EGS(1,I,J)
        199 CONTINUE
            R(I) = R(I) + CTR * Z(I)*CR
        210 T(I) = T(I) + CTT * Z(I)*CT
```

```
220 JLM = JLM + 1
    CM = CM * CI
230 SM =  - SM
    CM = CONJG(CM)
    SL =  - SL
240 ST = ST * CI
    FAC = 1.0
    IF (LAB .EQ. 1)  GO TO 280
    IF (LAB .GT. 6)  GO TO 250
    FAC = 1.0/ASQ
    GO TO 280
250 IF (LAB .EQ. 10)  GO TO 260
    FAC = 0.5
    GO TO 280
260 CONTINUE
    FAC = 0.5/ASQ
280 DO 300 II=1,ID
    RB=CZ
    TB=CZ
    DO 275 I=1,NG
    RB=RB+R(I)*EDW(1,I)
275 TB=TB+T(I)*EDW(2,I)
    RA(II) = FAC * RB
    TA(II) = FAC * TB
    DO 290 I = 1, NG
    DO 290 K = 1, 2
    EDW(K,I) = FDW(K,I)
    FDW(K,I)=GDW(K,I)
290 GDW(K,I)=HDW(K,I)
300 CONTINUE
    IF (IGC)  320, 320, 310
310 GP(2) =  - GP(2)
    AK3 =  - AK3
320 RETURN
330 IF (LAB.LE.13) EP6=EP6*EP6
    DO 440 L=1,LM
    SM=SL
    LL=L+L-1
    L1=L1+LL
    DO 430 M=1,LL
    MM=L1-M
    CYM=CYLM(JG,MM)
    MM=L1-LL-1+M
    CY=CYLM(JG,MM)
    IF (INC) 332,331,331
331 CTR=SL*SM
    CTT=RU
    GO TO 333
332 CTR=RU
    CTT=SL*SM
333 CONTINUE
    Z(1)=CY
    IF (LAB-11) 340,340,350
340 Z(2)=CY*CM
    Z(3)=Z(2)*CM
    GO TO 410
350 IF (LAB-13)360,365,370
360 Z(3)=CY*CM
    Z(5)=Z(3)*CM
    Z(6)=CYM*SM
    Z(2)=Z(6)*CM
    Z(4)=Z(2)*CM
    GO TO 410
365 Z(3)=CY*CM
    Z(5)=Z(3)*CM
    Z(2)=CYM
    Z(4)=CYM*CM
```

```
      Z(6)=Z(4)*CM
      GO TO 410
  370 IF (LAB-15) 375,390,390
  375 Z(2)=CY*CM
      DO 380 I=3,6
  380 Z(I)=Z(I-1)*CM
      GO TO 410
  390 Z(3)=CY*CM
      Z(12)=CYM*SM
      Z(2)=Z(12)*CM
      DO 400 I=5,11,2
      Z(I)=Z(I-2)*CM
  400 Z(I-1)=Z(I-3)*CM
  410 KLM=LXM(JLM)
      DO 420 I=1,NG
      CR=CZ
      CT=CZ
      DO 419 J=1,NLAY
      JJ=(J-1)*LMMAX
      IF (INC.EQ.-1) GO TO 418
      CS=TS(JJ+KLM)*RG(1,J,JGP)/RG(1,1,JGP)
      CR=CR+RG(2,J,JG)/RG(2,1,JG)*CS*EGS(1,I,J)
      CT=CT+RG(1,NLAY,JG)/RG(1,J,JG)*CS*EGS(2,I,J)
      GO TO 419
  418 CS=TS(JJ+KLM)*RG(2,NLAY,JGP)/RG(2,J,JGP)
      CR=CR+RG(1,NLAY,JG)/RG(1,J,JG)*CS*EGS(2,I,J)
      CT=CT+RG(2,J,JG)/RG(2,1,JG)*CS*EGS(1,I,J)
  419 CONTINUE
      R(I)=R(I)+CTR*Z(I)*CR
  420 T(I)=T(I)+CTT*Z(I)*CT
      JLM=JLM+1
      CM=CM*EP6
  430 SM=-SM
      CM=CONJG(CM)
      SL=-SL
  440 ST=ST*CI
      FAC=1.0/ACU
      IF ((LAB.GE.12).AND.(LAB.LE.14)) FAC=FAC/ASQ
      IF (LAB.EQ.15) FAC=FAC*0.5
      DO 470 II=1,ID
      RB=CZ
      TB=CZ
      DO 450 I=1,NG
      RB=RB+R(I)*EDW(1,I)
  450 TB=TB+T(I)*EDW(2,I)
      RA(II)=RB*FAC
      TA(II)=TB*FAC
      DO 460 I=1,NG
      DO 460 K=1,2
      EDW(K,I)=FDW(K,I)
      FDW(K,I)=GDW(K,I)
  460 GDW(K,I)=HDW(K,I)
  470 CONTINUE
      RETURN
      END
C-------------------------------------------------------------------
C  SUBROUTINE MPERTI COMPUTES REFLECTION AND TRANSMISSION MATRICES
C  FOR A COMPOSITE ATOMIC LAYER CONSISTING OF NLAY SUBPLANES (THE
C  COMPOSITE LAYER NEED NOT HAVE A BRAVAIS LATTICE OR BE PLANAR). THE
C  REVERSE SCATTERING PERTURBATION (RSP) METHOD IS USED, BUT CAN BE
C  BYPASSED IN PART BY THE USE OF A VERSION OF THE BEEBY MATRIX
C  INVERSION METHOD, WHEN RSP CONVERGES POORLY (IF NO RSP AT ALL IS TO
C  BE USED, SUBROUTINE MTINV MUST BE SELECTED INSTEAD OF MPERTI).
C  MPERTI ACCEPTS UP TO THREE DIFFERENT CHEMICAL ELEMENTS AND UP TO
C  THREE DIFFERENT ORIGIN (REGISTRY) SHIFTS. IT SYMMETRIZES THE OUTPUT
C  AS REQUESTED THROUGH SYM AND INTERNALLY ORDERS THE SUBPLANES
C  ACCORDING TO INCREASING DISTANCE FROM THE SURFACE.
```

```
C   NOTE: REAL AND IMAGINARY PARTS OF THE MUFFIN-TIN CONSTANT ARE ASSUMED
C   CONSTANT THROUGHOUT LAYER (I.E. BETWEEN OUTERMOST NUCLEAR PLANES).
C   DIMENSIONS SET FOR (NTAU,LM)= AT MOST (3,10)
C     RA1,TA1,...,RC2,TC2= OUTPUT REFLECTION AND TRANSMISSION MATRICES
C       (LETTERS R AND T STAND FOR REFLECTION AND TRANSMISSION, RESP.,
C       LETTERS A,B,C REFER TO THE THREE ORIGIN SHIFTS S1,S2,S3, RESP.,
C       NUMBERS 1 AND 2 REFER TO INCIDENCE TOWARDS +X AND -X, RESP.).
C     N= NO. OF BEAMS IN CURRENT BEAM SET (MUST BE EQUAL TO TOTAL NO. OF
C       BEAMS AT CURRENT ENERGY).
C     NM,NP= DIMENSIONS OF MATRICES (.GE.N)
C     AMULT,CYLM= WORKING SPACE.
C     PQ= LIST OF RECIPROCAL LATTICE VECTORS G (BEAMS).
C     SYM= LIST OF SYMMETRY CODE NUMBERS OF BEAMS.
C     NT= TOTAL NO. OF BEAMS IN MAIN PROGRAM AT CURRENT ENERGY.
C     FLMS= LATTICE SUMS FROM SUBROUTINE FMAT.
C     NL= NO. OF SUBLATTICES CONSIDERED IN SLIND AND FMAT.
C     KLM= (2*LMAX+1)*(2*LMAX+2)/2
C     LX,LXI,LT,LXM= PERMUTATIONS OF (L,M) SEQUENCE, FROM SUBROUTINE
C       LXGENT.
C     LMMAX= (LMAX+1)**2.
C     KLM=(2*LMAX+1)*(2*LMAX+2)/2.
C     XEV,TAU,TAUG,TAUGM= WORKING SPACE.
C     LEV= (LMAX+1)*(LMAX+2)/2.
C     LEV2= 2*LEV.
C     LMT= NTAU*LMMAX.
C     LTAUG= (NTAU+NINV)*LMMAX.
C     CLM= CLEBSCH-GORDON COEFFICIENTS, FROM SUBROUTINE CELMG.
C     NLM= DIMENSION OF CLM (SEE MAIN PROGRAM).
C     POS= INPUT ATOMIC POSITIONS IN UNIT CELL (ONE ATOM PER SUBPLANE).
C     POSS,MGH,DRL,SDRL,NUGH,NGEQ,NGOL,TEST,GH,RG,TS,TG,VT= WORKING SPACES
C     NLAY= NO. OF SUBPLANES IN LAYER.
C     NLAY2= NLAY*(NLAY-1)/2.
C     LMG= 2*NLAY2*LMMAX.
C     LMN= NLAY*LMMAX.
C     LM2N= 2*LMN.
C     CAA= CLEBSCH-GORDON COEFFICIENTS, FROM SUBROUTINE CAAA.
C     NCAA= DIMENSION OF CAA (SEE MAIN PROGRAM).
C     TH= WORKING SPACE.
C     LMNI= NINV*LMMAX.
C     LMNI2= 2*LMNI.
C   IN COMMON BLOCKS:
C     E,VPI= CURRENT COMPLEX ENERGY.
C     AK2,AK3= PARALLEL COMPONENTS OF PRIMARY INCIDENT K-VECTOR.
C     /MFB/ PASSES DATA TO SUBROUTINE MFOLT.
C     LMAX= LARGEST VALUE OF L.
C     EPS1,LITER: NOT USED.
C     SO1,SO2,SO3= SPACE FOR UP TO 3 DIFFERENT OVERLAYER REGISTRIES.
C     SS1,SS2,SS3,SS4= SPACE FOR UP TO 4 DIFFERENT SUBSTRATE LAYER
C       REGISTRIES (SS4 NOT USED).
C     NA=OFFSET OF CURRENT BEAM SET IN LIST PQ (NORMALLY 0).
C     NS= OFFSET FOR FIRST INDEX OF MATRIX ELEMENTS IN RA1,...,TC2
C       (NORMALLY 0).
C     ID= NO. OF ORIGIN SHIFTS (I.E. REGISTRIES) TO BE CONSIDERED.
C     LAY: INDICATES WHICH SYMMETRY CODES SYM(LAY,I) TO BE USED AND WHICH
C       SET OF ORIGIN SHIFTS IS CHOSEN (NORMALLY LAY=1 FOR OVERLAYERS,
C       LAY=2 FOR SUBSTRATE LAYERS).
C     LM=LMAX+1.
C     NTAU= NO. OF CHEMICAL ELEMENTS TO BE USED.
C     TST= CUTOFF PARAMETER FOR RECIPROCAL LATTICE SUM IN SUBROUTINE GHMAT
C     TV= AREA OF UNIT CELL.
C     DCUT= CUTOFF RADIUS FOR DIRECT LATTICE SUM IN GHD.
C     EPS= CONVERGENCE CRITERION FOR RS PERTURBATION.
C     NPERT= LIMIT ON RS PERTURBATION ORDER.
C     NOPT=1: COMPUTE ALL OUTPUT MATRICES.
C     NOPT=2: LAYER IS SYMMETRICAL IN +-X: COMPUTE ONLY OUTPUT MATRICES
C       WITH NUMBER 1 IN THEIR NAME. BY SYMMETRY RA2=RA1,TA2=TA1, ETC. (RA2
C       TA2, ETC. ARE NOT WRITTEN ONTO).
```

```
C      NOPT=3: COMPUTE ONLY A REFLECTION VECTOR (NOT MATRIX) FOR (00)-BEAM
C      INCIDENCE (FROM BOTH SIDES +-X).
C      NEW=-1: INPUT FOR SKIPPING COMPUTATION OF THOSE QUANTITIES THAT HAVE
C      NOT CHANGED SINCE THE PREVIOUS CALL TO MPERTI (I.E. THE PREVIOUS
C      CALL TO MPERTI WAS FOR THE SAME ENERGY, BUT A DIFFERENT GEOMETRY).
C      NOTE: IF USING NEW=-1, DO NOT OVERWRITE CYLM,TAU,AMULT,TH OR GH IN
C      MAIN PROGRAM, SINCE THEIR OLD VALUES WILL BE REUSED.
C      NEW=+1: NO REUSE OF OLD VALUES OF CYLM,TAU,AMULT,TH OR GH WILL OCCUR
C      (USE NEW=+1 WHEN ENERGY HAS CHANGED SINCE THE LAST CALL TO MPERTI).
C      LPS= CHEMICAL ELEMENT ASSIGNMENT FOR EACH SUBPLANE: LPS(I)=J MEANS
C      LAYER NO. I USES ATOMIC T-MATRIX ELEMENTS STORED IN TSF(J,K),K=1,
C      LMAX+1.
C      LPSS= WORKING SPACE.
C      INV(I)=1 MEANS SUBPLANE NO. I IS TO BE TREATED BY BEEBY-TYPE
C      INVERSION.
C      INV(I)=0 MEANS SUBPLANE NO. I IS TO BE TREATED BY RS PERTURBATION.
C      NOTE: SUBPLANES TO BE INVERTED MUST BE GROUPED TOGETHER IN SPACE,
C      I.E. NO SUBPLANE TO BE TREATED BY RSP MAY BE INTERCALATED BETWEEN
C      THEM (THIS MUST STILL BE TRUE AFTER REORDERING BY SRTLAY).
C      NINV= NO. OF SUBPLANES TO BE TREATED BY BEEBY-TYPE INVERSION.
C      TSF= ATOMIC T-MATRIX ELEMENTS (UP TO 3 CHEMICAL ELEMENTS, UP TO
C      LMAX=9).
C      /SL/: DATA PASSED TO SUBROUTINE GHMAT (SEE SUBROUTINE SLIND).
       SUBROUTINE MPERTI(RA1,TA1,RA2,TA2,RB1,TB1,RB2,TB2,RC1,TC1,
      1RC2,TC2,N,NM,NP,AMULT,CYLM,PQ,
      2SYM,NT,FLMS,NL,LX,LXI,LT,LXM,LMMAX,KLM,
      3XEV,LEV,LEV2,TAU,LMT,TAUG,TAUGM,LTAUG,CLM,NLM,POS,POSS,
      4MGH,NLAY,DRL,SDRL,NUGH,NGEQ,NGOL,NLAY2,TEST,
      5GH,LMG,RG,TS,LMN,TG,LM2N,VT,CAA,NCAA,TH,LMNI,LMNI2)
       INTEGER  SYM
       COMPLEX XEV,RA1,TA1,RA2,TA2,CYLM,AMULT,XA,YA,ST,CF,CT,CI,RU,CZ,
      1XB,YB,AM,FLMS,SU
       COMPLEX CSQRT,CEXP
       COMPLEX TH(LMNI,LMNI2)
       COMPLEX TAU(LMT,LEV),TSF(3,10),GH(LMG,LMMAX),RG(3,NLAY,N)
       COMPLEX TS(LMN),TG(2,LM2N),VT(LMMAX),TAUG(LTAUG),TAUGM(LTAUG)
       COMPLEX RC1(NM,NP),TC1(NM,NP),RC2(NM,NP),TC2(NM,NP)
       COMPLEX RB1(NM,NP),TB1(NM,NP),RB2(NM,NP),TB2(NM,NP)
       DIMENSION YB(4),XB(4),S3(2),S4(2),SS3(2),SS4(2)
       DIMENSION  RA1(NM,NP),TA1(NM,NP),RA2(NM,NP),TA2(NM,NP),AMULT(N)
       DIMENSION  XEV(LEV,LEV2),CYLM(NT,LMMAX),PQ(2,NT)
       DIMENSION CLM(NLM),GP(2),SO1(2),SO2(2),SO3(2),SS1(2),CAA(NCAA)
       DIMENSION  SS2(2),  S1(2),  S2(2),   FLMS(NL,KLM)
       DIMENSION  SYM(2,NT),LX(LMMAX),LXI(LMMAX),LT(LMMAX),LXM(LMMAX)
       DIMENSION POS(NLAY,3),POSS(NLAY,3),LPS(12),LPSS(12)
       DIMENSION MGH(NLAY,NLAY),DRL(NLAY2,3),SDRL(NLAY2,3)
       DIMENSION NUGH(NLAY2),NGEQ(NLAY2),NGOL(NLAY2)
       DIMENSION RBR1(2),RBR2(2),TEST(NLAY2),INV(12)
       DIMENSION ARA1(2),ARA2(2),ARB1(2),ARB2(2)
       COMMON   E, AK2,AK3, VPI
       COMMON /MFB/ GP,L1,LAY1,S1,S2,S3,S4
       COMMON  /MS/LMAX, EPS1, LITER, SO1,SO2,SO3, SS1, SS2,SS3,SS4
       COMMON  /MPT/ NA,NS,ID,LAY,LM,NTAU,TST,TV,DCUT,
      1EPS,NPERT,NOPT,NEW,LPS,LPSS,INV,NINV,TSF
       COMMON /SL/ ARA1,ARA2,ARB1,ARB2,RBR1,RBR2,NL1,NL2
C
C
       IF (NINV.LT.NLAY) GO TO 20
       WRITE(6,10)
10     FORMAT(37H MPERTI DOES NOT ACCEPT NINV.GE.NLAY )
       STOP
20     CONTINUE
       PI = 3.14159265
       L1=LM
       LAY1=LAY
       L2M=2*LMAX+1
       LMS=L2M*L2M
```

```
          LOD=LMMAX-LEV
          LEE=(LMAX/2+1)**2
          LOE=((LMAX-1)/2+1)**2
          CI = CMPLX(0.0,1.0)
          CZ = CMPLX(0.0,0.0)
          RU = CMPLX(1.0,0.0)
          ASQ = 1.41421356
          ACU=1.7320508
C
C   SELECT ORIGIN SHIFTS (REGISTRIES) TO BE USED
          IF (LAY-1) 40,25,40
   25     DO 30 I=1,2
          S3(I)=SO3(I)
          S2(I)=SO2(I)
   30     S1(I)=SO1(I)
          GO TO 52
   40 DO 50 I = 1, 2
          S3(I)=SS3(I)
          S1(I) = SS1(I)
   50 S2(I) = SS2(I)
C
C   CHECK WHETHER TAU NEEDS TO BE RECOMPUTED SINCE LAST CALL TO MPERTI
   52     IF (NEW.EQ.-1) GO TO 55
C   COMPUTE TAU FOR EACH CHEMICAL ELEMENT
          CALL TAUMAT(TAU,LMT,NTAU,XEV,LEV,LEV2,LOD,TSF,LMMAX,LMAX,
         1FLMS,NL,KLM,LM,CLM,NLM,LXI)
C
C   SORT SUBPLANES BY INCREASING POSITION ALONG +X. FIND WHICH OLD GH'S
C   CAN BE KEPT, WHICH MUST BE RECOMPUTED AND WHICH ARE MUTUALLY
C   IDENTICAL
   55 CALL SRTLAY(POS,POSS,LPS,LPSS,MGH,NLAY,DRL,SDRL,NLAY2,
         1NUGH,NGEQ,NGOL,NEW)
C
C   DETERMINE WHETHER NEW INVERSION WILL BE NECESSARY (LINV=1) OR THE
C   OLD VALUES OF TH APE STILL VALID (LINV=0)
          LINV=0
          NLAY1=NLAY-1
          DO 56 IN=1,NLAY1
          IN1=IN+1
          DO 56 INP=IN1,NLAY
          M=MGH(IN,INP)
          IF ((INV(IN).EQ.1).AND.(INV(INP).EQ.1).AND.
         1((NUGH(M).EQ.1).OR.(NGEQ(M).NE.0))) LINV=1
   56     CONTINUE
C
C   COMPUTE (OR COPY) THE GH MATRICES (INTERLAYER PROPAGATION MATRICES
C   IN (L,M)-SPACE).
          CALL GHMAT(GH,LMG,LMMAX,MGH,NLAY,NUGH,NGEQ,NGOL,NLAY2,
         1TST,TEST,XEV,L2M,VT,LM,TG,LMS,DRL,TV,LXM,LEV,DCUT,CAA,NCAA)
C
C
          YA = CMPLX(2.0 * E, - 2.0 * VPI + 0.000001)
          YA = CSQRT(YA)
          DO 90 IG = 1, N
          JG = IG + NA
          BK2 = PQ(1,JG) + AK2
          BK3 = PQ(2,JG) + AK3
          C = BK2 * BK2 + BK3 * BK3
          XA = CMPLX(2.0 * E - C, - 2.0 * VPI + 0.000001)
          XA = CSQRT(XA)
C   GENERATE PROPAGATION FACTORS FOR UNSCATTERED WAVES
          RG(3,1,IG)=CEXP(CI*XA*(POSS(NLAY,1)-POSS(1,1)))
          DO 60 I=1,NLAY
          X=BK2*POSS(I,2)+BK3*POSS(I,3)
C   GENERATE PLANE-WAVE PROPAGATORS BETWEEN SUBPLANES FOR USE AS
C   PREFACTORS FOR GH AND TH AND IN MFOLT
          RG(1,I,IG)=CEXP(CI*(XA*POSS(I,1)+X))
```

```
   60 RG(2,I,IG)=CEXP(CI*(XA*POSS(I,1)-X))
C
C  IF NEW=1 (I.E. NEW ENERGY) COMPUTE NEW PREFACTORS AND NEW SPHERICAL
C  HARMONICS FOR THE BEAM DIRECTIONS IN SPHRM.
      IF (NEW.EQ.-1) GO TO 90
      B = 0.0
      CF = RU
      IF (C-1.0E-7)  80, 80, 70
   70 B = SQRT(C)
      CF = CMPLX(BK2/B,BK3/B)
   80 CT = XA/YA
      ST = B/YA
C  GENERATE PREFACTORS OF REFLECTION AND TRANSMISSION MATRIX ELEMENTS
      AMULT(IG) =  - 16.0 * PI * PI * CI/(TV * XA)
C
      CALL  SPHRM(LMAX,VT,LMMAX,CT, ST, CF)
C  STORE THE SPHERICAL HARMONICS
      DO 85 K = 1, LMMAX
      CYLM(IG,K) = VT(K)
   85 CONTINUE
   90 CONTINUE
C
C
      IF (LINV.EQ.0) GO TO 160
C  IF NECESSARY, PRODUCE AND INVERT MATRIX TH BY BEEBY METHOD, INCL.
C  MULTIPLICATION OF TH INTO TAU'S
      CALL THINV(TH,LMNI,LMNI2,GH,LMG,LMMAX,MGH,NLAY,TAU,LMT,LEV,
     1INV,LPSS)
C
C
C  START LOOP OVER INCIDENT BEAMS
  160 DO 430 JGP = 1, N
C  IF ONLY DIFFRACTION FROM (00) BEAM REQUIRED, SKIP FURTHER LOOPING
      IF (NOPT.EQ.3.AND.JGP.EQ.2) GO TO 440
      JGPS=JGP+NS
      GP(1) = PQ(1,JGP + NA)
      GP(2) = PQ(2,JGP + NA)
C  CHOOSE APPROPRIATE SYMMETRY NORMALIZATION FACTOR FAC
      LAB = SYM(LAY,JGP + NA)
      IF (LAB-10) 165,220,180
  165 IF (LAB-2)  190, 170, 170
  170 IF (LAB-7)  200, 210, 210
  180 IF (LAB-12) 221,185,185
  185 IF (LAB-15) 222,223,223
  190 FAC = 1.0
      GO TO 230
  200 FAC = ASQ
      GO TO 230
  210 FAC = 2.0
      GO TO 230
  220 FAC = 2.0 * ASQ
      GO TO 230
  221 FAC=ACU
      GO TO 230
  222 FAC=ASQ*ACU
      GO TO 230
  223 FAC=2.0*ACU
  230 CONTINUE
C
C  GENERATE INITIAL VALUES TAUG,TAUGM= TAU*Y(G+-) AND TH*Y(G+-) FOR RSP
C  COMPUTATION
      CALL TAUY(TAUG,TAUGM,LTAUG,TAU,LMT,LEV,CYLM,NT,LMMAX,LT,TH,
     1LMNI,LMNI2,NTAU,LOD,LEE,LOE,JGP,LINV,RG,NLAY,N,INV)
  280 CONTINUE
C
C
C  FIRST CONSIDER INCIDENCE TOWARDS +X
```

```
         INC=1
   285 CONTINUE
C   PERFORM REVERSE SCATTERING PERTURBATION COMPUTATION
         IF (INC.EQ.-1) GO TO 290
         CALL TPERTI(TS,LMN,TG,LM2N,LMMAX,TAU,LMT,LEV,GH,LMG,RG,N,MGH,
      1LPSS,NLAY,INC,TAUG,LTAUG,VT,EPS,NPERT,JGP,INV,NINV,TH,LMNI,LMNI2)
         GO TO 295
   290 CALL TPERTI(TS,LMN,TG,LM2N,LMMAX,TAU,LMT,LEV,GH,LMG,RG,N,MGH,
      1LPSS,NLAY,INC,TAUGM,LTAUG,VT,EPS,NPERT,JGP,INV,NINV,TH,LMNI,LMNI2)
   295 CONTINUE
C
C
C   START LOOP OVER SCATTERED BEAMS
   400 DO 410 JG = 1, N
C   COMPLETE COMPUTATION IN MFOLT, INCL. ORIGIN SHIFTS AND SYMMETRIZATION
         CALL   MFOLT (JG,NA,E,VPI,AK2,AK3,CYLM,N,LMMAX,PQ,SYM,NT,
      1TS,LMN,RG,NLAY,JGP,XEV,LXM,ID,YB,XB,INC,POSS)
C
         JGS=JG+NS
         AM = AMULT(JG) * FAC
         IF (INC.EQ.-1) GO TO 500
         IF (ID-2) 409,408,407
C   PUT FINAL MATRIX ELEMENTS IN PROPER LOCATIONS
   407 RC1(JGS,JGP)=YB(3)*AM
       TC1(JGS,JGP)=XB(3)*AM
   408 RB1(JGS,JGP)=YB(2)*AM
       TB1(JGS,JGP)=XB(2)*AM
   409 RA1(JGS,JGP) = YB(1)* AM
       TA1(JGS,JGP) = XB(1)* AM
         GO TO 410
   500 IF (ID-2) 509,508,507
   507 RC2(JGS,JGP)=YB(3)*AM
       TC2(JGS,JGP)=XB(3)*AM
   508 RB2(JGS,JGP)=YB(2)*AM
       TB2(JGS,JGP)=XB(2)*AM
   509 RA2(JGS,JGP)=YB(1)*AM
       TA2(JGS,JGP)=XB(1)*AM
   410 CONTINUE
         IF (INC.EQ.-1) GO TO 520
         IF (ID-2) 431,429,428
C   ADD UNSCATTERED PLANE WAVE TO TRANSMISSION
   428 TC1(JGPS,JGP)=TC1(JGPS,JGP)+RG(3,1,JGP)
   429 TB1(JGPS,JGP)=TB1(JGPS,JGP)+RG(3,1,JGP)
   431 TA1(JGPS,JGP)=TA1(JGPS,JGP)+RG(3,1,JGP)
C   IF +-X SYMMETRY APPLIES (RA2=RA1, ETC.), SKIP COMPUTATION OF RA2,ETC.
         IF (NOPT-2) 510,430,510
C   REVERSE INCIDENCE DIRECTION AND REPEAT COMPUTATIONS
   510 INC=-1
         GO TO 285
   520 IF (ID-2) 530,529,528
   528 TC2(JGPS,JGP)=TC2(JGPS,JGP)+RG(3,1,JGP)
   529 TB2(JGPS,JGP)=TB2(JGPS,JGP)+RG(3,1,JGP)
   530 TA2(JGPS,JGP)=TA2(JGPS,JGP)+RG(3,1,JGP)
   430 CONTINUE
   440 RETURN
         END
C-------------------------------------------------------------------
C   SUBROUTINE MSMF GENERATES REFLECTION AND TRANSMISSION MATRICES FOR A
C   SINGLE BRAVAIS-LATTICE LAYER BY PENDRY'S FORMULA. IT USES SYMMETRY
C   AS REQUESTED AND PRODUCES MATRICES FOR UP TO FOUR ORIGIN SHIFTS
C   TOWARDS A SYMMETRY AXIS OR PLANE. THERMAL VIBRATIONS ARE TAKEN INTO
C   ACCOUNT AND MAY BE ANISOTROPIC (APPROXIMATE TREATMENT).
C   RA,TA,...,RD,TD= REFLECTION AND TRANSMISSION MATRICES (LETTERS R AND
C    T STAND FOR REFLECTION AND TRANSMISSION, RESP.; LETTERS A,B,C,D
C    CORRESPOND TO THE FOUR POSSIBLE ORIGIN SHIFTS, S1,S2,S3,S4,RESP.).
C   N= NO. OF BEAMS IN CURRENT BEAM SET.
C   NM,NP= DIMENSIONS OF DIFFRACTION MATRICES (.GE.N).
```

```
C      AMULT,CYLM= WORKING SPACE.
C      PQ= LIST OF BEAMS G.
C      SYM= LIST OF SYMMETRY CODES FOR ALL BEAMS.
C      NT= TOTAL NO. OF BEAMS IN MAIN PROGRAM AT CURRENT ENERGY.
C      FLMS= LATTICE SUMS FROM SUBROUTINE FMAT.
C      FLM= WORKING SPACE.
C      V: SUBLATTICE INFORMATION FROM SUBROUTINE SLIND.
C      NL= NO. OF SUBLATTICES IN LATTICE SUM (CF. SLIND).
C      NA= OFFSET OF CURRENT BEAM SET IN LIST PQ.
C      NS= OFFSET OF THE FIRST INDEX OF THE MATRIX ELEMENTS IN RA,...,TD.
C      ID= NO. OF ORIGIN SHIFTS TO BE CONSIDERED.
C      NLL= NO. OF SUBLATTICES TO BE CONSIDERED IN CURRENT USE OF MSMF
C       (NLL=1 FOR OVERLAYER, NLL=NL FOR SUBSTRATE LAYER).
C      AF= TEMPERATURE-INDEPENDENT ATOMIC T-MATRIX ELEMENTS.
C      CAF= TEMPERATURE-DEPENDENT ATOMIC T-MATRIX ELEMENTS
C      LM= LMAX+1.
C      LX,LXI= PERMUTATIONS OF THE (L,M) SEQUENCE, FROM SUBROUTINE LXGENT.
C      LMMAX= (LMAX+1)**2.
C      KLM= (2*LMAX+1)*(2*LMAX+2)/2.
C      XEV,XOD= WORKING SPACE.
C      LEV= (LMAX+1)*(LMAX+2)/2.
C      LOD= LMAX*(LMAX+1)/2.
C      YLM,YLME,YLMO,IPLE,IPLO= WORKING SPACES.
C      CLM= CLEBSCH-GORDON COEFFICIENTS, FROM SUBROUTINE CELMG.
C      NLM= DIMENSION OF CLM (SEE MAIN PROGRAM).
C      LAY= INDEX FOR CHOICE OF SYMMETRY CODES SYM(LAY,I) AND FOR CHOICE OF
C       SET OF ORIGIN SHIFTS (NORMALLY LAY=1 FOR OVERLAYER, LAY=2 FOR
C       SUBSTRATE LAYERS).
C      IN COMMON BLOCKS:
C      E,VPI= CURRENT COMPLEX ENERGY.
C      AK2,AK3= PARALLEL COMPONENTS OF PRIMARY INCIDENT K-VECTOR.
C      TV= AREA OF UNIT CELL.
C      EMACH= MACHINE ACCURACY.
C      /MFB/: DATA PASSED TO SUBROUTINE MFOLD.
C      LMAX= LARGEST VALUE OF L.
C      EPS,LITER: NOT USED.
C      SO1,SO2,SO3= SPACE FOR UP TO 3 ORIGIN SHIFTS (REGISTRIES) OF AN
C      OVERLAYER, FIRST ID OF WHICH ARE USED IF LAY=1.
C      SS1,SS2,SS3,SS4= SPACE FOR UP TO 4 ORIGIN SHIFTS (REGISTRIES) OF A
C      SUBSTRATE LAYER, FIRST ID OF WHICH ARE USED IF LAY=2.
C      /BT/: TEMPERATURE DATA PASSED TO MFOLD.
       SUBROUTINE  MSMF(RA,TA,RB,TB,RC,TC,RD,TD,N,NM,NP,AMULT,CYLM,PQ,
      1SYM,NT,FLMS,FLM,V,NL,NA,NS,ID,NLL,AF,CAF,LM,LX,LXI,LMMAX,KLM,
      2XEV,XOD,LEV,LOD,YLM,YLME,YLMO,IPLE,IPLO,CLM,NLM,LAY)
       INTEGER  SYM, LX, LXI
       COMPLEX  AF, XEV, XOD, RA, TA, RB, TB, CYLM, YLM, YLME, YLMO,
      1AMULT, XA, YA, ST, CF, CT, CI, RU, CZ, XB, YB, AM
       COMPLEX  CAF, FLMS, FLM
       COMPLEX CSQRT,CEXP
       COMPLEX RC(NM,NP),TC(NM,NP),RD(NM,NP),TD(NM,NP)
       DIMENSION YB(4),XB(4),S3(2),S4(2),SS3(2),SS4(2)
       DIMENSION  RA(NM,NP),TA(NM,NP),RB(NM,NP),TB(NM,NP),AMULT(N)
       DIMENSION  XEV(LEV,LEV),XOD(LOD,LOD),CYLM(NT,LMMAX),PQ(2,NT)
       DIMENSION  YLM(LMMAX),CLM(NLM),AF(LM),YLME(LEV),YLMO(LOD)
       DIMENSION  IPLE(LEV),IPLO(LOD),V(NL,2), GP(2), SS1(2)
       DIMENSION  SS2(2), S1(2), S2(2), CAF(LM), FLMS(NL,KLM), FLM(KLM)
       DIMENSION SYM(2,NT),LX(LMMAX),LXI(LMMAX),SO1(2),SO2(2),SO3(2)
       COMMON  E, AK2,AK3, VPI, TV, EMACH
       COMMON /MFB/ GP,L1,LAY1,S1,S2,S3,S4
       COMMON /MS/LMAX,EPS,LITER,SO1,SO2,SO3,SS1,SS2,SS3,SS4
       COMMON  /BT/IT, T, TO, DRX, DRY, DRO
       PI = 3.14159265
       L1=LM
       LAY1=LAY
       LEV1=LEV+1
       CI = CMPLX(0.0,1.0)
       CZ = CMPLX(0.0,0.0)
```

```
      RU = CMPLX(1.0,0.0)
      ASQ = 1.41421356
      ACU=1.7320508
C   SELECT ORIGIN SHIFTS (REGISTRIES) TO BE USED
      IF (LAY-1)   40, 20, 40
   20 DO 30 I = 1, 2
      S3(I)=SO3(I)
      S1(I) = SO1(I)
   30 S2(I) = SO2(I)
      GO TO 60
   40 DO 50 I = 1, 2
      S4(I)=SS4(I)
      S3(I)=SS3(I)
      S1(I) = SS1(I)
   50 S2(I) = SS2(I)
   60 AK = SQRT(2.0 * E)
      YA = CMPLX(2.0 * E, - 2.0 * VPI + 0.000001)
      YA = CSQRT(YA)
      DO 90 IG = 1, N
      JG = IG + NA
      BK2 = PQ(1,JG) + AK2
      BK3 = PQ(2,JG) + AK3
      C = BK2 * BK2 + BK3 * BK3
      XA = CMPLX(2.0 * E - C, - 2.0 * VPI + 0.000001)
      XA = CSQRT(XA)
      B = 0.0
      CF = RU
      IF (C-1.0E-7)   80, 80, 70
   70 B = SQRT(C)
      CF = CMPLX(BK2/B,BK3/B)
   80 CT = XA/YA
      ST = B/YA
C   PREFACTOR FOR REFLECTION AND TRANSMISSION MATRIX ELEMENTS
      AMULT(IG) =  - 8.0 * PI * PI * CI/(AK * TV * XA)
C   COMPUTE AND STORE SPHERICAL HARMONICS FOR ALL BEAM DIRECTIONS
      CALL   SPHRM(LMAX,YLM,LMMAX,CT, ST, CF)
      DO 90 K = 1, LMMAX
   90 CYLM(IG,K) = YLM(K)
      DO 100 K = 1, KLM
  100 FLM(K) = CZ
C   PERFORM SUM OVER SUBLATTICES
      BK2 = PQ(1,1 + NA)
      BK3 = PQ(2,1 + NA)
      DO 110 JS = 1, NLL
      ABR = V(JS,1) * (BK2 * COS(V(JS,2)) + BK3 * SIN(V(JS,2)))
      XA = CEXP(ABR * CI)
      DO 110 I = 1, KLM
  110 FLM(I) = FLM(I) + FLMS(JS,I) * XA
C   COMPUTE INTRALAYER MULTIPLE-SCATTERING MATRIX X, SPLIT ACCORDING TO
C   L+M=EVEN (XEV) AND L+M=ODD (XOD)
      CALL   XM (FLM,XEV,XOD,LEV,LOD,AF,CAF,LM,LX,LXI,LMMAX,KLM,CLM,NLM)
      DO 120 J = 1, LEV
  120 XEV(J,J) = XEV(J,J) - RU
      DO 130 J = 1, LOD
  130 XOD(J,J) = XOD(J,J) - RU
C   PREPARE INVERSION OF 1-X (- SIGN IS PUT IN AMULT) BY GAUSSIAN
C   ELIMINATION
  140 CALL   ZGE (XEV, IPLE, LEV, LEV, EMACH)
      CALL   ZGE (XOD, IPLO, LOD, LOD, EMACH)
C   START LOOP OVER SCATTERED BEAMS
  160 DO 430 JGP = 1, N
      JGPS=JGP+NS
      GP(1) = PQ(1,JGP + NA)
      GP(2) = PQ(2,JGP + NA)
C   SELECT SYMMETRY NORMALIZATION FACTOR FAC
      LAB = SYM(LAY,JGP + NA)
      IF (LAB-10) 165,220,180
```

```
      165 IF (LAB-2)    190, 170, 170
      170 IF (LAB-7)    200, 210, 210
      180 IF (LAB-12) 221,185,185
      185 IF (LAB-15) 222,223,223
      190 FAC = 1.0
          GO TO 230
      200 FAC = ASQ
          GO TO 230
      210 FAC = 2.0
          GO TO 230
      220 FAC = 2.0 * ASQ
          GO TO 230
      221 FAC=ACU
          GO TO 230
      222 FAC=ASQ*ACU
          GO TO 230
      223 FAC=2.0*ACU
C   PREPARE QUANTITIES TO BE MULTIPLIED FROM LEFT BY INVERSE OF 1-X
      230 JLM = 1
          ST = RU
          DO 250 L = 1, LM
          LL = L + L - 1
          DO 240 JM = 1, LL
          YLM(JLM) = CYLM(JGP,JLM) * AF(L)/ST
      240 JLM = JLM + 1
      250 ST = ST * CI
C   CHANGE ORDERING OF (L,M) PAIRS
      260 DO 270 JLM = 1, LEV
          JJ = LX(JLM)
      270 YLME(JLM) = YLM(JJ)
          DO 280 JLM = LEV1,LMMAX
          JJ = LX(JLM)
      280 YLMO(JLM - LEV) = YLM(JJ)
C   COMPLETE INVERSION AND MULTIPLICATION BY BACK-SUBSTITUTION
          CALL   ZSU (XEV, IPLE, YLME, LEV,LEV, EMACH)
          CALL   ZSU (XOD, IPLO, YLMO, LOD,LOD, EMACH)
C   REORDER (L,M) PAIRS
          DO 290 JLM = 1, LEV
          JJ = LX(JLM)
      290 YLM(JJ) = YLME(JLM)
          DO 300 JLM = LEV1,LMMAX
          JJ = LX(JLM)
      300 YLM(JJ) = YLMO(JLM - LEV)
C   START LOOP OVER INCIDENT BEAMS
      400 DO 410 JG = 1, N
C   COMPLETE COMPUTATION IN MFOLD, INCL. ORIGIN SHIFTS, DEBYE-WALLER
C   FACTORS AND SYMMETRIZATION
          CALL   MFOLD (JG,NA,E,VPI,AK2,AK3,YLM,CYLM,N,LMMAX,PQ,SYM,NT,
         1ID,YB,XB)
          AM = AMULT(JGP) * FAC
          IF (ID-2) 409,408,405
      405 IF (ID-3) 407,407,406
C   PUT FINAL MATRIX ELEMENTS IN PROPER LOCATION
      406 RD(JGPS,JG) = YB(4)* AM
          TD(JGPS,JG) = XB(4)* AM
      407 RC(JGPS,JG)=YB(3)*AM
          TC(JGPS,JG)=XB(3)*AM
      408 RB(JGPS,JG)=YB(2)*AM
          TB(JGPS,JG)=XB(2)*AM
      409 RA(JGPS,JG) = YB(1)* AM
      410 TA(JGPS,JG) = XB(1)* AM
          IF (ID-2) 430,429,426
      426 IF (ID-3) 428,428,427
C   ADD UNSCATTERED PLANE WAVE TO TRANSMISSION
      427 TD(JGPS,JGP) = TD(JGPS,JGP) + RU
      428 TC(JGPS,JGP)=TC(JGPS,JGP)+RU
      429 TB(JGPS,JGP)=TB(JGPS,JGP)+RU
```

```
  430 TA(JGPS,JGP) = TA(JGPS,JGP) + RU
      RETURN
      END
C-----------------------------------------------------------------------
C  SUBROUTINE MTINV COMPUTES REFLECTION AND TRANSMISSION MATRICES FOR
C  AN ATOMIC LAYER CONSISTING OF NLAY SUBPLANES. A BEEBY-TYPE MATRIX
C  INVERSION IS USED (A PERTURBATION TREATMENT IS DONE IN SUBROUTINE
C  MPERTI).
C  SPECIFICATIONS NOT FOUND BELOW CAN BE FOUND IN THE SPECIFICATIONS
C  OF SUBROUTINE MPERTI, WHICH HAS THE SAME STRUCTURE AS MTINV.
C  (SUBPLANE SORTING BY SRTLAY IS DONE IN MTINV FOR SIMILARITY WITH
C  MPERTI, ALTHOUGH REORDERING IS NOT NECESSARY HERE.)
C  NOTE: REAL AND IMAGINARY PARTS OF MUFFIN-TIN CONSTANT ARE ASSUMED
C  CONSTANT THROUGHOUT LAYER (I.E. BETWEEN OUTERMOST NUCLEAR PLANES).
C   TAUG, TAUGM HAVE DIMENSION LMT=NTAU*LMMAX HERE.
C   TH(LMNI,LMNI2) IS SQUARE HERE WITH LMNI2=LMNI=NLAY*LMMAX.
C   IPL IS ADDITIONAL WORKING SPACE.
C   INV,NINV ARE NOT USED HERE (BUT NINV IS TO BE GIVEN THE SAME
C    VALUE AS NLAY IN THE INPUT).
C DIMENSIONS SET FOR (NTAU,LM)= AT MOST (3,10)
      SUBROUTINE MTINV(RA1,TA1,RA2,TA2,RB1,TB1,RB2,TB2,RC1,TC1,
     1RC2,TC2,N,NM,NP,AMULT,CYLM,PQ,
     2SYM,NT,FLMS,NL,LX,LXI,LT,LXM,LMMAX,KLM,
     3XEV,LEV,LEV2,TAU,LMT,TAUG,TAUGM,CLM,NLM,POS,POSS,
     4MGH,NLAY,DRL,SDRL,NUGH,NGEQ,NGOL,NLAY2,TEST,
     5GH,LMG,RG,TS,LMN,TG,LM2N,VT,CAA,NCAA,TH,LMNI,IPL)
      INTEGER  SYM
      COMPLEX XEV,RA1,TA1,RA2,TA2,CYLM,AMULT,XA,YA,ST,CF,CT,CI,RU,CZ,
     1XB,YB,AM,FLMS,SU
      COMPLEX CSQRT,CEXP
      COMPLEX TH(LMNI,LMNI)
      COMPLEX TAU(LMT,LEV),TSF(3,10),GH(LMG,LMMAX),RG(3,NLAY,N)
      COMPLEX TS(LMN),TG(2,LM2N),VT(LMMAX),TAUG(LMT),TAUGM(LMT)
      COMPLEX RC1(NM,NP),TC1(NM,NP),RC2(NM,NP),TC2(NM,NP)
      COMPLEX RB1(NM,NP),TB1(NM,NP),RB2(NM,NP),TB2(NM,NP)
      DIMENSION YB(4),XB(4),S3(2),S4(2),SS3(2),SS4(2)
      DIMENSION  RA1(NM,NP),TA1(NM,NP),RA2(NM,NP),TA2(NM,NP),AMULT(N)
      DIMENSION  XEV(LEV,LEV2),CYLM(NT,LMMAX),PQ(2,NT)
      DIMENSION CLM(NLM),GP(2),SO1(2),SO2(2),SO3(2),SS1(2),CAA(NCAA)
      DIMENSION  SS2(2),  S1(2), S2(2),   FLMS(NL,KLM)
      DIMENSION  SYM(2,NT),LX(LMMAX),LXI(LMMAX),LT(LMMAX),LXM(LMMAX)
      DIMENSION POS(NLAY,3),POSS(NLAY,3),LPS(12),LPSS(12)
      DIMENSION MGH(NLAY,NLAY),DRL(NLAY2,3),SDRL(NLAY2,3)
      DIMENSION NUGH(NLAY2),NGEQ(NLAY2),NGOL(NLAY2)
      DIMENSION RBR1(2),RBR2(2),TEST(NLAY2),INV(12),IPL(LMNI)
      DIMENSION ARA1(2),ARA2(2),ARB1(2),ARB2(2)
      COMMON  E, AK2,AK3, VPI
      COMMON /MFB/ GP,L1,LAY1,S1,S2,S3,S4
      COMMON  /MS/LMAX, EPS1, LITER, SO1,SO2,SO3, SS1, SS2,SS3,SS4
      COMMON /MPT/ NA,NS,ID,LAY,LM,NTAU,TST,TV,DCUT,
     1EPS,NPERT,NOPT,NEW,LPS,LPSS,INV,NINV,TSF
      COMMON /SL/ ARA1,ARA2,ARB1,ARB2,RBR1,RBR2,NL1,NL2
      PI = 3.14159265
      L1=LM
      LAY1=LAY
      L2M=2*LMAX+1
      LMS=L2M*L2M
      LOD=LMMAX-LEV
      LEE=(LMAX/2+1)**2
      LOE=((LMAX-1)/2+1)**2
      CI = CMPLX(0.0,1.0)
      CZ = CMPLX(0.0,0.0)
      RU = CMPLX(1.0,0.0)
      ASQ = 1.41421356
      ACU=1.7320508
      IF (LAY-1) 40,20,40
   20 DO 30 I=1,2
```

```
        S3(I)=S03(I)
        S2(I)=S02(I)
30      S1(I)=S01(I)
        GO TO 52
   40 DO 50 I = 1, 2
        S3(I)=SS3(I)
        S1(I) = SS1(I)
   50 S2(I) = SS2(I)
C
C
C  CHECK WHETHER TAU NEEDS RECOMPUTING SINCE LAST CALL TO MTINV
52     IF (NEW.EQ.-1) GO TO 55
C  COMPUTE TAU FOR EACH CHEMICAL ELEMENT
        CALL TAUMAT(TAU,LMT,NTAU,XEV,LEV,LEV2,LOD,TSF,LMMAX,LMAX,
      1FLMS,NL,KLM,LM,CLM,NLM,LXI)
C
C
C  SORT SUBPLANES ACCORDING TO INCREASING POSITION ALONG +X AXIS.
C  FIGURE OUT WHICH OLD GH'S CAN BE KEPT, WHICH MUST BE RECOMPUTED
C  AND WHICH ARE MUTUALLY IDENTICAL
   55 CALL SRTLAY(POS,POSS,LPS,LPSS,MGH,NLAY,DRL,SDRL,NLAY2,
      1NUGH,NGEQ,NGOL,NEW)
C
C
C  COMPUTE (OR COPY) THE GH MATRICES (INTERLAYER PROPAGATORS IN (LM)-
C  SPACE)
        CALL GHMAT(GH,LMG,LMMAX,MGH,NLAY,NUGH,NGEQ,NGOL,NLAY2,
      1TST, TEST,XEV,L2M,VT,LM,TG,LMS,DRL,TV,LXM,LEV,DCUT,CAA,NCAA)
C
C
        YA = CMPLX(2.0 * E, - 2.0 * VPI + 0.000001)
        YA = CSQRT(YA)
        DO 90 IG = 1, N
        JG = IG + NA
        BK2 = PQ(1,JG) + AK2
        BK3 = PQ(2,JG) + AK3
        C = BK2 * BK2 + BK3 * BK3
        XA = CMPLX(2.0 * E - C, - 2.0 * VPI + 0.000001)
        XA = CSQRT(XA)
        RG(3,1,IG)=CEXP(CI*XA*(POSS(NLAY,1)-POSS(1,1)))
        DO 60 I=1,NLAY
        X=BK2*POSS(I,2)+BK3*POSS(I,3)
C  GENERATE PLANE-WAVE PROPAGATORS FOR USE AS PREFACTORS FOR GH AND TH
        RG(1,I,IG)=CEXP(CI*(XA*POSS(I,1)+X))
   60 RG(2,I,IG)=CEXP(CI*(XA*POSS(I,1)-X))
C  IF NEW=1 (I.E. NEW ENERGY) COMPUTE NEW SPHERICAL HARMONICS FOR THE
C  BEAM DIRECTIONS IN SPHRM
        IF (NEW.EQ.-1) GO TO 90
        B = 0.0
        CF = RU
        IF (C-1.0E-7) 80, 80, 70
   70 B = SQRT(C)
        CF = CMPLX(BK2/B,BK3/B)
   80 CT = XA/YA
        ST = B/YA
C  GENERATE PREFACTOR OF REFLECTION AND TRANSMISSION MATRIX ELEMENTS
        AMULT(IG) =  - 16.0 * PI * PI * CI/(TV * XA)
        CALL  SPHRM(LMAX,VT,LMMAX,CT, ST, CF)
C  STORE THE SPHERICAL HARMONICS
        DO 85 K = 1, LMMAX
        CYLM(IG,K) = VT(K)
   85 CONTINUE
   90 CONTINUE
C
C
C  CREATE MATRIX TH TO BE INVERTED IN BEEBY'S SCHEME
        CALL  THMAT(TH,LMNI,LMNI,GH,LMG,LMMAX,MGH,NLAY,TAU,LMT,LEV,
```

```
      1LPSS)
C  PREPARE TH FOR INVERSION (GAUSSIAN ELIMINATION)
      CALL ZGE(TH,IPL,LMNI,LMNI,1.E-6)
C
C
C  START LOOP OVER INCIDENT BEAMS
  160 DO 430 JGP = 1, N
C  IF ONLY DIFFRACTION FROM (00) BEAM REQUIRED, SKIP FURTHER LOOPING
      IF (NOPT.EQ.3.AND.JGP.EQ.2) GO TO 440
      JGPS=JGP+NS
      GP(1) = PQ(1,JGP + NA)
      GP(2) = PQ(2,JGP + NA)
      LAB = SYM(LAY,JGP + NA)
C  CHOOSE APPROPRIATE SYMMETRY NORMALIZATION FACTOR FAC
      IF (LAB-10) 165,220,180
  165 IF (LAB-2)  190, 170, 170
  170 IF (LAB-7)  200, 210, 210
  180 IF (LAB-12) 221,185,185
  185 IF (LAB-15) 222,223,223
  190 FAC = 1.0
      GO TO 230
  200 FAC = ASQ
      GO TO 230
  210 FAC = 2.0
      GO TO 230
  220 FAC = 2.0 * ASQ
      GO TO 230
  221 FAC=ACU
      GO TO 230
  222 FAC=ASQ*ACU
      GO TO 230
  223 FAC=2.0*ACU
  230 CONTINUE
C
C
C  GENERATE QUANTITIES TAUG,TAUGM INTO WHICH INVERSE OF TH WILL BE
C  MULTIPLIED
      CALL TAUY(TAUG,TAUGM,LMT,TAU,LMT,LEV,CYLM,NT,LMMAX,LT,TH,
     1LMNI,LMNI,NTAU,LOD,LEE,LOE,JGP,0,RG,NLAY,N,INV)
  280 CONTINUE
C
C
C  FIRST CONSIDER INCIDENCE TOWARDS +X
      INC=1
  285 CONTINUE
C  INCLUDE APPROPRIATE PLANE-WAVE PROPAGATING FACTORS
      DO 315 I=1,NLAY
      IN=(I-1)*LMMAX
      LP=(LPSS(I)-1)*LMMAX
      IF (INC.EQ.-1) GO TO 290
      ST=RG(1,I,JGP)
      GO TO 295
  290 ST=RU/RG(2,I,JGP)
  295 CONTINUE
      IF (INC.EQ.-1) GO TO 305
      DO 300 K=1,LMMAX
  300 TS(IN+K)=TAUG(LP+K)*ST
      GO TO 315
  305 DO 310 K=1,LMMAX
  310 TS(IN+K)=TAUGM(LP+K)*ST
  315 CONTINUE
C  DO INVERSION WITH MULTIPLICATION
      CALL ZSU(TH,IPL,TS,LMNI,LMNI,1.E-6)
C  INCLUDE FURTHER PLANE-WAVE PROPAGATING FACTORS
      DO 335 I=1,NLAY
      IN=(I-1)*LMMAX
      IF (INC.EQ.-1) GO TO 320
```

```
      ST=RU/RG(1,I,JGP)
      GO TO 325
320   ST=RG(2,I,JGP)
325   CONTINUE
      DO 330 K=1,LMMAX
330   TS(IN+K)=TS(IN+K)*ST
335   CONTINUE
C
C
C  START LOOP OVER SCATTERED BEAMS
  400 DO 410 JG = 1, N
C  COMPLETE COMPUTATION IN MFOLT, INCLUDING REGISTRY SHIFTS AND
C  SYMMETRIZATION
      CALL  MFOLT (JG,NA,E,VPI,AK2,AK3,CYLM,N,LMMAX,PQ,SYM,NT,
     1TS,LMN,RG,NLAY,JGP,XEV,LXM,ID,YB,XB,INC,POSS)
      JGS=JG+NS
      AM = AMULT(JG) * FAC
      IF (INC.EQ.-1) GO TO 500
      IF (ID-2) 409,408,407
C  PUT FINAL MATRIX ELEMENTS IN PROPER LOCATION
  407 RC1(JGS,JGP)=YB(3)*AM
      TC1(JGS,JGP)=XB(3)*AM
  408 RB1(JGS,JGP)=YB(2)*AM
      TB1(JGS,JGP)=XB(2)*AM
  409 RA1(JGS,JGP) = YB(1)* AM
      TA1(JGS,JGP) = XB(1)* AM
      GO TO 410
  500 IF (ID-2) 509,508,507
  507 RC2(JGS,JGP)=YB(3)*AM
      TC2(JGS,JGP)=XB(3)*AM
  508 RB2(JGS,JGP)=YB(2)*AM
      TB2(JGS,JGP)=XB(2)*AM
  509 RA2(JGS,JGP)=YB(1)*AM
      TA2(JGS,JGP)=XB(1)*AM
  410 CONTINUE
      IF (INC.EQ.-1) GO TO 520
      IF (ID-2) 431,429,428
C  ADD UNSCATTERED PLANE WAVE TO TRANSMISSION
  428 TC1(JGPS,JGP)=TC1(JGPS,JGP)+RG(3,1,JGP)
  429 TB1(JGPS,JGP)=TB1(JGPS,JGP)+RG(3,1,JGP)
  431 TA1(JGPS,JGP)=TA1(JGPS,JGP)+RG(3,1,JGP)
      IF (NOPT-2) 510,430,510
  510 INC=-1
      GO TO 285
  520 IF (ID-2) 530,529,528
  528 TC2(JGPS,JGP)=TC2(JGPS,JGP)+RG(3,1,JGP)
  529 TB2(JGPS,JGP)=TB2(JGPS,JGP)+RG(3,1,JGP)
  530 TA2(JGPS,JGP)=TA2(JGPS,JGP)+RG(3,1,JGP)
  430 CONTINUE
  440 RETURN
      END
C----------------------------------------------------------------------
C  SUBROUTINE PRPGAT PERFORMS ONE PROPAGATION THROUGH ONE LAYER FOR RFS
C  SUBROUTINES LIKE RFS02 AND RFS03. A LAYER ASYMMETRICAL IN +-X CAN BE
C  HANDLED, AS WELL AS INDEPENDENT BEAM SETS.
C  FOR QUANTITIES NOT EXPLAINED BELOW SEE SUBROUTINES RFS03, ETC.
C  RA,TA,RB,TB= INPUT DIFFRACTION MATRICES FOR CURRENT LAYER (R FOR
C   REFLECTION, T FOR TRANSMISSION, A,B FOR INCIDENCE TOWARDS +-X,
C   RESP.).
C  N= TOTAL NO. OF BEAMS AT CURRENT ENERGY.
C  NM= LARGEST NO. OF BEAMS IN ANY CURRENT BEAM SET (BUT =N FOR
C   OVERLAYER).
C  AW= WORKING SPACE.
C  I= INDEX OF INTERLAYER SPACING TO WHICH CURRENT CALL TO PRPGAT WILL
C   LEAD.
C  L1,L2= CURRENT CHOICE OF PLANE-WAVE PROPAGATORS, REFERRING TO
C   SECOND INDEX OF MATRIX PK. L1 IS TO DESCRIBE PROPAGATION IN
```

```
C       DIRECTION TOWARDS INTERLAYER SPACING I, L2 IN OPPOSITE DIRECTION.
C       CRIT= CRITERION FOR PENETRATION CONVERGENCE (SET IN CALLING ROUTINE)
C       IR= OUTPUT FLAG FOR PENETRATION CONVERGENCE.
C       IA=+-1: INDICATES PROPAGATION TOWARDS +-X.
C       BNORM= OUTPUT MEASURE OF CURRENT WAVEFIELD AMPLITUDE AT CURRENT
C       LAYER.
        SUBROUTINE PRPGAT(RA,TA,RB,TB,N,NM,ANEW,ND,NSL,NL,AW,I,PK,L1,L2,
       1CRIT,IR,IA,BNORM)
        COMPLEX RA(N,NM),TA(N,NM),ANEW(N,ND),AW(N,2),PK(N,20),CZ
        COMPLEX RB(N,NM),TB(N,NM)
        DIMENSION NSL(NL)
        CZ=CMPLX(0.0,0.0)
        BNORM=0.0
        NA=0
        DO 4 NN=1,NL
        NB=NSL(NN)
        DO 1 K=1,NB
        KNA=K+NA
C    PROPAGATE WAVEFIELD TO CURRENT LAYER FROM THE TWO NEAREST LAYERS
        AW(K,1)=PK(KNA,L1)*ANEW(KNA,I-IA)
   1    AW(K,2)=PK(KNA,L2)*ANEW(KNA,I)
        DO 10 J=1,NB
        JNA=J+NA
        ANEW(JNA,I)=CZ
C    SELECT FORMULA ACCORDING TO PROPAGATION DIRECTION
        IF (IA.EQ.-1) GO TO 8
        DO 2 K=1,NB
C    TRANSMIT AND REFLECT AND ADD WAVES
   2    ANEW(JNA,I)=ANEW(JNA,I)+TA(JNA,K)*AW(K,1)+RB(JNA,K)*AW(K,2)
        GO TO 3
   8    DO 9 K=1,NB
   9    ANEW(JNA,I)=ANEW(JNA,I)+TB(JNA,K)*AW(K,1)+RA(JNA,K)*AW(K,2)
C    GET MEASURE OF NEW WAVEFIELD MAGNITUDE
   3    BNORM=BNORM+REAL(ANEW(JNA,I))**2+AIMAG(ANEW(JNA,I))**2
  10    CONTINUE
   4    NA=NA+NB
C    CHECK AGAINST PENETRATION LIMIT, EXCEPT WHEN EMERGING (IN WHICH CASE
C    IR=1)
        IF ((I.LT.ND-1).OR.(IR.EQ.1)) GO TO 7
   5    WRITE(6,6)
   6    FORMAT(23H0***THIS ORDER TOO DEEP)
        IR=1
   7    CONTINUE
C    CHANGE I TO INDICATE TO WHICH NEW INTERLAYER SPACING THE NEXT CALL
C    TO PRPGAT WILL HAVE TO LEAD
        I=I+IA
C    CHECK ON PENETRATION CONVERGENCE
        IF (BNORM.LE.CRIT) IR=1
        RETURN
        END
C-------------------------------------------------------------------------
C    SUBROUTINE PSTEMP INCORPORATES THE THERMAL VIBRATION EFFECTS IN THE
C    PHASE SHIFTS, THROUGH A DEBYE-WALLER FACTOR. ISOTROPIC VIBRATION
C    AMPLITUDES ARE ASSUMED.
C       PPP= CLEBSCH-GORDON COEFFICIENTS FROM SUBROUTINE CPPP.
C       N3= NO. OF INPUT PHASE SHIFTS.
C       N2= DESIRED NO. OF OUTPUT TEMPERATURE-DEPENDENT PHASE SHIFTS.
C       N1= N2+N3-1.
C       DRO= 4TH POWER OF RMS ZERO-TEMPERATURE VIBRATION AMPLITUDES.
C       DR= ISOTROPIC RMS VIBRATION AMPLITUDE AT REFERENCE TEMPERATURE TO.
C       TO= ARBITRARY REFERENCE TEMPERATURE FOR DR.
C       TEMP= ACTUAL TEMPERATURE.
C       E = CURRENT ENERGY (REAL NUMBER).
C       PHS= INPUT PHASE SHIFTS.
C       DEL= OUTPUT (COMPLEX) PHASE SHIFTS.
C    DIMENSIONS ARE SET FOR N3.LE.8, N2.LE.13, N1.LE.20.
        SUBROUTINE  PSTEMP (PPP, N1, N2, N3, DRO, DR, TO, TEMP, E, PHS,
```

```
      1DEL)
       COMPLEX  BJ, DEL, SUM, CTAB
       COMPLEX  Z, FF, FN, FN1, FN2, CI, CS, CC, CL
       COMPLEX CSIN,CCOS,CEXP,CLOG
       DIMENSION  PPP(N1,N2,N3), PHS(N3), BJ(20), DEL(N2), SUM(13)
       DIMENSION  CTAB(8)
       PI = 3.14159265
       CI = CMPLX(0.0,1.0)
       DO 170 J = 1, N2
  170 DEL(J) = (0.0,0.0)
       ALFA = DR * DR * TEMP/TO
       ALFA = 0.166667 * SQRT(ALFA * ALFA + DRO)
       FALFE =   - 4.0 * ALFA * E
       IF (ABS(FALFE)-1.0E-3)  180, 200, 200
  180 DO 190 J = 1, N3
  190 DEL(J) = CMPLX(PHS(J),0.0)
       GO TO 360
C
COMMENT BJ(N1) IS LOADED WITH SPHERICAL BESSEL FUNCTIONS OF
C      THE FIRST KIND; ARGUMENT Z
  200 Z = FALFE * CI
       LLL = N1 - 1
       CS = CSIN(Z)
       CC = CCOS(Z)
       BJ(1) = CS/Z
       LP = INT(3.5 * CABS(Z))
       IF (LLL-LP)  240, 240, 210
  210 FF = CMPLX(1.0,0.0)
       CL = FF
       LPP = LP + 1
       DO 220 MM = 1, LPP
       CL = CL + (2.0,0.0)
  220 FF = FF * Z/CL
       FL = FLOAT(LP)
       DO 230 J = LPP, LLL
       FL = FL + 1.0
       B = 1.0/(12.0 * FL + 42.0)
       FN = 1.0 - B * Z * Z
       B = 1.0/(8.0 * FL + 20.0)
       FN = 1.0 - B * Z * Z * FN
       B = 1.0/(4.0 * FL + 6.0)
       FN = 1.0 - B * Z * Z * FN
       BJ(J + 1) = FF * FN
       CL = CL + (2.0,0.0)
  230 FF = FF * Z/CL
       LLL = LP
       IF (LP)  240, 270, 240
  240 BJ(2) = (BJ(1) - CC)/Z
       FN1 = BJ(2)
       FN2 = BJ(1)
       AM = 3.0
       IF (LP-1)  250, 270, 250
  250 ILL = LLL + 1
       DO 260 IA = 3, ILL
       FN = AM * FN1/Z - FN2
       FN2 = FN1
       FN1 = FN
       AM = AM + 2.0
  260 BJ(IA) = FN
  270 CS = CMPLX(1.0,0.0)
       FL = 1.0
       DO 280 I = 1, N1
       BJ(I) = EXP(FALFE) * FL * CS * BJ(I)
       FL = FL + 2.0
  280 CS = CS * CI
C
       FL = 1.0
```

```
      DO 290 I = 1, N3
      CTAB(I) = (CEXP(2.0 * PHS(I) * CI) - (1.0,0.0)) * FL
  290 FL = FL + 2.0
C
      ITEST = 0
      LLLMAX = N2
      FL = 1.0
      DO 350 LLL = 1, N2
      SUM(LLL) = CMPLX(0.0,0.0)
      DO 300 L = 1, N3
      LLMIN = IABS(L - LLL) + 1
      LLMAX = L + LLL - 1
      DO 300 LL = LLMIN, LLMAX
  300 SUM(LLL) = SUM(LLL) + PPP(LL,LLL,L) * CTAB(L) * BJ(LL)
      DEL(LLL) = - CI * CLOG(SUM(LLL) + (1.0,0.0))/(2.0,0.0)
      ABSDEL = CABS(DEL(LLL))
      IL = LLL - 1
      IF (ABSDEL-1.0E-2)  320, 310, 310
  310 ITEST = 0
      GO TO 350
  320 IF (ITEST-1)  340, 330, 340
  330 LLLMAX = LLL
      GO TO 360
  340 ITEST = 1
  350 FL = FL + 2.0
  360 RETURN
      END
C-------------------------------------------------------------------
C  SUBROUTINE READCL READS IN DATA RELEVANT TO A COMPOSITE LAYER.
C  INPUT FROM MAIN PROGRAM:
C    NLAY=NO. OF SUBPLANES IN THE COMPOSITE LAYER
C    LMMAX= NO. OF SPHERICAL HARMONICS USED: (LMAX+1)**2
C    FOR OTHER QUANTITIES SEE BELOW
      SUBROUTINE READCL(NLAY,NLAY2,FPOS,VPOS,NVAR,NTAU,LPS,NINV,INV,
     1LMMAX,LMG,LMN,LM2N,LMT,LTAUG,LMNI,LMNI2)
      DIMENSION FPOS(NLAY,3),VPOS(20,3),LPS(12),INV(12)
  160 FORMAT(3F7.4)
  200 FORMAT(20I3)
  215 FORMAT(26H COMPOSITE LAYER VECTOR : ,3F7.4)
  216 FORMAT(8H NTAU = ,1I3)
  217 FORMAT(43H PHASE SHIFT ASSIGNMENT IN COMPOSITE LAYER ,20I3)
  218 FORMAT(1I3,32H SUBPLANES IN BEEBY INVERSION : ,20I3)
  220 FORMAT(39H VARYING POSITION IN COMPOSITE LAYER : ,3F7.4)
C  NLAY2= NO. OF DISTINCT PAIRS OF SUBPLANES
      NLAY2=NLAY*(NLAY-1)/2
      DO 480 I=1,NLAY
C  FPOS: 3-D VECTORS POINTING FROM AN ARBITRARY REFERENCE POINT TO ONE
C  ATOM IN EACH SUBPLANE (THE REFERENCE POINT WILL BE USED TO PRODUCE
C  A PAIR OF DIFFERENT REFERENCE POINTS IN MPERTI AND MTINV)(ANGSTROM)
      READ(5,160)(FPOS(I,J),J=1,3)
      WRITE(6,215)(FPOS(I,J),J=1,3)
      DO 480 J=1,3
  480 FPOS(I,J)=FPOS(I,J)/0.529
C  NVAR= NO. OF VECTORS VPOS TO BE READ IN
      READ(5,200)NVAR
      DO 490 I=1,NVAR
C  VPOS: FURTHER VECTORS ANALOGOUS TO FPOS GIVING DESIRED VARIATIONS
C  OF THE INTERNAL GEOMETRY OF THE COMPOSITE LAYER (VPOS CAN BE USED
C  IN VARIOUS WAYS IN THE MAIN PROGRAM)(ANGSTROM)
      READ(5,160)(VPOS(I,J),J=1,3)
      WRITE(6,220)(VPOS(I,J),J=1,3)
      DO 490 J=1,3
  490 VPOS(I,J)=VPOS(I,J)/0.529
C  NTAU= NO. OF CHEMICAL ELEMENTS IN THE COMPOSITE LAYER
      READ(5,200)NTAU
      WRITE(6,216)NTAU
C  LPS GIVES CHEMICAL IDENTITY FOR EACH SUBPLANE, REFERRING TO ORDER
```

```
C   OF INPUT OF PHASE SHIFTS IN SUBROUTINE READIN. LPS(I)=K MEANS I-TH
C   SUBPLANE HAS K-TH SET OF PHASE SHIFTS IN INPUT SEQUENCE
        READ(5,200)(LPS(I),I=1,NLAY)
        WRITE(6,217)(LPS(I),I=1,NLAY)
C   NINV= NO. OF SUBPLANES TO BE TREATED BY BEEBY INVERSION IN MPERTI
C   OR MTINV
        READ(5,200)NINV
C   INV GIVES FOR EACH SUBPLANE WHETHER IT IS TO BE TREATED BY BEEBY
C   INVERSION OR NOT (NOT NEEDED FOR MTINV). INV(I)=(0),1 MEANS (NO)
C   INVERSION REQUIRED (NOTE RESTRICTIONS DESCRIBED IN BOOK)
        READ(5,200)(INV(I),I=1,NLAY)
        WRITE(6,218)NINV,(INV(I),I=1,NLAY)
C   LMG= DIMENSION OF GH
        LMG=NLAY2*LMMAX*2
C   LMN= DIMENSION OF TS
        LMN=NLAY*LMMAX
C   LM2N= DIMENSION OF TG
        LM2N=2*LMN
C   LMT= DIMENSION OF TAU
        LMT=LMMAX*NTAU
C   LTAUG= DIMENSION OF TAUG AND TAUGM IN MPERTI
        LTAUG=(NTAU+NINV)*LMMAX
C   LMNI= DIMENSION OF TH
        LMNI=NINV*LMMAX
C   LMNI2= SECOND DIMENSION OF TH IN MTINV (IN MPERTI LMNI2=LMNI)
        LMNI2=2*LMNI
        RETURN
        END
C------------------------------------------------------------------------
C   SUBROUTINE READIN READS IN MOST OF THE INPUT FOR THE LEED PROGRAMS.
C   INPUT FROM MAIN PROGRAM:
C   NL=NL1*NL2= NO. OF SUBLATTICES DUE TO SUPERLATTICE
C   KNBS= NO. OF BEAM SETS TO BE READ IN
C   KNT= TOTAL NO. OF BEAMS TO BE READ IN
C   NPUN= NO. OF BEAMS FOR WHICH INTENSITIES ARE TO BE PUNCHED OUT
C   NPSI= NO. OF ENERGIES AT WHICH PHASE SHIFTS WILL BE READ IN
C   FOR FURTHER QUANTITIES SEE BELOW
        SUBROUTINE READIN(TVA,RAR1,RAR2,ASA,TVB,ASB,STEP,NSTEP,IDEG,
       1NL,V,VL,JJS,KNBS,KNB,KNT,SPQF,KSYM,SPQ,TST,TSTS,NPUN,NPU,THETA,FI,
       2LMMAX,NPSI,ES,PHSS,L1)
        DIMENSION ARA1(2),ARA2(2),RAR1(2),RAR2(2),SS1(2),SS2(2),SS3(2),
       1SS4(2),S1(2),S2(2),S3(2),S4(2),ASA(3),ARB1(2),ARB2(2),RBR1(2),
       2RBR2(2),SO1(2),SO2(2),SO3(2),ASB(3)
        DIMENSION V(NL,2),JJS(NL,IDEG),KNB(KNBS),SPQF(2,KNT),KSYM(2,KNT),
       1SPQ(2,KNT),NPU(NPUN),ES(NPSI),PHSS(NPSI,30)
        COMPLEX VL(NL,2)
        COMMON /SL/ARA1,ARA2,ARB1,ARB2,RBR1,RBR2,NL1,NL2
        COMMON /MS/LMAX,EPS,LITER,SO1,SO2,SO3,SS1,SS2,SS3,SS4
        COMMON /ADS/ASL,FR,ASE,VPIS,VPIO,VO,VV
        COMMON /TEMP/IT1,IT2,IT3,TI,TF,DT,TO,DRPER1,DRPAR1,DRO1,DRPER2,
       1DRPAR2,DRO2,DRPER3,DRPAR3,DRO3
  130 FORMAT(5F9.4)
  140 FORMAT(25HOPARAMETERS FOR INTERIOR:)
  160 FORMAT(3F7.4)
  161 FORMAT(3F7.2)
  170 FORMAT(10H SURF VECS,2(5X,2F8.4))
  171 FORMAT(15X,2F8.4)
  180 FORMAT(4H ASA,3X,3F7.4)
  181 FORMAT(4H ASB,3X,3F7.4)
  190 FORMAT(3H SO,4X,2F7.4)
  200 FORMAT(26I3)
  210 FORMAT(6H FR = ,F7.4,7H ASE = ,F7.4,8H STEP = ,F7.4,9H NSTEP = ,
     11I3)
  220 FORMAT(3H SS,3X,2F7.4)
  230 FORMAT(24HOPARAMETERS FOR SURFACE:)
  250 FORMAT(1H0,9X,3HPQ1,4X,3HPQ2,8X,3HSYM)
  260 FORMAT(1H )
```

```
      270 FORMAT(2F7.4,4I3)
      280 FORMAT(1H ,1I4,F10.3,F7.3,4X,4I3)
      281 FORMAT(7H TST = ,1F7.4)
      283 FORMAT(I3,2F6.3,I3)
      284 FORMAT(17H PUNCHED BEAMS    ,26I3)
      285 FORMAT(13H THETA   FI = ,2F7.2)
      290 FORMAT(3H VO,4X,F7.2,2HVV,4X,F7.2)
      295 FORMAT(7H EPS = ,1F7.4,10H  LITER = ,1I3)
      305 FORMAT(7H IT1 = ,1I3,7H  IT2 = ,1I3,7H IT3 = ,1I3)
      325 FORMAT(8H THDB = ,1F9.4,6H AM = ,1F9.4,8H FPER = ,1F9.4,8H FPAR =
     1,1F9.4,7H DRO = ,1F9.4)
      340 FORMAT(10F7.4)
      350 FORMAT(1H0,5X,12HPHASE SHIFTS)
      355 FORMAT(8H LMAX = ,1I3)
      360 FORMAT(1H0,4HE = ,1F7.4,13H  1ST ELEMENT,3X,10F8.4)
      361 FORMAT(1H ,11X,13H  2ND ELEMENT,3X,10F8.4)
      362 FORMAT(1H ,11X,13H  3RD ELEMENT,3X,10F8.4)
          PI = 3.14159265
          WRITE (6,140)
C   ARA1 AND ARA2 ARE TWO 2-D BASIS VECTORS OF THE SUBSTRATE LAYER
C   LATTICE. THEY SHOULD BE EXPRESSED IN TERMS OF THE PLANAR CARTESIAN
C   Y- AND Z-AXES (X-AXIS IS PERPENDICULAR TO SURFACE)(ANGSTROM)
          READ(5,160)(ARA1(I),I=1,2)
          READ(5,160)(ARA2(I),I=1,2)
          WRITE(6,170)(ARA1(I),I=1,2)
          WRITE(6,171)(ARA2(I),I=1,2)
          DO 460 I=1,2
          ARA1(I)=ARA1(I)/0.529
      460 ARA2(I)=ARA2(I)/0.529
          TVA=ABS(ARA1(1)*ARA2(2)-ARA1(2)*ARA2(1))
          ATV=2.0*PI/TVA
C   RAR1 AND RAR2 ARE THE RECIPROCAL-LATTICE BASIS VECTORS CORRESPONDING
C   TO ARA1 AND ARA2
          RAR1(1)=ARA2(2)*ATV
          RAR1(2)=-ARA2(1)*ATV
          RAR2(1)=-ARA1(2)*ATV
          RAR2(2)=ARA1(1)*ATV
C   SS1,SS2,SS3,SS4 ARE 4 POSSIBLE DIFFERENT REGISTRIES FOR SUBSTRATE
C   LAYERS, TO BE GIVEN IN TERMS OF ARA1,ARA2
          READ(5,160)(SS1(I),I=1,2)
          READ(5,160)(SS2(I),I=1,2)
          READ(5,160)(SS3(I),I=1,2)
          READ(5,160)(SS4(I),I=1,2)
          WRITE(6,220)(SS1(I),I=1,2)
          WRITE(6,220)(SS2(I),I=1,2)
          WRITE(6,220)(SS3(I),I=1,2)
          WRITE(6,220)(SS4(I),I=1,2)
          DO 461 I=1,2
          S1(I)=SS1(1)*ARA1(I)+SS1(2)*ARA2(I)
          S2(I)=SS2(1)*ARA1(I)+SS2(2)*ARA2(I)
          S3(I)=SS3(1)*ARA1(I)+SS3(2)*ARA2(I)
      461 S4(I)=SS4(1)*ARA1(I)+SS4(2)*ARA2(I)
          DO 463 I=1,2
          SS1(I)=S1(I)
          SS2(I)=S2(I)
          SS3(I)=S3(I)
      463 SS4(I)=S4(I)
C   ASA IS THE SUBSTRATE INTERLAYER VECTOR (LINKING REFERENCE POINTS
C   ON SUCCESSIVE LAYERS)(ANGSTROM)
          READ(5,160)(ASA(I),I=1,3)
          WRITE(6,180)(ASA(I),I=1,3)
          DO 466 I=1,3
      466 ASA(I)=ASA(I)/0.529
          WRITE (6,230)
C   ARB1,ARB2,RBR1,RBR2,SO1,SO2,SO3 ARE EQUIVALENT TO ARA1,ARA2,RAR1,
C   RAR2,SS1,SS2,SS3 BUT FOR AN OVERLAYER (NOTE THAT SO1,SO2,SO3 ARE
C   TO BE EXPRESSED IN TERMS OF THE SUBSTRATE VECTORS ARA1,ARA2)
```

```
      READ(5,160)(ARB1(I),I=1,2)
      READ(5,160)(ARB2(I),I=1,2)
      WRITE(6,170)(ARB1(I),I=1,2)
      WRITE(6,171)(ARB2(I),I=1,2)
      DO 467 I=1,2
      ARB1(I)=ARB1(I)/0.529
  467 ARB2(I)=ARB2(I)/0.529
      TVB=ABS(ARB1(1)*ARB2(2)-ARB1(2)*ARB2(1))
      ATV=2.0*PI/TVB
      RBR1(1) = ARB2(2) * ATV
      RBR1(2) =  - ARB2(1) * ATV
      RBR2(1) =  - ARB1(2) * ATV
      RBR2(2) = ARB1(1) * ATV
      READ(5,160)(SO1(I),I=1,2)
      READ(5,160)(SO2(I),I=1,2)
      READ(5,160)(SO3(I),I=1,2)
      WRITE(6,190)(SO1(I),I=1,2)
      WRITE(6,190)(SO2(I),I=1,2)
      WRITE(6,190)(SO3(I),I=1,2)
      DO 468 I=1,2
      S1(I)=SO1(1)*ARA1(I)+SO1(2)*ARA2(I)
      S2(I)=SO2(1)*ARA1(I)+SO2(2)*ARA2(I)
  468 S3(I)=SO3(1)*ARA1(I)+SO3(2)*ARA2(I)
      DO 470 I=1,2
      SO1(I)=S1(I)
      SO2(I)=S2(I)
  470 SO3(I)=S3(I)
C  ASB IS A VECTOR POINTING FROM THE TOP LAYER TO THE TOPMOST SUBSTRATE
C  LAYER (LINKING REFERENCE POINTS ON THE TWO LAYERS)(ANGSTROM)
      READ(5,160)(ASB(I),I=1,3)
      WRITE(6,181)(ASB(I),I=1,3)
      DO 472 I=1,3
  472 ASB(I)=ASB(I)/0.529
C  FR IS THE FRACTION OF THE TOP LAYER TO SUBSTRATE SPACING THAT IS
C  ASSIGNED TO THE SUBSTRATE (OVER WHICH THE SUBSTRATE MUFFIN-TIN
C  CONSTANT AND DAMPING APPLY).
C  ASE IS THE SPACING BETWEEN THE SURFACE (WHERE MUFFIN-TIN CONSTANT
C  AND DAMPING SET IN) AND THE TOP-LAYER NUCLEI (ANGSTROM).
C  STEP IS AN INCREMENTAL DISTANCE FOR VARYING E.G. THE TOP-LAYER
C  SPACING (ANGSTROM)
      READ (5,160) FR, ASE, STEP
C  NSTEP IS A LOOP LIMIT FOR GEOMETRY VARIATIONS
      READ (5,200)  NSTEP
      WRITE (6,210)  FR, ASE, STEP, NSTEP
      ASE=ASE/0.529
      STEP=STEP/0.529
C  SLIND COMPARES SUBSTRATE LATTICE AND SUPERLATTICE FOR THE BENEFIT
C  OF FMAT
      CALL    SLIND (V, VL, JJS, NL, IDEG, 1.0E-5)
      WRITE (6,250)
      KNTT= 0
C  READ IN KNBS SETS OF BFAMS
      DO 580 J = 1, KNBS
      WRITE (6,260)
C  READ IN KNB(J) BEAMS IN J-TH SET
      READ (5,200)  KNB(J)
      N = KNB(J)
      DO 570 K = 1, N
      KK = K + KNTT
C  READ IN A BEAM (AS A RECIPROCAL LATTICE VECTOR IN TERMS OF RAR1,RAR2
C  I.E. IN THE FORM (0,0),(1,0),(0,1),ETC.) AND ITS SYMMETRY PROPERTY
C  (BOTH FOR THE SUPERLATTICE REGION AND THE SUBSTRATE REGION)
      READ (5,270) SPQF(1,KK), SPQF(2,KK), (KSYM(I,KK),I=1,2)
      DO 560 I = 1, 2
  560 SPQ(I,KK)=SPQF(1,KK)*RAR1(I)+SPQF(2,KK)*RAR2(I)
      WRITE (6,280) KK, SPQF(1,KK), SPQF(2,KK), (KSYM(I,KK),I=1,2)
  570 CONTINUE
```

```
      580 KNTT= KNTT+ KNB(J)
C   TST GIVES THE CRITERION FOR THE SELECTION OF BEAMS AT ANY ENERGY:
C   TO BE SELECTED A BEAM MAY NOT DECAY TO LESS THAN A FRACTION TST OF
C   ITS INITIAL VALUE ON TRAVELING FROM ANY LAYER TO THE NEXT
          READ(5,160)TST
          WRITE(6,281)TST
          TSTS=TST
          TST=ALOG(TST)/(AMIN1(ASB(1),ASA(1)))
          TST=TST*TST
C   PUNCH A CARD WITH NPUN
          WRITE(7,283)NPUN
C   NPU GIVES THOSE BEAMS FOR WHICH INTENSITIES ARE TO BE PUNCHED OUT.
C   NPU(K)=J MEANS PUNCH OUTPUT IS DESIRED FOR THE J-TH BEAM IN THE
C   INPUT LIST
          READ(5,200)(NPU(K),K=1,NPUN)
          DO 600 K=1,NPUN
          N=NPU(K)
C   PUNCH A CARD FOR EACH BEAM FOR WHICH PUNCH OUTPUT IS DESIRED (THIS
C   PASSES THE BEAM IDENTITY (0,0),(1,0),(0,1),ETC. TO THE DECK OF CARDS)
      600 WRITE(7,283)N,(SPQF(I,N),I=1,2),KSYM(1,N)
          WRITE(6,284)(NPU(K),K=1,NPUN)
C   (THETA,FI) IS THE DIRECTION OF INCIDENCE. THETA=0 AT NORMAL
C   INCIDENCE. FI=0 FOR INCIDENCE ALONG Y-AXIS IN (YZ) SURFACE PLANE
C   (X-AXIS IS PERPENDICULAR TO SURFACE)(DEGREE)
          READ (5,161)   THETA, FI
          WRITE(6,285) THETA,FI
          THETA=THETA*PI/180.0
          FI=FI*PI/180.0
C   VO IS MUFFIN-TIN CONSTANT OF OVERLAYER WITH RESPECT TO THAT OF
C   SUBSTRATE (EV). VV IS VACUUM LEVEL WITH RESPECT TO SUBSTRATE MUFFIN-
C   TIN CONSTANT (EV)
          READ (5,161)   VO, VV
          WRITE (6,290)   VO, VV
          VO=VO/27.18
          VV=VV/27.18
C   EPS,LITER ARE CONVERGENCE CRITERIA FOR LAYER DOUBLING
          READ (5,160)   EPS
          READ (5,200)   LITER
          WRITE(6,295)EPS,LITER
C   IT1,IT2,IT3=(0),1 MEANS: DO (NOT) INCLUDE THERMAL VIBRATION EFFECTS,
C   FOR UP TO 3 DIFFERENT TYPES OF ATOMS
          READ(5,200)IT1,IT2,IT3
          WRITE(6,305)IT1,IT2,IT3
C   THDB= DEBYE TEMPERATURE (KELVIN).
C   AM= ATOMIC MASS (IN NUCLEON UNITS).
C   FPER,FPAR ARE ENHANCEMENT FACTORS FOR THE MEAN SQUARE VIBRATION
C   AMPLITUDES PERPENDICULAR AND PARALLEL TO THE SURFACE, RESP.
C   DRO IS THE RMS ZERO-TEMPERATURE VIBRATION AMPLITUDE (ANGSTROM).
C   1,2,3 REFER TO UP TO 3 DIFFERENT TYPES OF ATOMS.
C   NOTE: NEVER INPUT ZERO VALUES FOR THDB AND AM
          READ(5,130)THDB1,AM1,FPER1,FPAR1,DRO1
          READ(5,130)THDB2,AM2,FPER2,FPAR2,DRO2
          READ(5,130)THDB3,AM3,FPER3,FPAR3,DRO3
          WRITE(6,325)THDB1,AM1,FPER1,FPAR1,DRO1
          WRITE(6,325)THDB2,AM2,FPER2,FPAR2,DRO2
          WRITE(6,325)THDB3,AM3,FPER3,FPAR3,DRO3
C   (TI,TF) IS A TEMPERATURE RANGE AND DT A TEMPERATURE INCREMENT FOR
C   A LOOP IN THE MAIN PROGRAM. BLANKS FOR TF AND DT WILL PRODUCE A
C   COMPUTATION AT ONLY TI (KELVIN)
          READ (5,130)   TI, TF, DT
          TO=TI
          IF (TO.LT.0.001) TO=0.001
C   DRPER AND DRPAR ARE RMS VIBRATION AMPLITUDES PERPENDICULAR AND
C   PARALLEL TO THE SURFACE, RESP., EVALUATED AT REFERENCE TEMPERATURE
C   TO=MAX(TI,0.001 K)
          DR2=1.546E3*TO/(AM1*THDB1*THDB1)
          DRPER1=SQRT(FPER1*DR2)
```

```
      DRPAR1=SQRT(FPAR1*DR2)
      DRO1=(DRO1/0.529)**4
      DR2=1.546E3*TO/(AM2*THDB2*THDB2)
      DRPER2=SQRT(FPER2*DR2)
      DRPAR2=SQRT(FPAR2*DR2)
      DRO2=(DRO2/0.529)**4
      DR2=1.546E3*TO/(AM3*THDB3*THDB3)
      DRPER3=SQRT(FPER3*DR2)
      DRPAR3=SQRT(FPAR3*DR2)
      DRO3=(DRO3/0.529)**4
C   LMAX= HIGHEST L-VALUE TO BE CONSIDERED
      READ(5,200)LMAX
      WRITE(6,355)LMAX
C   LMMAX= NO. OF SPHERICAL HARMONICS TO BE CONSIDERED
      LMMAX=(LMAX+1)**2
C   NEL= NO. OF CHEMICAL ELEMENTS FOR WHICH PHASE SHIFTS ARE TO BE READ
C   IN (.LE.3)
      READ(5,200)NEL
      L1=LMAX+1
      L2=L1+1
      L3=L1+L1
      L4=L3+1
      L5=L3+L1
      DO 660 I = 1, NPSI
C   ES= ENERGIES (HARTREES) AT WHICH PHASE SHIFTS ARE INPUT. LINEAR
C   INTERPOLATION OF THE PHASE SHIFTS WILL OCCUR FOR ACTUAL ENERGIES
C   FALLING BETWEEN THE VALUES OF ES (AND LINEAR EXTRAPOLATION ABOVE
C   THE HIGHEST ES)
      READ (5,160)  ES(I)
C   PHSS STORES THE INPUT PHASE SHIFTS (RADIAN)
      READ(5,340)(PHSS(I,L),L=1,L1)
      IF (NEL-2) 660,650,650
  650 READ (5,340)  (PHSS(I,L),L=L2,L3)
      IF (NEL-2) 660,660,655
  655 READ(5,340)(PHSS(I,L),L=L4,L5)
  660 CONTINUE
      WRITE (6,350)
      DO 670 I = 1, NPSI
      WRITE(6,280)
      WRITE(6,360)(ES(I),(PHSS(I,L),L=1,L1))
      IF (NEL-2) 670,665,665
  665 WRITE(6,361)(PHSS(I,L),L=L2,L3)
      IF (NEL-2) 670,670,668
  668 WRITE(6,362)(PHSS(I,L),L=L4,L5)
  670 CONTINUE
      RETURN
      END
C---------------------------------------------------------------------
C   SUBROUTINE RFS02 HAS THE SAME PURPOSE AND STRUCTURE AS SUBROUTINE
C   RFS03, BUT IT APPLIES TO SUBSTRATES WITH A 1- OR 2-LAYER PERIODICITY
C   OF THE LAYER DIFFRACTION MATRICES, AND A 1- OR 2-LAYER PERIODICITY
C   OF THE INTERLAYER VECTORS (TWO ALTERNATING INTERLAYER VECTORS ARE
C   USED).
C   FOR SPECIFICATIONS NOT EXPLAINED BELOW SEE SUBROUTINE RFS03.
C   ORGANIZATIONAL DIAGRAM OF LAYER ARRANGEMENT:
C
C   LAYER NO.    I    INTERL.VECTOR    INTERL.PROPAG'S    LAYER DIFFR.MATR'S
C
C   0(SURF.)    ------------------------------------------
C                1        ASE               1,2
C   1(OVERLAY.)-----------------------------------------------ROP,TOP,ROM,TOM
C                2        ASB               3,4
C   2(SUBSTR.)  -------------------------------------------------RA1,TA1
C                3        AS1               5,6
C   3          --------------------------------------------------RA2,TA2
C                4        AS2               7,8
C   4          --------------------------------------------------RA1,TA1
```

```
C                    5              AS1                  5,6
C   5       -----------------------------------------  RA2,TA2
C                    6              AS2                  7,8
C   ETC.
      SUBROUTINE RFSO2(ROP,TOP,ROM,TOM,RA1,TA1,RA2,TA2,N,
     1NSL,NL,NM,WV,PQ,PK,AW,ANEW,ND,ASB,AS1,AS2)
      DIMENSION RA1(N,NM),TA1(N,NM),RA2(N,NM),TA2(N,NM)
      DIMENSION ROP(N,N),TOP(N,N),AS1(3),AS2(3)
      DIMENSION NSL(NL),WV(N),MSL(1),ASB(3)
      DIMENSION PQ(2,N),ANEW(N,ND),AW(N,2),AS(3),PK(N,8)
      COMPLEX RA1,TA1,RA2,TA2,WV,ANEW,AW,PK
      COMPLEX ROP,TOP,EL,EK,EKP,CZ,CI,ROM(N,N),TOM(N,N)
      COMPLEX CSQRT,CEXP
      COMMON /ADS/ ASL,FR,ASE,VPIS,VPIO,VO,VV
      COMMON /X4/ E,VPI,BK2,BK3,AS
      CRIT=0.003
      CZ=CMPLX(0.0,0.0)
      CI=CMPLX(0.0,1.0)
      DO 1 J=1,N
      AK2=BK2+PQ(1,J)
      AK3=BK3+PQ(2,J)
      A=2.0*E-AK2*AK2-AK3*AK3
      EK=CSQRT(CMPLX(A,-2.0*VPIS+0.000001))
      EL=CSQRT(CMPLX(A-2.0*VO,-2.0*VPIO+0.000001))
      EKP=ASB(1)*(FR*EK+(1.0-FR)*EL)
      PK(J,3)=CEXP(CI*(EKP+AK2*ASB(2)+AK3*ASB(3)))
      PK(J,4)=CEXP(CI*(EKP-AK2*ASB(2)-AK3*ASB(3)))
      PK(J,1)=CEXP(CI*EL*ASE)
      PK(J,2)=PK(J,1)
C   TWO SETS OF DIFFERENT INTERLAYER PROPAGATORS ARE PRODUCED FOR THE
C   SUBSTRATE, USING THE BULK INTERLAYER VECTORS AS1 AND AS2
      EKP=AK2*AS1(2)+AK3*AS1(3)
      PK(J,5)=CEXP(CI*(EK*AS1(1)+EKP))
      PK(J,6)=CEXP(CI*(EK*AS1(1)-EKP))
      EKP=AK2*AS2(2)+AK3*AS2(3)
      PK(J,7)=CEXP(CI*(EK*AS2(1)+EKP))
1     PK(J,8)=CEXP(CI*(EK*AS2(1)-EKP))
      DO 2 I=1,N
      WV(I)=CZ
      DO 2 J=1,ND
2     ANEW(I,J)=CZ
      ANEW(1,1)=CMPLX(1.0,0.0)
      MSL(1)=N
      IO=0
      ANORM1=1.0E-6
3     ANORM2=0.0
      IO=IO+1
      IR=0
      I=2
      CALL PRPGAT(ROP,TOP,ROM,TOM,N,N,ANEW,ND,MSL,1,AW,I,PK,1,4,
     1CRIT,IR,1,BNORM)
      L1=3
C   PROPAGATE THE WAVEFIELD THROUGH THE NEXT TWO LAYERS AND REPEAT UNTIL
C   CONVERGENCE IN PENETRATION
50    CALL PRPGAT(RA1,TA1,RA1,TA1,N,NM,ANEW,ND,NSL,NL,AW,I,PK,L1,6,
     1CRIT,IR,1,BNORM)
      IF (IR.EQ.1) GO TO 52
51    CALL PRPGAT(RA2,TA2,RA2,TA2,N,NM,ANEW,ND,NSL,NL,AW,I,PK,5,8,
     1CRIT,IR,1,BNORM)
      L1=7
      IF (IR) 50,50,52
52    CONTINUE
      I=I-1
      IMAX=I-1
      WRITE(6,20)IMAX,BNORM
20    FORMAT(8H IMAX = ,I2,5X,8HBNORM = ,E12.4)
      DO 21 IG=1,N
```

```
21     ANEW(IG,I+1)=CZ
       L3=7
       IF (I.EQ.1) GO TO 59
       II=MOD(I,2)+1
       GO TO (58,57),II
57     CALL PRPGAT(RA2,TA2,RA2,TA2,N,NM,ANEW,ND,NSL,NL,AW,I,PK,8,5,
      1CRIT,IR,-1,BNORM)
       IF (I.EQ.2) L3=3
58     CALL PRPGAT(RA1,TA1,RA1,TA1,N,NM,ANEW,ND,NSL,NL,AW,I,PK,6,L3,
      1CRIT,IR,-1,BNORM)
       IF (I-1) 59,59,57
59     CALL PRPGAT(ROP,TOP,ROM,TOM,N,N,ANEW,ND,MSL,1,AW,I,PK,4,1,
      1CRIT,IR,-1,BNORM)
       DO 29 IG=1,N
       EK=ANEW(IG,1)*PK(IG,2)
       AB=CABS(EK)
       ANORM2=ANORM2+AB*AB
       WV(IG)=WV(IG)+EK
29     ANEW(IG,1)=CZ
       ANORM1=ANORM1+ANORM2
       WRITE(6,32)ANORM1
32     FORMAT(9H ANORM = ,E12.4,/)
       IF (ANORM2/ANORM1-0.001) 39,39,36
36     IF (IO-6) 3,37,37
37     WRITE(6,38) IO
38     FORMAT(/,24H ***NO CONVERGENCE AFTER,
      11I4,11H ITERATIONS)
39     RETURN
       END
C---------------------------------------------------------------------
C  SUBROUTINE RFSO2V HAS THE SAME PURPOSE AS SUBROUTINE RFSO2, BUT
C  ADDITIONNALLY THE TOPMOST SUBSTRATE INTERLAYER SPACING (HERE CALLED
C  ASC) IS ALLOWED TO VARY.
C  NOTE: PK HAS A LARGER DIMENSION THAN IN RFSO2
C  ORGANIZATIONAL DIAGRAM OF LAYER ARRANGEMENT:
C
C  LAYER NO.   I   INTERL.VECTOR   INTERL.PROPAG'S   LAYER DIFFR.MATR'S
C
C  O(SURF.)    -------------------------------------
C              1      ASE               1,2
C  1(OVERLAY.)-------------------------------------------- ROP,TOP,ROM,TOM
C              2      ASB               3,4
C  2(SUBSTR.) ------------------------------------------- RA1,TA1
C              3      ASC               9,10
C  3          ------------------------------------------- RA2,TA2
C              4      AS2               7,8
C  4          ------------------------------------------- RA1,TA1
C              5      AS1               5,6
C  5          ------------------------------------------- RA2,TA2
C              6      AS2               7,8
C  ETC.
       SUBROUTINE RFSO2V(ROP,TOP,ROM,TOM,RA1,TA1,RA2,TA2,N,
      1NSL,NL,NM,WV,PQ,PK,AW,ANEW,ND,ASB,ASC,AS1,AS2)
       DIMENSION RA1(N,NM),TA1(N,NM),RA2(N,NM),TA2(N,NM)
       DIMENSION ROP(N,N),TOP(N,N),AS1(3),AS2(3)
       DIMENSION NSL(NL),WV(N),MSL(1),ASB(3),ASC(3)
       DIMENSION PQ(2,N),ANEW(N,ND),AW(N,2),AS(3),PK(N,10)
       COMPLEX RA1,TA1,RA2,TA2,WV,ANEW,AW,PK
       COMPLEX ROP,TOP,EL,EK,EKP,CZ,CI,ROM(N,N),TOM(N,N)
       COMPLEX CSQRT,CEXP
       COMMON /ADS/ ASL,FR,ASE,VPIS,VPIO,VO,VV
       COMMON /X4/ E,VPI,BK2,BK3,AS
       CRIT=0.003
       CZ=CMPLX(0.0,0.0)
       CI=CMPLX(0.0,1.0)
       DO 1 J=1,N
       AK2=BK2+PQ(1,J)
```

```
        AK3=BK3+PQ(2,J)
        A=2.0*E-AK2*AK2-AK3*AK3
        EK=CSQRT(CMPLX(A,-2.0*VPIS+0.000001))
        EL=CSQRT(CMPLX(A-2.0*VO,-2.0*VPIO+0.000001))
        EKP=ASB(1)*(FR*EK+(1.0-FR)*EL)
        PK(J,3)=CEXP(CI*(EKP+AK2*ASB(2)+AK3*ASB(3)))
        PK(J,4)=CEXP(CI*(EKP-AK2*ASB(2)-AK3*ASB(3)))
        PK(J,1)=CEXP(CI*EL*ASE)
        PK(J,2)=PK(J,1)
C     THREE SETS OF DIFFERENT INTERLAYER PROPAGATORS ARE PRODUCED FOR THE
C     SUBSTRATE, USING THE BULK INTERLAYER VECTORS AS1 AND AS2, AND THE
C     TOPMOST ONE ASC
        EKP=AK2*AS1(2)+AK3*AS1(3)
        PK(J,5)=CEXP(CI*(EK*AS1(1)+EKP))
        PK(J,6)=CEXP(CI*(EK*AS1(1)-EKP))
        EKP=AK2*AS2(2)+AK3*AS2(3)
        PK(J,7)=CEXP(CI*(EK*AS2(1)+EKP))
        PK(J,8)=CEXP(CI*(EK*AS2(1)-EKP))
        EKP=AK2*ASC(2)+AK3*ASC(3)
        PK(J,9)=CEXP(CI*(EK*ASC(1)+EKP))
1       PK(J,10)=CEXP(CI*(EK*ASC(1)-EKP))
        DO 2 I=1,N
        WV(I)=CZ
        DO 2 J=1,ND
2       ANEW(I,J)=CZ
        ANEW(1,1)=CMPLX(1.0,0.0)
        MSL(1)=N
        IO=0
        ANORM1=1.0E-6
3       ANORM2=0.0
        IO=IO+1
        IR=0
        I=2
        CALL PRPGAT(ROP,TOP,ROM,TOM,N,N,ANEW,ND,MSL,1,AW,I,PK,1,4,
       1CRIT,IR,1,BNORM)
        L1=3
        L2=10
        L3=9
C     PROPAGATE THE WAVEFIELD THROUGH THE NEXT TWO LAYERS AND REPEAT UNTIL
C     CONVERGENCE IN PENETRATION
50      CALL PRPGAT(RA1,TA1,RA1,TA1,N,NM,ANEW,ND,NSL,NL,AW,I,PK,L1,L2,
       1CRIT,IR,1,BNORM)
        IF (IR.EQ.1) GO TO 52
51      CALL PRPGAT(RA2,TA2,RA2,TA2,N,NM,ANEW,ND,NSL,NL,AW,I,PK,L3,8,
       1CRIT,IR,1,BNORM)
        L1=7
        L2=6
        L3=5
        IF (IR) 50,50,52
52      CONTINUE
        I=I-1
        IMAX=I-1
        WRITE(6,20)IMAX,BNORM
20      FORMAT(8H IMAX = ,I2,5X,8HBNORM = ,E12.4)
        DO 21 IG=1,N
21      ANEW(IG,I+1)=CZ
        L1=5
        L2=6
        L3=7
        IF (I.EQ.1) GO TO 59
        II=MOD(I,2)+1
        GO TO (58,57),II
57      IF (I.EQ.3) L1=9
        CALL PRPGAT(RA2,TA2,RA2,TA2,N,NM,ANEW,ND,NSL,NL,AW,I,PK,8,L1,
       1CRIT,IR,-1,BNORM)
58      IF (I.NE.2) GO TO 60
        L2=10
```

```
         L3=3
60       CALL PRPGAT(RA1,TA1,RA1,TA1,N,NM,ANEW,ND,NSL,NL,AW,I,PK,L2,L3,
        1CRIT,IR,-1,BNORM)
         IF (I-1) 59,59,57
59       CALL PRPGAT(ROP,TOP,ROM,TOM,N,N,ANEW,ND,MSL,1,AW,I,PK,4,1,
        1CRIT,IR,-1,BNORM)
         DO 29 IG=1,N
         EK=ANEW(IG,1)*PK(IG,2)
         AB=CABS(EK)
         ANORM2=ANORM2+AB*AB
         WV(IG)=WV(IG)+EK
29       ANEW(IG,1)=CZ
         ANORM1=ANORM1+ANORM2
         WRITE(6,32)ANORM1
32       FORMAT(9H ANORM = ,E12.4,/)
         IF (ANORM2/ANORM1-0.001) 39,39,36
36       IF (IO-6) 3,37,37
37       WRITE(6,38) IO
38       FORMAT(/,24H ***NO CONVERGENCE AFTER,
        11I4,11H ITERATIONS)
39       RETURN
         END
C-------------------------------------------------------------------
C   SUBROUTINE RFS03 PERFORMS THE RENORMALIZED FORWARD SCATTERING (RFS)
C   PERTURBATION EXPANSION, GIVEN LAYER DIFFRACTION MATRICES, FOR ONE
C   OVERLAYER (WITH SUPERLATTICE, IF DESIRED) ON A SUBSTRATE THAT CAN
C   BE DESCRIBED BY A SINGLE CONSTANT INTERLAYER VECTOR AND A 1- OR 3-
C   LAYER PERIODICITY OF THE LAYER DIFFRACTION MATRICES. THE OVERLAYER
C   MAY BE ASYMMETRICAL IN +-X, WHILE SUBSTRATE LAYERS ARE ASSUMED
C   SYMMETRICAL IN +-X. THE OVERLAYER MAY BE CHOSEN EQUAL TO A SUBSTRATE
C   LAYER, REPRESENTING THE CASE OF A CLEAN SURFACE (RELAXED OR NOT).
C   ORGANIZATIONAL DIAGRAM OF LAYER ARRANGEMENT:
C
C   LAYER NO.    I    INTERL.VECTOR    INTERL.PROPAG'S    LAYER DIFFR.MATR'S
C
C   0(SURF.)   ------------------------------------------
C                1        ASE                1,2
C   1(OVERLAY.)-------------------------------------------- ROP,TOP,ROM,TOM
C                2        ASB                3,4
C   2(SUBSTR.) --------------------------------------------- RA1,TA1
C                3        ASA                5,6
C   3          --------------------------------------------- RA2,TA2
C                4        ASA                5,6
C   4          --------------------------------------------- RA3,TA3
C                5        ASA                5,6
C   5          --------------------------------------------- RA1,TA1
C                6        ASA                5,6
C   ETC.
C    ROP,TOP,ROM,TOM: OVERLAYER DIFFRACTION MATRICES (R FOR REFLECTION,
C     T FOR TRANSMISSION, P FOR INCIDENCE TOWARDS +X, M FOR INCIDENCE
C     TOWARDS -X).
C    RA1,TA1,RA2,TA2,RA3,TA3: SUBSTRATE LAYER DIFFRACTION MATRICES
C    (R FOR REFLECTION, T FOR TRANSMISSION, 1 FOR LAYERS 2,5,8,...,
C    2 FOR LAYERS 3,6,9,..., 3 FOR LAYERS 4,7,10,..).
C    N= NO. OF BEAMS USED AT CURRENT ENERGY (INCL. ALL BEAM SETS).
C    NSL: GIVES NO. OF BEAMS IN EACH SET OF BEAMS.
C    NL= NO. OF BEAM SETS.
C    NM= 2ND DIMENSION OF SUBSTRATE DIFFRACTION MATRICES (= LARGEST
C    VALUE OF NSL).
C    WV= OUTPUT REFLECTED BEAM AMPLITUDES.
C    PQ: LIST OF BEAMS.
C    PK,AW,ANEW: WORKING SPACES.
C    ND= NO.OF THE LAYER TO WHICH PENETRATION INTO SURFACE IS ALLOWED
C    (DIMENSIONS ANEW).
C    ASB=INTERLAYER VECTOR BETWEEN OVERLAYER AND TOP SUBSTRATE LAYER.
C    ASA= SUBSTRATE INTERLAYER VECTOR.
C   IN COMMON BLOCKS:
```

```
C     ASL: NOT USED.
C     FR=FRACTION OF SPACING BETWEEN OVERLAYER AND TOP SUBSTRATE LAYER
C     THAT IS ALLOTTED TO THE SUBSTRATE.
C     ASE= SPACING BETWEEN SURFACE AND OVERLAYER NUCLEI.
C     VPIS,VPIO= OPTICAL POTENTIAL IN SUBSTRATE AND OVERLAYER, RESP.
C     VO,VV= LEVEL OF CONSTANT POTENTIAL IN OVERLAYER AND VACUUM, RESP.,
C     REFERRED TO LEVEL OF CONSTANT POTENTIAL IN SUBSTRATE.
C     E= CURRENT ENERGY.
C     VPI: NOT USED.
C     BK2,BK3: PARALLEL COMPONENTS OF INCIDENT K-VECTOR.
C     AS: NOT USED.
      SUBROUTINE RFSO3(ROP,TOP,ROM,TOM,RA1,TA1,RA2,TA2,RA3,TA3,N,
     1NSL,NL,NM,WV,PQ,PK,AW,ANEW,ND,ASB,ASA)
      DIMENSION RA1(N,NM),TA1(N,NM),RA2(N,NM),TA2(N,NM)
      DIMENSION RA3(N,NM),TA3(N,NM),ROP(N,N),TOP(N,N)
      DIMENSION NSL(NL),WV(N),MSL(1),ASB(3)
      DIMENSION PQ(2,N),ANEW(N,ND),AW(N,2),ASA(3),PK(N,8),AS(3)
      COMPLEX RA1,TA1,RA2,TA2,RA3,TA3,WV,ANEW,AW,PK
      COMPLEX ROP,TOP,EL,EK,EKP,CZ,CI,ROM(N,N),TOM(N,N)
      COMPLEX CSQRT,CEXP
      COMMON /ADS/ ASL,FR,ASE,VPIS,VPIO,VO,VV
      COMMON /X4/ E,VPI,BK2,BK3,AS
C     SET CRITERION FOR PENETRATION CONVERGENCE
      CRIT=0.003
      CZ=CMPLX(0.0,0.0)
      CI=CMPLX(0.0,1.0)
      DO 1 J=1,N
      AK2=BK2+PQ(1,J)
      AK3=BK3+PQ(2,J)
      A=2.0*E-AK2*AK2-AK3*AK3
      EK=CSQRT(CMPLX(A,-2.0*VPIS+0.000001))
      EL=CSQRT(CMPLX(A-2.0*VO,-2.0*VPIO+0.000001))
      EKP=ASB(1)*(FR*EK+(1.0-FR)*EL)
C     PREPARE PLANE-WAVE PROPAGATORS FOR PROPAGATION FROM OVERLAYER TO
C     NEXT LAYER, AND VICE VERSA
      PK(J,3)=CEXP(CI*(EKP+AK2*ASB(2)+AK3*ASB(3)))
      PK(J,4)=CEXP(CI*(EKP-AK2*ASB(2)-AK3*ASB(3)))
C     SAME FOR PROPAGATION FROM SURFACE TO OVERLAYER
      PK(J,1)=CEXP(CI*EL*ASE)
      PK(J,2)=PK(J,1)
      EKP=AK2*ASA(2)+AK3*ASA(3)
C     SAME FOR PROPAGATION BETWEEN SUBSTRATE LAYERS
      PK(J,5)=CEXP(CI*(EK*ASA(1)+EKP))
1     PK(J,6)=CEXP(CI*(EK*ASA(1)-EKP))
C
C     INITIALIZE ANEW (RUNNING WAVEFIELD)
      DO 2 I=1,N
      WV(I)=CZ
      DO 2 J=1,ND
2     ANEW(I,J)=CZ
      ANEW(1,1)=CMPLX(1.0,0.0)
C     MSL TO GIVE NO. OF BEAMS AT OVERLAYER (ANALOGOUS TO NSL FOR SUBSTRATE
C     LAYERS)
      MSL(1)=N
      IO=0
      ANORM1=1.0E-6
C
C
C     START ITERATION OVER ORDERS OF PERTURBATION (NO. OF BACK-SCATTERINGS)
3     ANORM2=0.0
      IO=IO+1
C     IR WILL BE SET =1 BY PRPGAT TO INDICATE PENETRATION CONVERGENCE
      IR=0
C     I INDICATES INTERLAYER SPACING TO WHICH NEXT CALL TO PRPGAT WILL
C     LEAD (I=1 IS NEAR SURFACE). PRPGAT INCREASES I BY IA=+-1, DEPENDING
C     ON DIRECTION OF PROPAGATION
      I=2
```

```
C
C   PROPAGATE WAVEFIELD THROUGH OVERLAYER, PICKING UP REFLECTIONS FROM
C   EMERGING WAVES, IF PRESENT
        CALL PRPGAT(ROP,TOP,ROM,TOM,N,N,ANEW,ND,MSL,1,AW,I,PK,1,4,
       1CRIT,IR,1,BNORM)
        L1=3
C
C   PROPAGATE WAVEFIELD THROUGH NEXT 3 LAYERS AND REPEAT FOR SUBSEQUENT
C   3 LAYERS,ETC., STOPPING WHEN PENETRATION CONVERGENCE IS ACHIEVED
50      CALL PRPGAT(RA1,TA1,RA1,TA1,N,NM,ANEW,ND,NSL,NL,AW,I,PK,L1,6,
       1CRIT,IR,1,BNORM)
        IF (IR.EQ.1) GO TO 52
51      CALL PRPGAT(RA2,TA2,RA2,TA2,N,NM,ANEW,ND,NSL,NL,AW,I,PK,5,6,
       1CRIT,IR,1,BNORM)
        IF (IR.EQ.1) GO TO 52
        CALL PRPGAT(RA3,TA3,RA3,TA3,N,NM,ANEW,ND,NSL,NL,AW,I,PK,5,6,
       1CRIT,IR,1,BNORM)
        L1=5
        IF (IR) 50,50,52
52      CONTINUE
        I=I-1
        IMAX=I-1
C   PRINT OUT LAYER REACHED AND MEASURE OF WAVE MAGNITUDE AT THAT LAYER
        WRITE(6,20)IMAX,BNORM
20      FORMAT(8H IMAX = ,I2,5X,8HBNORM = ,E12.4)
C
C
C   PREPARE PROPAGATION BACK TO SURFACE BY SETTING AMPLITUDES OF WAVES
C   COMING FROM DEEP IN SURFACE TO ZERO
        DO 21 IG=1,N
21      ANEW(IG,I+1)=CZ
        L3=5
C   FIGURE OUT AT WHICH LAYER TO START PROPAGATION TO SURFACE
        IF (I.EQ.1) GO TO 59
        II=MOD(I,3)+1
        GO TO (57,56,58),II
56      CALL PRPGAT(RA3,TA3,RA3,TA3,N,NM,ANEW,ND,NSL,NL,AW,I,PK,6,5,
       1CRIT,IR,-1,BNORM)
57      CALL PRPGAT(RA2,TA2,RA2,TA2,N,NM,ANEW,ND,NSL,NL,AW,I,PK,6,5,
       1CRIT,IR,-1,BNORM)
        IF (I.EQ.2) L3=3
58      CALL PRPGAT(RA1,TA1,RA1,TA1,N,NM,ANEW,ND,NSL,NL,AW,I,PK,6,L3,
       1CRIT,IR,-1,BNORM)
        IF (I-1) 59,59,56
59      CALL PRPGAT(ROP,TOP,ROM,TOM,N,N,ANEW,ND,MSL,1,AW,I,PK,4,1,
       1CRIT,IR,-1,BNORM)
C
C   DO FINAL BIT OF PROPAGATION FROM OVERLAYER TO SURFACE AND ACCUMULATE
C   REFLECTED BEAM AMPLITUDES
        DO 29 IG=1,N
        EK=ANEW(IG,1)*PK(IG,2)
        AB=CABS(EK)
        ANORM2=ANORM2+AB*AB
        WV(IG)=WV(IG)+EK
29      ANEW(IG,1)=CZ
        ANORM1=ANORM1+ANORM2
C   PRINT OUT MEASURE OF TOTAL REFLECTED BEAM AMPLITUDES
        WRITE(6,32)ANORM1
32      FORMAT(9H ANORM = ,E12.4,/)
C   CHECK FOR CONVERGENCE OVER ORDERS OF PERTURBATION, WHICH ARE LIMITED
C   TO 6
        IF (ANORM2/ANORM1-0.001) 39,39,36
36      IF (IO-6) 3,37,37
37      WRITE(6,38) IO
38      FORMAT(/,24H ***NO CONVERGENCE AFTER,
       11I4,11H ITERATIONS)
39      RETURN
```

```
      END
C-----------------------------------------------------------------------
C   SUBROUTINE RFS2OV HAS THE SAME PURPOSE AS SUBROUTINE RFSO2, BUT
C   TWO STACKED OVERLAYERS ARE ASSUMED (THE OVERLAYERS MUST HAVE
C   IDENTICAL SUPERLATTICES).
C   NOTE: PK HAS A LARGER DIMENSION THAN IN RFSO2
C   ORGANIZATIONAL DIAGRAM OF LAYER ARRANGEMENT:
C
C   LAYER NO.     I   INTERL.VECTOR   INTERL.PROPAG'S   LAYER DIFFR.MATR'S
C
C   0(SURF.)     ------------------------------------------
C                1       ASE               1,2
C   1(OVERLAY.)--------------------------------------------- ROP,TOP,ROM,TOM
C                2       ASB               3,4
C   2(OVERLAY.)--------------------------------------------- RTP,TTP,RTM,TTM
C                3       ASC               9,10
C   3(SUBSTR.) --------------------------------------------- RA1,TA1
C                4       AS2               7,8
C   4          --------------------------------------------- RA2,TA2
C                5       AS1               5,6
C   5          --------------------------------------------- RA1,TA1
C                6       AS2               7,8
C   ETC.
      SUBROUTINE RFS2OV(ROP,TOP,ROM,TOM,RTP,TTP,RTM,TTM,RA1,TA1,RA2,TA2,
     1N,NSL,NL,NM,WV,PQ,PK,AW,ANEW,ND,ASB,ASC,AS1,AS2)
      DIMENSION RA1(N,NM),TA1(N,NM),RA2(N,NM),TA2(N,NM)
      DIMENSION ROP(N,N),TOP(N,N),AS1(3),AS2(3)
      DIMENSION NSL(NL),WV(N),MSL(1),ASB(3),ASC(3)
      DIMENSION PQ(2,N),ANEW(N,ND),AW(N,2),AS(3),PK(N,10)
      COMPLEX RA1,TA1,RA2,TA2,WV,ANEW,AW,PK
      COMPLEX ROP,TOP,EL,EK,EKP,CZ,CI,ROM(N,N),TOM(N,N)
      COMPLEX RTP(N,N),TTP(N,N),RTM(N,N),TTM(N,N)
      COMPLEX CSQRT,CEXP
      COMMON /ADS/ ASL,FR,ASE,VPIS,VPIO,VO,VV
      COMMON /X4/ E,VPI,BK2,BK3,AS
      CRIT=0.003
      CZ=CMPLX(0.0,0.0)
      CI=CMPLX(0.0,1.0)
      DO 1 J=1,N
      AK2=BK2+PQ(1,J)
      AK3=BK3+PQ(2,J)
      A=2.0*E-AK2*AK2-AK3*AK3
      EK=CSQRT(CMPLX(A,-2.0*VPIS+0.000001))
      EL=CSQRT(CMPLX(A-2.0*VO,-2.0*VPIO+0.000001))
      EKP=ASB(1)*(FR*EK+(1.0-FR)*EL)
      PK(J,3)=CEXP(CI*(EKP+AK2*ASB(2)+AK3*ASB(3)))
      PK(J,4)=CEXP(CI*(EKP-AK2*ASB(2)-AK3*ASB(3)))
      PK(J,1)=CEXP(CI*EL*ASE)
      PK(J,2)=PK(J,1)
C   THREE SETS OF DIFFERENT INTERLAYER PROPAGATORS ARE PRODUCED FOR THE
C   SUBSTRATE, USING THE BULK INTERLAYER VECTORS AS1 AND AS2, AND THE
C   TOPMOST ONE ASC
      EKP=AK2*AS1(2)+AK3*AS1(3)
      PK(J,5)=CEXP(CI*(EK*AS1(1)+EKP))
      PK(J,6)=CEXP(CI*(EK*AS1(1)-EKP))
      EKP=AK2*AS2(2)+AK3*AS2(3)
      PK(J,7)=CEXP(CI*(EK*AS2(1)+EKP))
      PK(J,8)=CEXP(CI*(EK*AS2(1)-EKP))
      EKP=AK2*ASC(2)+AK3*ASC(3)
      PK(J,9)=CEXP(CI*(EK*ASC(1)+EKP))
1     PK(J,10)=CEXP(CI*(EK*ASC(1)-EKP))
      DO 2 I=1,N
      WV(I)=CZ
      DO 2 J=1,ND
2     ANEW(I,J)=CZ
      ANEW(1,1)=CMPLX(1.0,0.0)
      MSL(1)=N
```

```
      IO=0
      ANORM1=1.0E-6
3     ANORM2=0.0
      IO=IO+1
      IR=0
      I=2
      CALL PRPGAT(ROP,TOP,ROM,TOM,N,N,ANEW,ND,MSL,1,AW,I,PK,1,4,
     1CRIT,IR,1,BNORM)
      CALL PRPGAT(RTP,TTP,RTM,TTM,N,N,ANEW,ND,MSL,1,AW,I,PK,3,10,
     1CRIT,IR,1,BNORM)
      IF (IR) 49,49,52
49    CONTINUE
      L1=9
C  PROPAGATE THE WAVEFIELD THROUGH THE NEXT TWO LAYERS AND REPEAT UNTIL
C  CONVERGENCE IN PENETRATION
50    CALL PRPGAT(RA1,TA1,RA1,TA1,N,NM,ANEW,ND,NSL,NL,AW,I,PK,L1,8,
     1CRIT,IR,1,BNORM)
      IF (IR.EQ.1) GO TO 52
51    CALL PRPGAT(RA2,TA2,RA2,TA2,N,NM,ANEW,ND,NSL,NL,AW,I,PK,7,6,
     1CRIT,IR,1,BNORM)
      L1=5
      IF (IR) 50,50,52
52    CONTINUE
      I=I-1
      IMAX=I-1
      WRITE(6,20)IMAX,BNORM
20    FORMAT(8H IMAX = ,I2,5X,8HBNORM = ,E12.4)
      DO 21 IG=1,N
21    ANEW(IG,I+1)=CZ
      L3=5
      IF (I.EQ.1) GO TO 60
      IF (I.EQ.2) GO TO 59
      II=MOD(I,2)+1
      GO TO (57,58),II
57    CONTINUE
      CALL PRPGAT(RA2,TA2,RA2,TA2,N,NM,ANEW,ND,NSL,NL,AW,I,PK,6,7,
     1CRIT,IR,-1,BNORM)
58    IF (I.EQ.3) L3=9
      CALL PRPGAT(RA1,TA1,RA1,TA1,N,NM,ANEW,ND,NSL,NL,AW,I,PK,8,L3,
     1CRIT,IR,-1,BNORM)
      IF (I-2) 60,59,57
59    CALL PRPGAT(RTP,TTP,RTM,TTM,N,N,ANEW,ND,MSL,1,AW,I,PK,10,3,
     1CRIT,IR,-1,BNORM)
60    CALL PRPGAT(ROP,TOP,ROM,TOM,N,N,ANEW,ND,MSL,1,AW,I,PK,4,1,
     1CRIT,IR,-1,BNORM)
      DO 29 IG=1,N
      EK=ANEW(IG,1)*PK(IG,2)
      AB=CABS(EK)
      ANORM2=ANORM2+AB*AB
      WV(IG)=WV(IG)+EK
29    ANEW(IG,1)=CZ
      ANORM1=ANORM1+ANORM2
      WRITE(6,32)ANORM1
32    FORMAT(9H ANORM = ,E12.4,/)
      IF (ANORM2/ANORM1-0.001) 39,39,36
36    IF (IO-6) 3,37,37
37    WRITE(6,38) IO
38    FORMAT(/,24H ***NO CONVERGENCE AFTER,
     11I4,11H ITERATIONS)
39    RETURN
      END
C-------------------------------------------------------------------
C  SUBROUTINE RINT COMPUTES THE REFLECTED BEAM INTENSITIES FROM THE
C  (COMPLEX) REFLECTED AMPLITUDES. INCLUDED ARE ANGULAR PREFACTORS AND A
C  SYMMETRY-RELATED PREFACTOR (SO THAT THE RESULTING INTENSITY IS
C  DIRECTLY COMPARABLE WITH EXPERIMENT). RINT PRINTS OUT NON-ZERO
C  INTENSITIES AND PUNCHES OUT (IF REQUESTED) THE INTENSITIES OF THE
```

```
C     BEAMS SPECIFIED IN NPUC.
C     N= NO. OF BEAMS AT CURRENT ENERGY.
C     WV= INPUT REFLECTED AMPLITUDES.
C     AT= OUTPUT REFLECTED INTENSITIES.
C     PQ= LIST OF RECIPROCAL-LATTICE VECTORS G (BEAMS).
C     PQF= SAME AS PQ, BUT IN UNITS OF THE RECIPROCAL-LATTICE CONSTANTS.
C     SYM= LIST OF SYMMETRY CODES FOR ALL BEAMS.
C     VV= VACUUM LEVEL ABOVE SUBSTRATE MUFFIN-TIN CONSTANT.
C     THETA,FI= POLAR AND AZIMUTHAL ANGLES OF INCIDENT BEAM.
C     MPU= NO. OF BEAM INTENSITIES TO BE PUNCHED.
C     NPUC: INDICATES WHICH BEAM INTENSITIES ARE TO BE PUNCHED.
C     EEV= CURRENT ENERGY IN EV ABOVE VACUUM LEVEL.
C     A= STRUCTURAL PARAMETER OR OTHER IDENTIFIER TO BE PUNCHED ON CARDS.
C     NPNCH=0: NO PUNCH DESIRED.
C     NPNCH.NE.0: PUNCH DESIRED.
C     IN COMMON BLOCKS:
C     E= CURRENT ENERGY IN HARTREES ABOVE SUBSTRATE MUFFIN-TIN CONSTANT.
C     VPI= IMAGINARY PART OF CURRENT ENERGY.
C     CK2,CK3= PARALLEL COMPONENTS OF PRIMARY INCIDENT K-VECTOR.
C     AS: NOT USED.
      SUBROUTINE RINT(N,WV,AT,PQ,PQF,SYM,VV,THETA,FI,MPU,NPUC,EEV,AP,
     1NPNCH)
      INTEGER  SYM
      DIMENSION  WV(N),  PQ(2,N),  PQF(2,N),  AT(N)
      DIMENSION  AS(3),  SYM(2,N),NPUC(MPU),ATP(30)
      COMPLEX WV
      COMMON  /X4/E,  VPI,  CK2,  CK3,   AS
   10 FORMAT(3H E=,1F10.4,7H THETA=,1F10.4,4H FI=,1F10.4,/)
   20 FORMAT(35H  PQ1     PQ2      REF INT     SYM)
   30 FORMAT(2F7.3,1E13.5,I7)
   35 FORMAT(1F7.2,1F7.4,4E14.5,/,5(5E14.5,/))
      AK = SQRT(AMAX1(2.0 * E - 2.0 * VV,0.0))
      BK2 = AK * SIN(THETA) * COS(FI)
      BK3 = AK * SIN(THETA) * SIN(FI)
C C IS K(PERP) IN VACUUM FOR INCIDENT BEAM
      C = AK * COS(THETA)
      TH1=THETA*180.0/3.14159265
      FI1=FI*180.0/3.14159265
      WRITE(6,10)E,TH1,FI1
      WRITE(6,20)
      DO 120 J = 1, N
      AK2 = PQ(1,J) + BK2
      AK3 = PQ(2,J) + BK3
      A = 2.0 * E - 2.0 * VV - AK2 * AK2 - AK3 * AK3
      AT(J) = 0.0
C     SKIP PRINT OUTPUT FOR NON-EMERGING BEAMS
      IF (A)  120, 120, 40
C     A IS K(PERP) IN VACUUM FOR SCATTERED BEAMS
   40 A = SQRT(A)
C     USE APPROPRIATE SYMMETRY PREFACTOR
      LAB = SYM(1,J)
      IF (LAB-10) 45,100,60
   45 IF (LAB-2)  70, 50, 50
   50 IF (LAB-7)  80,90,90
   60 IF (LAB-12) 101,65,65
   65 IF (LAB-15) 102,103,103
   70 FAC = 1.0
      GO TO 110
   80 FAC = 2.0
      GO TO 110
   90 FAC = 4.0
      GO TO 110
  100 FAC = 8.0
      GO TO 110
  101 FAC=3.0
      GO TO 110
  102 FAC=6.0
```

```
            GO TO 110
    103 FAC=12.0
    110 WR = REAL(WV(J))
        WI = AIMAG(WV(J))
C   AT IS REFLECTED INTENSITY (FOR UNIT INCIDENT CURRENT)
        AT(J) = (WR * WR + WI * WI) * A/(FAC * C)
        WRITE (6,30)  PQF(1,J),PQF(2,J),AT(J),SYM(1,J)
    120 CONTINUE
C   PUNCH IF REQUESTED
        IF (NPNCH.EQ.0) RETURN
        DO 140 K=1,MPU
        NPC=NPUC(K)
C   PUNCH ONLY FOR BEAMS SELECTED IN INPUT (NPU)
        IF (NPC.EQ.0) GO TO 130
        ATP(K)=AT(NPC)
        GO TO 140
    130 ATP(K)=0.0
    140 CONTINUE
        WRITE(7,35)EEV,AP,(ATP(K),K=1,MPU)
        RETURN
        END
C-----------------------------------------------------------------------
C   SUBROUTINE SB COMPUTES SPHERICAL BESSEL FUNCTIONS.
C   X= COMPLEX ARGUMENT OF BESSEL FUNCTIONS.
C   HH= OUTPUT COMPLEX BESSEL FUNCTIONS.
C   N3= LMAX+1.
        SUBROUTINE  SB (X, HH, N3)
        COMPLEX  A, B, C, HH, X
        DIMENSION  HH(N3)
    330 FORMAT(1H0,11HSB INFINITE)
        A = (0.0,1.0)
        C = X
        B = A * C
        F = CABS(X)
        IF (F-1.0E-38)  1540, 1540, 1520
   1520 HH(1) = CEXP(B) * ( - A)/C
        HH(2) = CEXP(B) * (1.0/( - C) - A/C**2)
        DO 1530 J = 3, N3
        HH(J) = (2.0 * (J - 2) + 1.0)/C * HH(J - 1) - HH(J - 2)
   1530 CONTINUE
        GO TO 1550
   1540 PRINT 330
   1550 RETURN
        END
C-----------------------------------------------------------------------
C   SUBROUTINE SH COMPUTES SPHERICAL HARMONICS.
C   NHARM= LMAX+1.
C   Z= COMPLEX ARGUMENT COS(THETA).
C   FI= AZIMUTHAL ANGLE.
C   Y= OUTPUT COMPLEX SPHERICAL HARMONICS.
        SUBROUTINE  SH (NHARM, Z, FI, Y)
        COMPLEX  A, AA, ANOR, BB, BB1, Q1A, Y, YSTAR, YY, Z, ZNW
        DIMENSION  Y(NHARM,NHARM)
        AM = 1.0
        A = (0.0,1.0) * FI
        RZ = REAL(Z)
        ZNW = CSQRT((1.0,0.0) - Z * Z)
        ANORA = 0.2820948
        YY = (1.0,0.0)
        DO 1510 L = 1, NHARM
        BM = FLOAT(L) - 1.0
        Q1A = CEXP(BM * A)
        ANOR = ANORA * Q1A * ( - 1.0)**(L + 1)
        ANORA = ANORA * SQRT(1.0 + 0.5/(BM + 1.0))/(2.0 * BM + 1.0)
        IF((ABS(ABS(RZ) - 1.0) .LT. 1.0E-10) .AND. L .EQ. 1) GO TO 1480
        YY = ZNW**INT(BM + 0.1)
   1480 AA = AM * YY
```

```
      AM = AM * (2.0 * BM + 1.0)
      BB = AM * Z * YY
      DO 1500 LL = L, NHARM
      BN = FLOAT(LL)
      Y(LL,L) = AA * ANOR
      IF (L .EQ. 1)  GO TO 1490
      YSTAR = Y(LL,L)/(Q1A * Q1A)
      Y(L - 1,LL) = YSTAR
 1490 BB1 = BB
      BB = BB1 * Z + (BB1 * Z - AA) * (BN + BM)/(BN - BM + 1.0)
      AA = BB1
 1500 ANOR = ANOR * SQRT((1.0 + 1.0/(BN - 0.5)) * ((BN - BM)/(BN + BM)))
 1510 CONTINUE
      RETURN
      END
C----------------------------------------------------------------------
COMMENT SLIND SETS UP A MATRIX JJS(JS,J) CONTAINING DETAILS OF
C       HOW THE SUBLATTICES JS ARE TRANSFORMED INTO ONE
C       ANOTHER BY ROTATIONS THROUGH J*2.0*PI/IDEG. V(JS,2)
C       CONTAINS THE ADDING VECTORS DEFINING THE SUBLATTICES
C       JS IN POLAR FORM
C   AUTHOR: PENDRY.
C   V= VECTORS POINTING TO DIFFERENT SUBLATTICES (IN POLAR COORDINATES).
C   VL= WORKING SPACE.
C   JJS= OUTPUT RELATIONSHIP OF SUBLATTICES UNDER ROTATION.
C   NL= NO. OF SUBLATTICES.
C   IDEG= DEGREE OF SYMMETRY OF LATTICE: IDEG-FOLD ROTATION AXIS.
C   EPSD= SMALL NUMBER.
C   IN COMMON BLOCKS:
C   BR1,BR2= BASIS VECTORS OF SUBSTRATE LAYER LATTICE.
C   AR1,AR2,RAR1,RAR2= BASIS VECTORS OF SUPERLATTICE IN DIRECT AND
C     RECIPROCAL SPACE.
C   NL1,NL2= SUPERLATTICE CHARACTERIZATION (SEE MAIN PROGRAM).
      SUBROUTINE  SLIND (V, VL, JJS, NL, IDEG, EPSD)
      COMPLEX  VL, VLA, VLB, CI
      COMPLEX CEXP
      DIMENSION  V(NL,2), JJS(NL,IDEG), VL(NL,2), AR1(2), AR2(2)
      DIMENSION  BR1(2), BR2(2), RAR1(2), RAR2(2)
      COMMON  /SL/BR1, BR2, AR1, AR2, RAR1, RAR2, NL1, NL2
C
      CI = CMPLX(0.0,1.0)
      PI = 3.14159265
C SET UP VECTORS V DEFINING SUBLATTICES AND QUANTITIES VL FOR LATER
C REFERENCE.
      I = 1
      S1 = 0.0
      DO 560 J = 1, NL1
      S2 = 0.0
      DO 550 K = 1, NL2
      ADR1 = S1 * BR1(1) + S2 * BR2(1)
      ADR2 = S1 * BR1(2) + S2 * BR2(2)
      V(I,1) = SQRT(ADR1 * ADR1 + ADR2 * ADR2)
      V(I,2) = 0.0
      IF (V(I,1))  540, 540, 530
  530 V(I,2) = ATAN2(ADR2,ADR1)
  540 VL(I,1) = CEXP(CI * (ADR1 * RAR1(1) + ADR2 * RAR1(2)))
      VL(I,2) = CEXP(CI * (ADR1 * RAR2(1) + ADR2 * RAR2(2)))
      I = I + 1
  550 S2 = S2 + 1.0
  560 S1 = S1 + 1.0
C
C ROTATE EACH VECTOR V AND FIND TO WHICH V IT BECOMES EQUIVALENT IN
C TERMS OF THE QUANTITIES VL. THIS EQUIVALENCE MEANS BELONGING TO THE
C SAME SUBLATTICE.
      AINC = 2.0 * PI/FLOAT(IDEG)
      DO 590 I = 1, NL
      ADR = V(I,1)
```

```
      ANG = V(I,2)
      DO 590 K = 1, IDEG
      ANG = ANG + AINC
      CANG = COS(ANG)
      SANG = SIN(ANG)
      A = RAR1(1) * CANG + RAR1(2) * SANG
      B = RAR2(1) * CANG + RAR2(2) * SANG
      VLA = CEXP(CI * ADR * A)
      VLB = CEXP(CI * ADR * B)
      DO 570 J = 1, NL
      TEST = CABS(VLA - VL(J,1)) + CABS(VLB - VL(J,2))
      IF (TEST-5.0*EPSD)  580, 580, 570
  570 CONTINUE
  580 JJS(I,K) = J
  590 CONTINUE
      RETURN
      END
C----------------------------------------------------------------------
C    SUBROUTINE SPHRM COMPUTES SPHERICAL HARMONICS.
C    AUTHOR: PENDRY.
C    LMAX= LARGEST VALUE OF L.
C    YLM= OUTPUT COMPLEX SPHERICAL HARMONICS.
C    LMMAX= (LMAX+1)**2.
C    CT= COS(THETA) (COMPLEX).
C    ST= SIN(THETA) (COMPLEX).
C    CF= CEXP(I*FI).
      SUBROUTINE  SPHRM(LMAX,YLM,LMMAX,CT, ST, CF)
      COMPLEX   YLM
      COMPLEX   CT, ST, CF, SF, SA
      DIMENSION  FAC1(10),FAC3(10),FAC2(100),YLM(LMMAX)
      PI = 3.14159265
      LM = 0
      CL = 0.0
      A = 1.0
      B = 1.0
      ASG = 1.0
      LL = LMAX + 1
      DO 550 L = 1, LL
      FAC1(L) = ASG * SQRT((2.0 * CL + 1.0) * A/(4.0 * PI * B * B))
      FAC3(L) = SQRT(2.0 * CL)
      CM =  - CL
      LN = L + L - 1
      DO 540 M = 1, LN
      LO = LM + M
      FAC2(LO) = SQRT((CL + 1.0 + CM) * (CL + 1.0 - CM)/((2.0 * CL + 3.
     1O) * (2.0 * CL + 1.0)))
  540 CM = CM + 1.0
      CL = CL + 1.0
      A = A * 2.0 * CL * (2.0 * CL - 1.0)/4.0
      B = B * CL
      ASG =  - ASG
  550 LM = LM + LN
      LM = 1
      CL = 1.0
      ASG =  - 1.0
      SF = CF
      SA = CMPLX(1.0,0.0)
      YLM(1) = CMPLX(FAC1(1),0.0)
      DO 560 L = 1, LMAX
      LN = LM + L + L + 1
      YLM(LN) = FAC1(L + 1) * SA * SF * ST
      YLM(LM + 1) = ASG * FAC1(L + 1) * SA * ST/SF
      YLM(LN - 1) =  - FAC3(L + 1) * FAC1(L + 1) * SA * SF * CT/CF
      YLM(LM + 2) = ASG * FAC3(L + 1) * FAC1(L + 1) * SA * CT * CF/SF
      SA = ST * SA
      SF = SF * CF
      CL = CL + 1.0
```

```
      ASG =  - ASG
  560 LM = LN
      LM = 1
      LL = LMAX - 1
      DO 580 L = 1, LL
      LN = L + L - 1
      LM2 = LM + LN + 4
      LM3 = LM - LN
      DO 570 M = 1, LN
      LO = LM2 + M
      LP = LM3 + M
      LQ = LM + M + 1
      YLM(LO) =  - (FAC2(LP) * YLM(LP) - CT * YLM(LQ))/FAC2(LQ)
  570 CONTINUE
  580 LM = LM + L + L + 1
      RETURN
      END
C---------------------------------------------------------------------------
C  SUBROUTINE SRTLAY IS USED BY SUBROUTINES MPERTI AND MTINV TO REORDER
C  THE SUBPLANES OF A COMPOSITE LAYER ACCORDING TO INCREASING POSITION
C  ALONG THE +X AXIS (REORDERING THE CHEMICAL ELEMENT ASSIGNMENT AS
C  WELL). IF NEW=-1 IT FINDS WHICH RESULTS FOR GH WILL BE AVAILABLE BY
C  SIMPLY COPYING OLD ONES (OF A PREVIOUS CALL TO MPERTI OR MTINV),
C  WHICH ONES NEED TO BE COMPUTED AFRESH AND WHICH ONES CAN BE OBTAINED
C  BY COPYING OTHER NEW ONES.
C  POS= INPUT SUBPLANE POSITIONS (ATOMIC POSITIONS IN UNIT CELL).
C  POSS= OUTPUT REORDERED POS.
C  LPS= INPUT CHEMICAL ELEMENT ASSIGNMENT.
C  LPSS= OUTPUT REORDERED LPS.
C  MGH= OUTPUT: INDICATES ORGANIZATION OF GH IN MPERTI OR MTINV.
C  MGH(I,J)=K MEANS GH(I,J) (I,J= SUBPLANE INDICES) IS TO BE FOUND IN
C  K-TH POSITION IN COLUMNAR MATRIX GH.
C  NLAY= NO. OF SUBPLANES.
C  DRL= OUTPUT INTERPLANAR VECTORS.
C  SDRL= STORAGE FOR OLD DRL FROM PREVIOUS CALL TO MPERTI OR MTINV.
C  NLAY2= NLAY*(NLAY-1)/2.
C  NUGH= OUTPUT: NUGH(K)=1 MEANS THE GH(I,J) (I,J= SUBPLANE INDICES)
C  FOR WHICH MGH(I,J)=K MUST BE COMPUTED AFRESH. NUGH(K)=0 MEANS NO
C  COMPUTATION NECESSARY FOR THAT GH(I,J).
C  NGEQ= OUTPUT: NGEQ(K)=L MEANS THE GH(I,J) FOR WHICH MGH(I,J)=K CAN
C  BE COPIED FROM THE NEWLY CREATED GH(M,N) FOR WHICH MGH(M,N)=L.
C  NGEQ(K)=0 MEANS NO COPYING TO BE DONE FOR THAT GH(I,J).
C  NGOL= OUTPUT: SAME AS NGEQ, BUT FOR COPYING FROM OLD EXISTING VALUES
C  OF GH(M,N).
      SUBROUTINE SRTLAY(POS,POSS,LPS,LPSS,MGH,NLAY,DRL,SDRL,NLAY2,
     1NUGH,NGEQ,NGOL,NEW)
      DIMENSION POS(NLAY,3),POSS(NLAY,3),POSA(3),DRL(NLAY2,3),
     1SDRL(NLAY2,3)
      DIMENSION LPS(NLAY),LPSS(NLAY),MGH(NLAY,NLAY),NUGH(NLAY2),
     1NGEQ(NLAY2),NGOL(NLAY2)
      DO 1 I=1,NLAY
      LPSS(I)=LPS(I)
      DO 1 J=1,3
    1 POSS(I,J)=POS(I,J)
C  ANALYSE ORDER OF SUBPLANE POSITIONS ALONG X-AXIS AND REORDER IN
C  ASCENDING POSITION ALONG +X-AXIS (EQUALLY POSITIONED SUBPLANES ARE
C  NOT PERMUTED)
      NLAY1=NLAY-1
      DO 7 I=1,NLAY1
      II=I+1
      KM=I
      PM=POSS(I,1)
      DO 2 K=II,NLAY
      IF (POSS(K,1).GE.PM) GO TO 2
      PM=POSS(K,1)
      KM=K
    2 CONTINUE
```

```
        IF (KM.EQ.I) GO TO 7
        DO 3 J=1,3
3       POSA(J)=POSS(KM,J)
        LPSA=LPSS(KM)
        DO 5 KK=II,KM
        K=KM+II-KK
        DO 4 J=1,3
4       POSS(K,J)=POSS(K-1,J)
C   REORDER CHEMICAL ASSIGNMENTS CORRESPONDINGLY
5       LPSS(K)=LPSS(K-1)
        DO 6 J=1,3
6       POSS(I,J)=POSA(J)
        LPSS(I)=LPSA
7       CONTINUE
C   GENERATE INTERPLANAR VECTORS DRL
        IN=1
        DO 14 I=1,NLAY
        DO 14 J=I,NLAY
        MGH(I,J)=0
        MGH(J,I)=0
        IF (I-J) 8,14,8
8       DO 9 K=1,3
9       DRL(IN,K)=POSS(I,K)-POSS(J,K)
C   ASSIGN ORGANIZATION OF COLUMNAR MATRIX GH
        MGH(I,J)=IN
        MGH(J,I)=IN+NLAY2
        NUGH(IN)=1
        NGEQ(IN)=0
        NGOL(IN)=0
        IF (IN.EQ.1) GO TO 13
C   FIND EQUAL INTERPLANAR VECTORS, RECORD FINDINGS IN NGEQ
        IN1=IN-1
        DO 12 M=1,IN1
        DO 10 M1=1,3
10      POSA(M1)=ABS(DRL(IN,M1)-DRL(M,M1))
        IF ((POSA(1)+POSA(2)+POSA(3))-0.001) 11,11,12
11      NGEQ(IN)=M
        NUGH(IN)=0
        GO TO 13
12      CONTINUE
13      CONTINUE
        IN=IN+1
14      CONTINUE
C   IF OLD VALUES OF GH NOT RELEVANT (DIFFERENT ENERGY) SKIP NEXT
C   SECTIONS
        IF (NEW+1) 19,15,19
C   COMPARE PRESENT WITH OLD INTERPLANAR VECTORS
15      DO 18 IN=1,NLAY2
        NGOL(IN)=0
        DO 16 K=1,3
16      POSA(K)=ABS(DRL(IN,K)-SDRL(IN,K))
        IF (POSA(1)+POSA(2)+POSA(3)-0.001) 17,17,24
C   OLD GH CAN BE LEFT IN ITS PLACE
17      NUGH(IN)=0
        NGEQ(IN)=0
        NGOL(IN)=0
        GO TO 18
24      DO 23 IO=1,NLAY2
        DO 21 K=1,3
21      POSA(K)=ABS(DRL(IN,K)-SDRL(IO,K))
        IF (POSA(1)+POSA(2)+POSA(3)-0.001) 22,22,23
C   OLD GH CAN BE USED, BUT NEEDS COPYING INTO OTHER PLACE. HOWEVER, DO
C   NOT ALLOW COPYING FROM A PLACE THAT COPYING WILL ALREADY HAVE
C   OVERWRITTEN
22      IF (NGOL(IO).NE.0) GO TO 23
        NUGH(IN)=0
        NGEQ(IN)=0
```

```
        NGOL(IN)=IO
        GO TO 18
 23     CONTINUE
 18     CONTINUE
C  STORE PRESENT INTERPLANAR VECTORS FOR LATER COMPARISON
 19     DO 20 IN=1,NLAY2
        DO 20 J=1,3
 20     SDRL(IN,J)=DRL(IN,J)
        RETURN
        END
C----------------------------------------------------------------------
C  SUBROUTINE SUBREF USES LAYER DOUBLING TO GENERATE REFLECTION AND
C  TRANSMISSION MATRICES FOR A STACK OF LAYERS OF SUFFICIENT THICKNESS
C  TO ACHIEVE CONVERGENCE TO THE SEMI-INFINITE CRYSTAL LIMIT. THE
C  INDIVIDUAL ATOMIC LAYERS MUST BE REPRESENTED BY ALTERNATING
C  REFLECTION AND TRANSMISSION MATRICES (OR EQUAL ONES), SO THAT AN
C  ABABA...STACKING SEQUENCE CAN BE HANDLED. THE INTERLAYER VECTORS MUST
C  BE THE SAME BETWEEN ALL SUCCESSIVE LAYERS. THE INDIVIDUAL LAYERS MUST
C  HAVE EQUAL DIFFRACTION MATRICES FOR INCIDENCE ON EITHER SIDE OF THEM.
C  RA,TA,RB,TB= INPUT REFLECTION AND TRANSMISSION MATRICES FOR
C   INDIVIDUAL LAYERS (RA,TA IS FOR LAYER CLOSEST TO THE SURFACE).
C   THESE MATRICES WILL BE OVERWRITTEN AND FINALLY ON OUTPUT REPRESENT
C   REFLECTION AND TRANSMISSION BY A STACK OF LAYERS FOR INCIDENCE
C   TOWARDS +X (RA,TA) AND FOR INCIDENCE TOWARDS -X (RB,TB).
C  N= NO. OF BEAMS USED IN CURRENT BEAM SET.
C  S1,S2,S3,S4,PP,XS,INT= WORKING SPACES.
C  NP= DIMENSION FOR WORKING SPACES (NP.GE.N).
C  NA= OFFSET OF CURRENT BEAM SET IN LIST PQ.
C  NT= TOTAL NO. OF BEAMS IN MAIN PROGRAM AT CURRENT ENERGY.
C  PQ= LIST OF RECIPROCAL LATTICE VECTORS G (BEAMS).
C  IN COMMON BLOCKS:
C  LMAX,SO1,SO2,SO3,SS1,SS2,SS3,SS4: NOT USED.
C  EPS= CONVERGENCE CRITERION FOR LAYER DOUBLING
C  LITER= LIMIT ON NUMBER OF DOUBLINGS.
C  E,VPI: COMPLEX CURRENT ENERGY.
C  AK2,AK3: PARALLEL COMPONENTS OF PRIMARY INCIDENT K-VECTOR.
C  AS= INTERLAYER VECTOR (AS(1).GT.0.0).
C  NOTE: DO NOT IN GENERAL ASSIGN THE SAME STORAGE AREA FOR RA AND RB,
C  AND FOR TA AND TB, IN THE CALLING PROGRAM, EVEN THOUGH RA=RB, TA=TB
C  MAY HOLD FOR THE INDIVIDUAL LAYERS: THESE EQUALITIES WILL IN GENERAL
C  NO LONGER HOLD AFTER DOUBLING.
        SUBROUTINE  SUBREF (RA, TA, RB, TB, N, S1, S2, S3, S4, PP, XS,
       1INT, NP, NA, NT, PQ)
        INTEGER   INT
        COMPLEX   RA, TA, RB, TB, PP, XS, IU, XX
        COMPLEX   S1, S2, S3, S4
        COMPLEX CSQRT,CEXP
        DIMENSION  RA(N,N), TA(N,N), RB(N,N), TB(N,N), PP(2,NP), XS(NP)
        DIMENSION  S1(NP,NP), S2(NP,NP), S3(NP,NP), S4(NP,NP), PQ(2,NT)
        DIMENSION  AS(3),SS1(2),SS2(2),INT(NP),SS3(2),SS4(2)
        DIMENSION  SO1(2),SO2(2),SO3(2)
        COMMON   /MS/LMAX,EPS,LITER,SO1,SO2,SO3,SS1,SS2,SS3,SS4
        COMMON   /X4/E, VPI, AK2, AK3,  AS
 10     FORMAT(4H X =,E15.4)
 20     FORMAT(24H NO CONV IN SUBREF AFTER,I3,2X,9HITER, X =,E15.4)
 30     FORMAT(I4,16H  ITER IN SUBREF)
        IU = CMPLX(0.0,1.0)
        AK = 2.0 * E
        DO 790 I = 1, N
        BK2 = AK2 + PQ(1,I + NA)
        BK3 = AK3 + PQ(2,I + NA)
        XX = CMPLX(AK - BK2 * BK2 - BK3 * BK3, - 2.0 * VPI + 0.000001)
        XX = CSQRT(XX)
        X = BK2 * AS(2) + BK3 * AS(3)
C  GENERATE PLANE-WAVE PROPAGATORS BETWEEN LAYERS
        PP(1,I) = CEXP(IU * (XX * AS(1) + X))
 790    PP(2,I) = CEXP( - IU * ( - XX * AS(1) + X))
```

```
      L = 0
      X = 0.0
C  START ITERATION OVER NUMBER OF DOUBLINGS. M=1 INDICATES PAIRING OF
C  SYMMETRICAL, BUT POSSIBLY DIFFERENT LAYERS
      M = 1
  800 XO = X
C  IF NO. OF BEAMS IS 1, USE SIMPLE FORMULA RATHER THAN MATRIX INVERSION
C  OF TLRTA
      IF (N-1)  840, 810, 840
  810 XX = 1.0/(CMPLX(1.0,0.0) - RA(1,1) * PP(2,1) * RB(1,1) * PP(1,1))
      IF (M-1)  830, 820, 830
  820 IU = RA(1,1) + TA(1,1) * PP(2,1) * RB(1,1) * PP(1,1) * TA(1,1) *
     1XX
      RB(1,1) = RB(1,1) + TB(1,1) * PP(1,1) * RA(1,1) * PP(2,1) * TB(1,
     11) * XX
      RA(1,1) = IU
      IU = TB(1,1) * PP(1,1) * TA(1,1) * XX
      TB(1,1) = TA(1,1) * PP(2,1) * TB(1,1) * XX
      TA(1,1) = IU
      GO TO 850
  830 IU = RA(1,1) + TB(1,1) * PP(2,1) * RA(1,1) * PP(1,1) * TA(1,1) *
     1XX
      RB(1,1) = RB(1,1) + TA(1,1) * PP(1,1) * RB(1,1) * PP(2,1) * TB(1,
     11) * XX
      RA(1,1) = IU
      IU = TA(1,1) * PP(1,1) * TA(1,1) * XX
      TB(1,1) = TB(1,1) * PP(2,1) * TB(1,1) * XX
      TA(1,1) = IU
      GO TO 850
C  PERFORM ONE DOUBLING STEP
  840 CALL   TLRTA (RA, TA, RB, TB, N, M, S1, S2, S3, S4, PP, XS, INT,
     1NP)
  850 X = 0.0
      DO 860 I = 1, N
      DO 860 J = 1, N
      XX = RA(I,J) * PP(2,I) * PP(1,J)
  860 X = X + ABS(REAL(XX)) + ABS(AIMAG(XX))
      WRITE (6,10)  X
      L = L + 1
C  CHECK NUMBER OF ITERATIONS
      IF (L-LITER)  880, 870, 870
  870 WRITE (6,20)  LITER, X
      GO TO 900
C  CHECK CONVERGENCE
  880 IF (ABS(X-XO)/X .LT. EPS)  GO TO 890
C  M=2 INDICATES PAIRING OF EQUAL, BUT POSSIBLY ASYMMETRICAL (IN +-X)
C  LAYERS OR SLABS
      M = 2
      GO TO 800
  890 WRITE (6,30)  L
  900 RETURN
      END
C-------------------------------------------------------------------
C  SUBROUTINE TAUMAT COMPUTES THE MATRIX TAU (INTRA-SUBPLANE MULTIPLE
C  SCATTERING IN (L,M)-REPRESENTATION) FOR A SINGLE BRAVAIS-LATTICE
C  LAYER FOR EACH OF SEVERAL CHEMICAL ELEMENTS. THE MATRIX ELEMENTS ARE
C  ORDERED THUS: THE (L,M) SEQUENCE IS THE 'SYMMETRIZED' ONE (CF. BOOK).
C  TAU IS SPLIT INTO BLOCK-DIAGONAL PARTS CORRESPONDING TO EVEN L+M
C  AND ODD L+M OF DIMENSIONS LEV AND LOD, RESP. IN THE MATRIX TAU THE
C  PART FOR L+M= ODD IS LEFT-JUSTIFIED AND THE MATRIX REDUCED IN WIDTH
C  FOR STORAGE EFFICIENCY. THE TAU MATRICES FOR DIFFERENT CHEMICAL
C  ELEMENTS ARE ARRANGED UNDER EACH OTHER IN COLUMNAR FASHION: AT THE
C  TOP IS THE TAU BASED ON THE ATOMIC T-MATRIX ELEMENTS TSF(1,L),
C  FOLLOWED BY THE TAU'S BASED ON TSF(2,L), TSF(3,L), ETC.
C  FOR QUANTITIES NOT EXPLAINED BELOW SEE MPERTI OR MTINV.
C   TAU= OUTPUT MATRIX (SEE ABOVE).
C   LMT= NTAU*LMMAX.
```

```
C     NTAU= NO. OF CHEMICAL ELEMENTS TO BE USED.
C     X= WORKING SPACE.
C     TSF= ATOMIC T-MATRIX ELEMENTS.
C     FLMS= LATTICE SUMS FROM SUBROUTINE FMAT.
C     NL=NO. OF SUBLATTICES CONSIDERED IN FMAT (ONLY THE FIRST SUBLATTICE
C        SUM IS USED BY TAUMAT).
C     CLM= CLEBSCH-GORDON COEFFICIENTS FROM SUBROUTINE CELMG.
C     NLM= DIMENSION OF CLM (SEE MAIN PROGRAM).
C     LXI= PERMUTATION OF (L,M) SEQUENCE FROM SUBROUTINE LXGENT.
C     IN COMMON BLOCKS:
C     E,VPI= CURRENT COMPLEX ENERGY.
C     NOTE: DIMENSIONS ARE SET FOR UP TO 3 CHEMICAL ELEMENTS AND LMAX.LE.9.
      SUBROUTINE TAUMAT(TAU,LMT,NTAU,X,LEV,LEV2,LOD,TSF,
     1LMMAX,LMAX,FLMS,NL,KLM,LM,CLM,NLM,LXI)
      DIMENSION CLM(NLM),LXI(LMMAX)
      COMPLEX AK,CZ,TAU(LMT,LEV),X(LEV,LEV2),TSF(3,10),
     1FLMS(NL,KLM),DET
      COMMON E,AK2,AK3,VPI
      CZ=(0.0,0.0)
      AK=-0.5/CSQRT(CMPLX(2.0*E,-2.0*VPI+0.000001))
      DO 13 IT=1,NTAU
      DO 13 IL=1,2
      DO 1 I=1,LEV
      DO 1 J=1,LEV2
1     X(I,J)=CZ
      LL=LOD
      IF (IL.EQ.2) LL=LEV
C     GENERATE MATRIX 1-X FOR L+M= ODD (IL=1), LATER FOR L+M= EVEN (IL=2)
      CALL XMT(IL,FLMS,NL,X,LEV,LL,TSF,NTAU,IT,LM,LXI,LMMAX,
     1KLM,CLM,NLM,NST)
C
C     PREPARE QUANTITIES INTO WHICH INVERSE OF 1-X WILL BE MULTIPLIED
      IF (IL-2) 6,2,2
2     IS=0
      LD1=0
      L=0
3     LD=LD1+1
      LD1=LD+L
      DO 4 I=LD,LD1
4     X(I,I+LEV)=AK*TSF(IT,L+1)
      L=L+2
      IF (L-LMAX) 3,3,5
5     IS=IS+1
      L=1
      IF (IS-1) 3,3,10
C
6     IS=0
      LD1=0
      L=1
7     LD=LD1+1
      LD1=LD+L-1
      DO 8 I=LD,LD1
8     X(I,I+LOD)=AK*TSF(IT,L+1)
      L=L+2
      IF (L-LMAX) 7,7,9
9     IS=IS+1
      L=2
      IF (IS-1) 7,7,10
10    LL2=LL+LL
C
C     PERFORM INVERSION AND MULTIPLICATION
      CALL CXMTXT(X,LEV,LL,LL,LL2,MARK,DET,-1)
C
      LD=(IT-1)*LMMAX
      IF (IL.EQ.1) LD=LD+LEV
      DO 11 I=1,LL
      DO 11 J=1,LL
```

```
C  PUT RESULT IN TAU
11     TAU(LD+I,J)=X(I,J+LL)
13     CONTINUE
       RETURN
       END
C-----------------------------------------------------------------------
C  SUBROUTINE TAUY PRODUCES THE QUANTITIES TAU*YLM(G) FOR ALL CHEMICAL
C  ELEMENTS AND, IF BEEBY-TYPE INVERSION IS CALLED FOR IN THE RSP
C  CALCULATION, TH*YLM(G) FOR ALL RELEVANT LAYERS. THESE ARE USED
C  EITHER FOR THE BEEBY-TYPE INVERSION IN MTINV OR AS INITIAL VALUES FOR
C  THE RSP ITERATION IN MPERTI.
C  FOR QUANTITIES NOT DESCRIBED BELOW SEE CALLING ROUTINES (MPERTI OR
C  MTINV).
C   TAUG,TAUGM= OUTPUT RESULTING VECTORS FOR K(G+) AND K(G-),RESP.
C   LTAUG= (NTAU+NINV)*LMMAX (NINV TO BE 0 IN CALL FROM MTINV).
C   LT=PERMUTATION OF (LM) ORDER, FROM LXGENT.
C   TH= T-MATRIX FROM THINV, USED ONLY IF LINV=1 (I.E. IF INVERSION IS
C   REQUIRED IN MPERTI).
C   LMNI= NINV*LMMAX, USED ONLY IF LINV=1.
C   LMNI2= 2*LMNI, USED ONLY IF LINV=1.
C   JGP= CURRENT INCIDENT BEAM.
C   LINV=0: NO TH*YLM(G) TO BE PRODUCED.
C   LINV=1: DO PRODUCE THOSE QUANTITIES.
C   INV: INDICATOR OF WHICH SUBPLANES ARE INVOLVED IN BEEBY-TYPE
C   INVERSION IN MPERTI.
       SUBROUTINE TAUY(TAUG,TAUGM,LTAUG,TAU,LMT,LEV,CYLM,NT,LMMAX,LT,
      1TH,LMNI,LMNI2,NTAU,LOD,LEE,LOE,JGP,LINV,RG,NLAY,NG,INV)
       COMPLEX CZ,ST,SU,CF,CF1
       COMPLEX TAU(LMT,LEV),CYLM(NT,LMMAX),TH(LMNI,LMNI2)
       COMPLEX TAUG(LTAUG),TAUGM(LTAUG),RG(3,NLAY,NG)
       DIMENSION INV(NLAY),LT(LMMAX)
       CZ=(0.0,0.0)
C
C  PERFORM MATRIX PRODUCT TAU*YLM(G+-) FOR EACH CHEMICAL ELEMENT
       DO 275 I=1,NTAU
       IS=(I-1)*LMMAX
       DO 250 JLM=1,LEV
       ST=CZ
       DO 245 ILM=1,LEV
       KLP=LT(ILM)
       IF (ILM.GT.LEE) GO TO 240
       ST=ST+TAU(IS+JLM,ILM)*CYLM(JGP,KLP)
       GO TO 245
 240   ST=ST-TAU(IS+JLM,ILM)*CYLM(JGP,KLP)
 245   CONTINUE
       TAUG(IS+JLM)=ST
 250   TAUGM(IS+JLM)=ST
       IS=IS+LEV
       DO 270 JLM=1,LOD
       ST=CZ
       SU=CZ
       DO 265 ILM=1,LOD
       KLP=LT(ILM+LEV)
       CF=TAU(IS+JLM,ILM)*CYLM(JGP,KLP)
       IF (ILM.GT.LOE) GO TO 260
       ST=ST+CF
       SU=SU-CF
       GO TO 265
 260   ST=ST-CF
       SU=SU+CF
 265   CONTINUE
       TAUG(IS+JLM)=ST
 270   TAUGM(IS+JLM)=SU
 275   CONTINUE
C
C  FIND WHETHER SIMILAR PRODUCT FOR TH IS NEEDED, ELSE RETURN
       IF (LINV.EQ.0) RETURN
```

```
      IC=0
      DO 380 IN=1,NLAY
      IF (INV(IN).EQ.0) GO TO 380
      IC=IC+1
      ICL=(IC-1)*LMMAX
      IS=LMT+ICL
      DO 370 JLM=1,LMMAX
      ST=CZ
      SU=CZ
      DO 360 ILM=1,LMMAX
      CF=CZ
      CF1=CZ
      ID=0
C  SUM OVER LAYERS INVOLVED IN BEEBY INVERSION
      DO 280 INP=1,NLAY
      IF (INV(INP).EQ.0) GO TO 280
      ID=ID+1
      IDL=(ID-1)*LMMAX
      CF=CF+TH(ICL+JLM,IDL+ILM)*RG(1,INP,JGP)
      CF1=CF1+TH(ICL+JLM,IDL+ILM)/RG(2,INP,JGP)
280   CONTINUE
      KLP=LT(ILM)
      CF=CF*CYLM(JGP,KLP)
      CF1=CF1*CYLM(JGP,KLP)
      IF (ILM-LEV) 290,290,320
290   IF (ILM-LEE) 300,300,310
300   ST=ST+CF
      SU=SU+CF1
      GO TO 350
310   ST=ST-CF
      SU=SU-CF1
      GO TO 350
320   IF (ILM-LEV-LOE) 330,330,340
330   ST=ST+CF
      SU=SU-CF1
      GO TO 350
340   ST=ST-CF
      SU=SU+CF1
350   CONTINUE
360   CONTINUE
      TAUG(IS+JLM)=ST/RG(1,IN,JGP)
370   TAUGM(IS+JLM)=SU*RG(2,IN,JGP)
380   CONTINUE
      RETURN
      END
C------------------------------------------------------------------------
C  SUBROUTINE THINV PERFORMS THE BEEBY-TYPE INVERSION WHEN CALLED FOR
C  IN MPERTI. FOR QUANTITIES NOT EXPLAINED BELOW SEE MPERTI.
C  TH= OUTPUT INVERSE MATRIX.
C  LMNI= NINV*LMMAX.
C  LMNI2=2*LMNI.
C  INV: INDICATOR OF WHICH SUBPLANES ARE INVOLVED IN BEEBY-TYPE
C  INVERSION IN MPERTI.
      SUBROUTINE THINV(TH,LMNI,LMNI2,GH,LMG,LMMAX,MGH,NLAY,TAU,LMT,LEV,
     1INV,LPS)
      COMPLEX TH(LMNI,LMNI2),GH(LMG,LMMAX),TAU(LMT,LEV)
      COMPLEX RU,CZ,ST,SU
      DIMENSION INV(NLAY),LPS(NLAY),MGH(NLAY,NLAY)
      LOD=LMMAX-LEV
      RU=(1.0,0.0)
      CZ=(0.0,0.0)
C
C  INSERT UNITY IN DIAGONAL
      DO 20 I=1,LMNI
      DO 10 J=1,LMNI2
10    TH(I,J)=CZ
20    TH(I,I)=RU
```

```
            IC=0
            NLAY1=NLAY-1
            DO 120 IN=1,NLAY1
            IF (INV(IN).EQ.0) GO TO 120
            IC=IC+1
            ICL=(IC-1)*LMMAX
            ID=IC
            IN1=IN+1
            DO 110 INP=IN1,NLAY
            IF (INV(INP).EQ.0) GO TO 110
            ID=ID+1
            IDL=(ID-1)*LMMAX
            M=(MGH(IN,INP)-1)*LMMAX
            M1=(MGH(INP,IN)-1)*LMMAX
            LP=(LPS(IN)-1)*LMMAX
            LP1=(LPS(INP)-1)*LMMAX
            DO 60 I=1,LEV
            DO 40 J=1,LMMAX
            ST=CZ
            SU=CZ
C     INSERT -TAU*GH FOR EVEN L+M (IN FIRST INDEX)
            DO 30 K=1,LEV
            ST=ST+TAU(LP+I,K)*GH(M+K,J)
30          SU=SU+TAU(LP1+I,K)*GH(M1+K,J)
            TH(ICL+I,IDL+J)=-ST
40          TH(IDL+I,ICL+J)=-SU
60          CONTINUE
            LP=LP+LEV
            LP1=LP1+LEV
            M=M+LEV
            M1=M1+LEV
            ICV=ICL+LEV
            IDV=IDL+LEV
            DO 100 I=1,LOD
            DO 80 J=1,LMMAX
            ST=CZ
            SU=CZ
C     INSERT -TAU*GH FOR ODD L+M (IN FIRST INDEX)
            DO 70 K=1,LOD
            ST=ST+TAU(LP+I,K)*GH(M+K,J)
70          SU=SU+TAU(LP1+I,K)*GH(M1+K,J)
            TH(ICV+I,IDL+J)=-ST
80          TH(IDV+I,ICL+J)=-SU
100         CONTINUE
110         CONTINUE
120         CONTINUE
C
C     INSERT TAU'S IN RIGHT-HAND PART OF TH. INVERSE WILL MULTIPLY INTO
C     THESE TAU'S
            IC=0
            DO 150 IN=1,NLAY
            IF (INV(IN).EQ.0) GO TO 150
            IC=IC+1
            ICL=(IC-1)*LMMAX
            IDL=ICL+LMNI
            LP=(LPS(IN)-1)*LMMAX
            DO 130 I=1,LEV
            DO 130 J=1,LEV
130         TH(ICL+I,IDL+J)=TAU(LP+I,J)
            ICL=ICL+LEV
            IDL=IDL+LEV
            LP=LP+LEV
            DO 140 I=1,LOD
            DO 140 J=1,LOD
140         TH(ICL+I,IDL+J)=TAU(LP+I,J)
150         CONTINUE
C     INVERT AND MULTIPLY
```

```
      CALL CXMTXT(TH,LMNI,LMNI,LMNI,LMNI2,MARK,ST,-1)
C  SHIFT RESULT TO LEFT-HAND PART OF TH
      DO 160 I=1,LMNI
      DO 160 J=1,LMNI
160   TH(I,J)=TH(I,J+LMNI)
      RETURN
      END
C-----------------------------------------------------------------------
C  SUBROUTINE THMAT PRODUCES THE MATRIX THAT IS INVERTED IN THE BEEBY-
C  TYPE METHOD IN SUBROUTINE MTINV. FOR QUANTITIES NOT EXPLAINED
C  BELOW SEE MTINV.
C  TH= OUTPUT MATRIX, STILL TO BE INVERTED.
C  LMNI= NLAY*LMMAX.
C  LMNI2= LMNI.
      SUBROUTINE THMAT(TH,LMNI,LMNI2,GH,LMG,LMMAX,MGH,NLAY,TAU,LMT,LEV,
     1LPS)
      COMPLEX TH(LMNI,LMNI2),GH(LMG,LMMAX),TAU(LMT,LEV)
      COMPLEX RU,CZ,ST,SU
      DIMENSION LPS(NLAY),MGH(NLAY,NLAY)
      LOD=LMMAX-LEV
      RU=(1.0,0.0)
      CZ=(0.0,0.0)
C
C  INSERT UNITY IN DIAGONAL
      DO 20 I=1,LMNI
      DO 10 J=1,LMNI2
10    TH(I,J)=CZ
20    TH(I,I)=RU
      IC=0
      NLAY1=NLAY-1
      DO 120 IN=1,NLAY1
      IC=IC+1
      ICL=(IC-1)*LMMAX
      ID=IC
      IN1=IN+1
      DO 110 INP=IN1,NLAY
      ID=ID+1
      IDL=(ID-1)*LMMAX
      M=(MGH(IN,INP)-1)*LMMAX
      M1=(MGH(INP,IN)-1)*LMMAX
      LP=(LPS(IN)-1)*LMMAX
      LP1=(LPS(INP)-1)*LMMAX
      DO 60 I=1,LEV
      DO 40 J=1,LMMAX
      ST=CZ
      SU=CZ
C  INSERT -TAU*GH FOR EVEN L+M (IN FIRST INDEX)
      DO 30 K=1,LEV
      ST=ST+TAU(LP+I,K)*GH(M+K,J)
30    SU=SU+TAU(LP1+I,K)*GH(M1+K,J)
      TH(ICL+I,IDL+J)=-ST
40    TH(IDL+I,ICL+J)=-SU
60    CONTINUE
      LP=LP+LEV
      LP1=LP1+LEV
      M=M+LEV
      M1=M1+LEV
      ICV=ICL+LEV
      IDV=IDL+LEV
      DO 100 I=1,LOD
      DO 80 J=1,LMMAX
      ST=CZ
      SU=CZ
C  INSERT -TAU*GH FOR ODD L+M (IN FIRST INDEX)
      DO 70 K=1,LOD
      ST=ST+TAU(LP+I,K)*GH(M+K,J)
70    SU=SU+TAU(LP1+I,K)*GH(M1+K,J)
```

```
          TH(ICV+I,IDL+J)=-ST
 80       TH(IDV+I,ICL+J)=-SU
100       CONTINUE
110       CONTINUE
120       CONTINUE
          RETURN
          END
C-------------------------------------------------------------------------
C   SUBROUTINE TLRT HAS THE SAME PURPOSE AS TLRTA, BUT IT MAKES NO
C   ASSUMPTION ABOUT THE SYMMETRY OF THE LAYERS (WITH RESPECT TO
C   INCIDENCE TOWARDS +-X) OR ON THE EQUALITY OF THE LAYERS.
C   RA1,...,TB2= INPUT REFLECTION AND TRANSMISSION MATRICES OF THE
C      LAYERS (LETTERS R AND T STAND FOR REFLECTION AND TRANSMISSION,
C      RESP., LETTERS A AND B STAND FOR TWO LAYERS, B TO +X SIDE OF A,
C      NUMBERS 1 AND 2 STAND FOR INCIDENCE TOWARDS +X AND -X, RESP.).
C   ON OUTPUT ONLY 4, NOT 8, SUCH MATRICES ARE NEEDED: REFLECTION AND
C   TRANSMISSION OF PAIRED LAYERS FOR INCIDENCE TOWARDS +X OVERWRITE
C   RA1,TA1. THOSE FOR INCIDENCE TOWARDS -X OVERWRITE EITHER RA2,TA2
C   (IF II=1) OR RB1,TB1 (IF II=2: THIS CHOICE ALLOWS RA2,TA2 NOT TO
C   BE OVERWRITTEN).
C   N= NO. OF BEAMS USED.
C   II: SEE DESCRIPTION ABOVE.
C   S1,S2,S3,XS,INT= WORKING SPACES.
C   PP= PLANE-WAVE PROPAGATORS, FROM SUBROUTINE LIKE SUBREF.
C   NP= DIMENSION OF MATRICES (.GE.N).
          SUBROUTINE TLRT(RA1,TA1,RA2,TA2,RB1,TB1,RB2,TB2,N,II,S1,S2,S3,
         1PP,XS,INT,NP)
COMMENT   DO NOT CALL THIS SUBROUTINE WITH IDENTICALLY NAMED MATRICES
          COMPLEX RA1,TA1,RA2,TA2,RB1,TB1,RB2,TB2,XX,YY,CZ,RU,S1,S2,S3,PP,XS
          DIMENSION  RA1(N,N),  TA1(N,N),  RA2(N,N),  TA2(N,N),  S1(NP,NP)
          DIMENSION  S2(NP,NP), S3(NP,NP), PP(2,NP), XS(NP)
          DIMENSION  RB1(N,N),TB1(N,N),RB2(N,N),TB2(N,N)
          DIMENSION  INT(NP)
          CZ = CMPLX(0.0,0.0)
          RU = CMPLX(1.0,0.0)
 950 DO 980 I = 1, N
          DO 970 J = 1, N
          XX = CZ
          YY = CZ
          DO 960 K = 1, N
          XX = XX - RA2(I,K) * PP(2,K) * RB1(K,J) * PP(1,J)
 960 YY = YY - RB1(I,K) * PP(1,K) * RA2(K,J) * PP(2,J)
          S1(I,J) = XX
 970 S2(I,J) = YY
          S1(I,I) = S1(I,I) + RU
 980 S2(I,I) = S2(I,I) + RU
 990 CALL  ZGE (S1, INT, NP, N, 1.0E-6)
          DO 1010 I = 1, N
          DO 1000 J = 1, N
1000 XS(J) = TA1(J,I)
          CALL  ZSU (S1, INT, XS, NP, N, 1.0E-6)
          DO 1010 J = 1, N
1010 S3(J,I) = XS(J)
          CALL  ZGE (S2, INT, NP, N, 1.0E-6)
          DO 1030 I = 1, N
          DO 1020 J = 1, N
1020 XS(J) = TB2(J,I)
          CALL  ZSU (S2, INT, XS, NP, N, 1.0E-6)
          DO 1030 J = 1, N
1030 S1(J,I) = XS(J)
1140 DO 1160 I = 1, N
          DO 1160 J = 1, N
          XX = CZ
          YY = CZ
          DO 1150 K = 1, N
          XX = XX + TB1(I,K) * PP(1,K) * S3(K,J)
1150 YY = YY + RB1(I,K) * PP(1,K) * S3(K,J)
```

```
      TA1(I,J) = XX
 1160 S2(I,J) = YY
      DO 1200 I = 1, N
      DO 1200 J = 1, N
      XX = RA1(I,J)
      DO 1190 K = 1, N
 1190 XX = XX + TA2(I,K) * PP(2,K) * S2(K,J)
 1200 RA1(I,J) = XX
      DO 1220 I = 1, N
      DO 1220 J = 1, N
      XX = CZ
      YY = CZ
      DO 1210 K = 1, N
      XX = XX + TA2(I,K) * PP(2,K) * S1(K,J)
 1210 YY=YY+RA2(I,K)*PP(2,K)*S1(K,J)
      S2(I,J) = XX
 1220 S3(I,J)=YY
      IF (II-1) 1223,1223,1235
 1223 DO 1230 I = 1, N
      DO 1230 J = 1, N
      XX=RB2(I,J)
      DO 1225 K=1,N
 1225 XX=XX+TB1(I,K)*PP(1,K)*S3(K,J)
      RA2(I,J)=XX
 1230 TA2(I,J) = S2(I,J)
      GO TO 1240
 1235 DO 1239 I=1,N
      DO 1239 J=1,N
      XX=RB2(I,J)
      DO 1238 K=1,N
 1238 XX=XX+TB1(I,K)*PP(1,K)*S3(K,J)
 1239 RB1(I,J)=XX
      DO 1241 I=1,N
      DO 1241 J=1,N
 1241 TB1(I,J)=S2(I,J)
 1240 RETURN
      END
C-----------------------------------------------------------------------
C  SUBROUTINE TLRTA PERFORMS EACH LAYER DOUBLING STEP FOR SUBROUTINE
C  SUBREF BY MATRIX INVERSION. THE TWO LAYERS ARE ASSUMED TO BE EITHER
C  SYMMETRICAL IN +-X AND MUTUALLY DIFFERENT (II=1), OR ASYMMETRICAL IN
C  +-X AND MUTUALLY IDENTICAL (II=2).
C   R1,T1,R2,T2= INPUT REFLECTION AND TRANSMISSION MATRICES OF THE TWO
C    LAYERS. IF II=1, THE NUMBERS 1 AND 2 REFER TO THE FIRST AND SECOND
C    LAYER, RESP. (2 ON +X SIDE OF 1). IF II=2, THE NUMBERS 1 AND 2
C    REFER TO INCIDENCE TOWARDS +X AND -X, RESP. ON OUTPUT, REGARDLESS
C    OF II, THE NUMBERS 1 AND 2 REFER TO THE INCIDENCE DIRECTION.
C   N= NO. OF BEAMS USED.
C   II: SEE DESCRIPTION ABOVE.
C   S1,S2,S3,S4,XS,INT= WORKING SPACES.
C   PP= PLANE-WAVE PROPAGATORS, FROM SUBROUTINE SUBREF.
C   NP= DIMENSION OF MATRICES (.GE.N).
      SUBROUTINE  TLRTA (R1, T1, R2, T2, N, II, S1, S2, S3, S4, PP, XS,
     1INT, NP)
      COMPLEX  R1, T1, R2, T2, XX, YY, CZ, RU, S1, S2, S3, S4, PP, XS
      DIMENSION  R1(N,N), T1(N,N), R2(N,N), T2(N,N), S1(NP,NP)
      DIMENSION  S2(NP,NP), S3(NP,NP), S4(NP,NP), PP(2,NP), XS(NP)
      DIMENSION  INT(NP)
      CZ = CMPLX(0.0,0.0)
      RU = CMPLX(1.0,0.0)
      IF (II-1)  910, 910, 950
  910 DO 940 I = 1, N
      DO 930 J = 1, N
      XX = CZ
      YY=CZ
      DO 920 K = 1, N
      XX = XX - R1(I,K) * PP(2,K) * R2(K,J) * PP(1,J)
```

```
 920 YY = YY - R2(I,K) * PP(1,K) * R1(K,J) * PP(2,J)
     S1(I,J) = XX
 930 S2(I,J) = YY
     S1(I,I) = S1(I,I) + RU
 940 S2(I,I) = S2(I,I) + RU
     GO TO 990
 950 DO 980 I = 1, N
     DO 970 J = 1, N
     XX = CZ
     YY = CZ
     DO 960 K = 1, N
     XX = XX - R2(I,K) * PP(2,K) * R1(K,J) * PP(1,J)
 960 YY = YY - R1(I,K) * PP(1,K) * R2(K,J) * PP(2,J)
     S1(I,J) = XX
 970 S2(I,J) = YY
     S1(I,I) = S1(I,I) + RU
 980 S2(I,I) = S2(I,I) + RU
 990 CALL   ZGE (S1, INT, NP, N, 1.0E-6)
     DO 1010 I = 1, N
     DO 1000 J = 1, N
1000 XS(J) = T1(J,I)
     CALL  ZSU (S1, INT, XS, NP, N, 1.0E-6)
     DO 1010 J = 1, N
1010 S3(J,I) = XS(J)
     CALL   ZGE (S2, INT, NP, N, 1.0E-6)
     DO 1030 I = 1, N
     DO 1020 J = 1, N
1020 XS(J) = T2(J,I)
     CALL  ZSU (S2, INT, XS, NP, N, 1.0E-6)
     DO 1030 J = 1, N
1030 S4(J,I) = XS(J)
     IF (II-1)   1040, 1040, 1140
1040 DO 1060 I = 1, N
     DO 1060 J = 1, N
     XX = CZ
     YY = CZ
     DO 1050 K = 1, N
     XX = XX + T2(I,K) * PP(1,K) * S3(K,J)
1050 YY = YY + R2(I,K) * PP(1,K) * S3(K,J)
     S1(I,J) = XX
1060 S2(I,J) = YY
     DO 1080 I = 1, N
     DO 1080 J = 1, N
     XX = CZ
     DO 1070 K = 1, N
1070 XX = XX + R1(I,K) * PP(2,K) * S4(K,J)
1080 S3(I,J) = XX
     DO 1100 I = 1, N
     DO 1100 J = 1, N
     XX = R1(I,J)
     YY = R2(I,J)
     DO 1090 K = 1, N
     XX = XX + T1(I,K) * PP(2,K) * S2(K,J)
1090 YY = YY + T2(I,K) * PP(1,K) * S3(K,J)
     R1(I,J) = XX
1100 R2(I,J) = YY
     DO 1120 I = 1, N
     DO 1120 J = 1, N
     XX = CZ
     DO 1110 K = 1, N
1110 XX = XX + T1(I,K) * PP(2,K) * S4(K,J)
1120 S2(I,J) = XX
     DO 1130 I = 1, N
     DO 1130 J = 1, N
     T2(I,J)=S2(I,J)
1130 T1(I,J) = S1(I,J)
     GO TO 1240
```

```
1140 DO 1160 I = 1, N
     DO 1160 J = 1, N
     XX = CZ
     YY = CZ
     DO 1150 K = 1, N
     XX = XX + T1(I,K) * PP(1,K) * S3(K,J)
1150 YY = YY + R1(I,K) * PP(1,K) * S3(K,J)
     S1(I,J) = XX
1160 S2(I,J) = YY
     DO 1180 I = 1, N
     DO 1180 J = 1, N
     XX = CZ
     DO 1170 K = 1, N
1170 XX = XX + R2(I,K) * PP(2,K) * S4(K,J)
1180 S3(I,J) = XX
     DO 1200 I = 1, N
     DO 1200 J = 1, N
     XX = R1(I,J)
     YY = R2(I,J)
     DO 1190 K = 1, N
     XX = XX + T2(I,K) * PP(2,K) * S2(K,J)
1190 YY = YY + T1(I,K) * PP(1,K) * S3(K,J)
     R1(I,J) = XX
1200 R2(I,J) = YY
     DO 1220 I = 1, N
     DO 1220 J = 1, N
     XX = CZ
     DO 1210 K = 1, N
1210 XX = XX + T2(I,K) * PP(2,K) * S4(K,J)
1220 S2(I,J) = XX
     DO 1230 I = 1, N
     DO 1230 J = 1, N
     T1(I,J) = S1(I,J)
1230 T2(I,J) = S2(I,J)
1240 RETURN
     END
C-------------------------------------------------------------------
C    SUBROUTINE TPERTI SUPERVISES THE REVERSE SCATTERING PERTURBATION
C    ITERATION FOR SUBROUTINE MPERTI. WHEN REQUESTED, IT TAKES INTO
C    ACCOUNT THE BEEBY-TYPE INVERSION, WHICH MUST HAVE BEEN PREVIOUSLY
C    PERFORMED BY SUBROUTINE THINV.
C    FOR QUANTITIES NOT EXPLAINED BELOW SEE MPERTI.
C    TS= OUTPUT VECTOR T(LM,I,G).
C    LMN=NLAY*LMMAX.
C    TG= WORKING SPACE.
C    LM2N= 2*LMN.
C    RG= PLANE-WAVE PROPAGATION FACTORS FROM MPERTI.
C    NG= NO. OF BEAMS IN CURRENT BEAM SET.
C    MGH= INDEX TO POSITION OF PROPAGATION MATRICES IN GH.
C    LPS: INDICATES WHICH CHEMICAL ELEMENT IS IN WHICH SUBPLANE.
C    NLAY= NO. OF SUBPLANES IN LAYER TREATED BY MPERTI.
C    INC= CURRENT INCIDENCE DIRECTION (+-1 TOWARDS +-X).
C    TAUG= INPUT INITIAL VALUES, FROM TAUY.
C    LTAUG= (NTAU+NINV)*LMMAX.
C    V= WORKING SPACE.
C    EPS= SMALL NUMBER USED IN MONITORING CONVERGENCE.
C    NPERT= LIMIT ON NO. OF ITERATIONS.
C    JG= INDEX OF CURRENT INCIDENT BEAM.
C    INV: INDICATES WHICH SUBPLANES ARE INVOLVED IN BEEBY-TYPE INVERSION.
C    NINV= NO. OF SUBPLANES INVOLVED IN BEEBY-TYPE INVERSION.
C    TH= T-MATRIX INPUT FROM THINV.
C    LMNI= NINV*LMMAX.
C    LMNI2= 2*LMNI (LMNI2=LMNI ACTUALLY SUFFICIENT HERE)
     SUBROUTINE TPERTI(TS,LMN,TG,LM2N,LMMAX,TAU,LMT,LEV,GH,LMG,RG,NG,
    1MGH,LPS,NLAY,INC,TAUG,LTAUG,V,EPS,NPERT,JG,INV,NINV,TH,LMNI,LMNI2)
     DIMENSION MGH(NLAY,NLAY),LPS(NLAY),XS(15),X(15),INV(NLAY)
     COMPLEX CZ,TS(LMN),TG(2,LM2N),TAU(LMT,LEV),GH(LMG,LMMAX),
```

```
      1RG(3,NLAY,NG),TAUG(LTAUG),V(LMMAX),TH(LMNI,LMNI2)
       CZ=(0.0,0.0)
       IC=0
C   INITIALIZE WORKING SPACE TG AND ACCUMULATOR TS
       DO 1 K=1,NLAY
       LP=(LPS(K)-1)*LMMAX
       IF (INV(K).EQ.0) GO TO 2
       IC=IC+1
       LP=LMT+(IC-1)*LMMAX
2      KK=(K-1)*LMMAX
       KL=LM2N-KK-LMMAX
       DO 1 L=1,LMMAX
       TS(KK+L)=TAUG(LP+L)
       TG(1,KK+L)=TAUG(LP+L)
       TG(1,KL+L)=TAUG(LP+L)
       TG(2,KK+L)=CZ
1      TG(2,KL+L)=CZ
       KM=1
       KT=LMMAX
       KL=NLAY*LMMAX
       DO 5 K=KM,KT
       DO 5 I=1,2
       TG(I,K)=CZ
5      TG(I,KL+K)=CZ
       IF (INV(1).NE.1) GO TO 6
       KM=1
       KT=NINV*LMMAX
       DO 20 K=KM,KT
20     TG(1,K)=CZ
       GO TO 7
6      IF (INV(NLAY).NE.1) GO TO 7
       KM=LMN+1
       KT=LMN+NINV*LMMAX
       DO 3 K=KM,KT
3      TG(1,K)=CZ
C
C   START MONITORING CONVERGENCE OF TS
7      DO 8 I=1,NLAY
       X(I)=0.0
       KT=(I-1)*LMMAX
       DO 8 L=1,LMMAX
8      X(I)=X(I)+CABS(TS(L+KT))
       WRITE(6,13)(X(K),K=1,NLAY)
C
C
C   START FIRST ITERATION
       IPERT=0
       MNC=-INC
       IT=2
C   FIGURE OUT AT WHICH SUBPLANE PROPAGATION IN INCIDENCE DIRECTION
C   SHOULD START
9      NI=2
       IF ((INC.EQ.1.AND.INV(1).EQ.1).OR.(INC.EQ.-1.AND.INV(NLAY).EQ.1))
      1 NI=NINV+1
C   PROPAGATE THROUGH ONE SUBPLANE AT A TIME IN INCIDENCE DIRECTION
       DO 10 I=NI,NLAY
       ILAY=I
       IF (INC.EQ.-1) ILAY=NLAY-I+1
10     CALL TPSTPI(TG,LM2N,LMMAX,NLAY,ILAY,INC,INC,TAU,LMT,
      1LEV,GH,LMG,MGH,LPS,V,RG,NG,JG,IT,INV,TH,LMNI,LMNI2)
C
C   REPEAT OPERATIONS FOR PROPAGATION IN OPPOSITE DIRECTION
       NI=2
       IF ((INC.EQ.1.AND.INV(NLAY).EQ.1).OR.(INC.EQ.-1.AND.INV(1).EQ.1))
      1 NI=NINV+1
       DO 11 I=NI,NLAY
       ILAY=NLAY-I+1
```

```
      IF (INC.EQ.-1) ILAY=I
11    CALL TPSTPI(TG,LM2N,LMMAX,NLAY,ILAY,INC,MNC,TAU,LMT,
     1LEV,GH,LMG,MGH,LPS,V,RG,NG,JG,IT,INV,TH,LMNI,LMNI2)
      DO 12 K=1,NLAY
      XS(K)=X(K)
      X(K)=0.0
      KL=(K-1)*LMMAX
      KM=(2*NLAY-K)*LMMAX
      DO 12 L=1,LMMAX
C  ADD CURRENT ORDER OF PERTURBATION TO ACCUMULATOR
      TS(KL+L)=TS(KL+L)+TG(IT,KM+L)+TG(IT,KL+L)
12    X(K)=X(K)+CABS(TS(KL+L))
      IPERT=IPERT+1
C  LIMIT NUMBER OF ITERATIONS
      IF (IPERT.GT.NPERT) GO TO 15
      IT=3-IT
      DO 14 I=1,NLAY
C  TEST FOR CONVERGENCE
      IF ((ABS(XS(I)-X(I))/XS(I)).GT.EPS) GO TO 9
14    CONTINUE
C  PRINT QUANTITIES TESTED FOR CONVERGENCE
      WRITE(6,13)(X(K),K=1,NLAY)
13    FORMAT(16H X IN TPERTI =  ,8E13.5,/,7E13.5)
21    FORMAT(26H    CONV IN TPERTI AFTER  ,I3,12H ITERATIONS )
      WRITE(6,21)IPERT
      RETURN
15    WRITE(6,13)(X(K),K=1,NLAY)
      WRITE(6,16)NPERT
16    FORMAT(26H NO CONV IN TPERTI AFTER  ,I3,12H ITERATIONS )
      RETURN
      END
C-----------------------------------------------------------------------
C  SUBROUTINE TPSTPI PERFORMS A SINGLE ONE-SUBPLANE PROPAGATION OF THE
C  WAVEFIELD IN THE REVERSE SCATTERING PERTURBATION METHOD. SUBPLANES
C  TREATED BY THE BEEBY-TYPE INVERSION CAN BE DEALT WITH.
C  FOR QUANTITIES NOT EXPLAINED BELOW SEE TPERTI AND MPERTI.
C  TG= RUNNING T(G) VECTOR IN PERTURBATION EXPANSION.
C  LM2N= 2*NLAY*LMMAX.
C  ILAY= SUBPLANE TO WHICH CURRENT PROPAGATION LEADS.
C  INC= INCIDENCE DIRECTION (+-1 TOWARDS +-X).
C  IPR= PROPAGATION DIRECTION (+-1 TOWARDS +-X).
C  V= WORKING SPACE.
C  IT= ALTERNATELY 1 AND 2 IN EACH SUCCESSIVE ITERATION.
C  INV: INDICATES WHICH SUBPLANES ARE INVOLVED IN BEEBY-TYPE INVERSION.
      SUBROUTINE TPSTPI(TG,LM2N,LMMAX,NLAY,ILAY,INC,IPR,TAU,
     1LMT,LEV,GH,LMG,MGH,LPS,V,RG,NG,JG,IT,INV,TH,LMNI,LMNI2)
      DIMENSION MGH(NLAY,NLAY),LPS(NLAY),INV(NLAY)
      COMPLEX CZ,SS,ST,TG(2,LM2N),TAU(LMT,LEV),GH(LMG,LMMAX),
     1V(LMMAX),RG(3,NLAY,NG),TH(LMNI,LMNI2)
      CZ=(0.0,0.0)
C  ITP IS 2 OR 1, DEPENDING ON WHETHER THE NUMBER IT IS 1 OR 2
      ITP=3-IT
      DO 1 K=1,LMMAX
1     V(K)=CZ
C  FIGURE OUT LIMITS ON SUM OVER SUBPLANES
      IF (IPR.EQ.-1) GO TO 2
      NSH=(ILAY-1)*LMMAX
      IL=1
      IM=ILAY-1
      GO TO 3
2     NSH=(2*NLAY-ILAY)*LMMAX
      IL=ILAY+1
      IM=NLAY
C
C
C  IF DESTINATION SUBPLANE WAS TREATED BY BEEBY-TYPE INVERSION, USE
C  SPECIAL FORMULA
```

```
3       IF (INV(ILAY).EQ.1) GO TO 20
C
C
C   SUM OVER SUBPLANES
        DO 5 J=IL,IM
C   FIGURE OUT LOCATIONS OF NEEDED QUANTITIES TG
        JL=(J-1)*LMMAX
        JJ=(2*NLAY-J)*LMMAX
        IF (IPR.EQ.1) GO TO 12
        NGH=JL
        JL=JJ
        JJ=NGH
C   FIGURE OUT LOCATION OF NEEDED GH
12      NGH=(MGH(ILAY,J)-1)*LMMAX
C   GET APPROPRIATE PLANE-WAVE PROPAGATION FACTORS
        IF (INC.EQ.-1) GO TO 8
        SS=RG(1,J,JG)/RG(1,ILAY,JG)
        GO TO 9
8       SS=RG(2,ILAY,JG)/RG(2,J,JG)
9       CONTINUE
        DO 5 K=1,LMMAX
        ST=CZ
C   PERFORM PRODUCT GH*(TG+TG)
        DO 4 L=1,LMMAX
4       ST=ST+GH(NGH+K,L)*(TG(IT,JL+L)+TG(ITP,JJ+L))
5       V(K)=V(K)+SS*ST
C
C   FIND LOCATION OF NEEDED TAU
        MT=(LPS(ILAY)-1)*LMMAX
        MTO=MT+LEV
        NSHO=NSH+LEV
        LOD=LMMAX-LEV
        DO 10 K=1,LEV
        ST=CZ
C   PERFORM PRODUCT TAU*GH*(TG+TG) FOR EVEN L+M
        DO 6 L=1,LEV
6       ST=ST+TAU(MT+K,L)*V(L)
10      TG(IT,NSH+K)=ST
        DO 11 K=1,LOD
        ST=CZ
C   PERFORM PRODUCT TAU*GH*(TG+TG) FOR ODD L+M
        DO 7 L=1,LOD
7       ST=ST+TAU(MTO+K,L)*V(L+LEV)
11      TG(IT,NSHO+K)=ST
        RETURN
C
C
C   DIFFERENT FORMULA FOR CASE OF BEEBY-TYPE INVERSION
20      CONTINUE
        DO 30 K=1,LMMAX
30      TG(IT,NSH+K)=CZ
        IC=0
C   FIND LOCATION OF NEEDED TH'S
        DO 40 J=1,ILAY
        IF (INV(J).EQ.1) IC=IC+1
40      CONTINUE
        ICL=(IC-1)*LMMAX
C   SUM OVER SUBPLANES, EXCEPT THOSE TREATED BY BEEBY-TYPE INVERSION
        DO 130 J=IL,IM
        IF (INV(J).EQ.1) GO TO 130
C   FIND LOCATION OF NEEDED TG'S
        JL=(J-1)*LMMAX
        JJ=(2*NLAY-J)*LMMAX
        IF (IPR.EQ.1) GO TO 50
        NGH=JL
        JL=JJ
        JJ=NGH
```

```
50      CONTINUE
C   GET APPROPRIATE PLANE-WAVE PROPAGATION FACTORS
        IF (INC.EQ.-1) GO TO 60
        SS=RG(1,J,JG)/RG(1,ILAY,JG)
        GO TO 70
60      SS=RG(2,ILAY,JG)/RG(2,J,JG)
70      CONTINUE
        ID=0
C   SUM OVER SUBPLANES TREATED BY BEEBY-TYPE INVERSION TO OBTAIN
C   GH*(TG+TG)
        DO 120 IINV=1,NLAY
        IF (INV(IINV).EQ.0) GO TO 120
        ID=ID+1
        IDL=(ID-1)*LMMAX
        NGH=(MGH(IINV,J)-1)*LMMAX
        DO 90 K=1,LMMAX
        ST=CZ
        DO 80 L=1,LMMAX
80      ST=ST+GH(NGH+K,L)*(TG(IT,JL+L)+TG(ITP,JJ+L))
90      V(K)=ST
        DO 110 K=1,LMMAX
        ST=CZ
C   PRODUCT TH*GH*(TG+TG)
        DO 100 L=1,LMMAX
100     ST=ST+TH(ICL+K,IDL+L)*V(L)
110     TG(IT,NSH+K)=TG(IT,NSH+K)+SS*ST
120     CONTINUE
130     CONTINUE
        RETURN
        END
C-----------------------------------------------------------------------
C   SUBROUTINE TRANS PERFORMS THE ORIGIN (REGISTRY) SHIFT TO AN AXIS OR
C   PLANE OF SYMMETRY FOR SUBROUTINE MFOLD.
C    FDW= OUTPUT SHIFTED VALUES.
C    EDW= INPUT UNSHIFTED VALUES.
C    NG= NO. OF SYMMETRICAL BEAMS RELATED TO BEAM G (INCL. G).
C    G= BEAM G AND ITS SYMMETRICAL COUNTERPARTS.
C    GP= BEAM G(PRIME).
C    S= ORIGIN SHIFT.
        SUBROUTINE  TRANS (FDW, EDW, NG, G, GP, S)
        COMPLEX EDW,FDW,TRL,CI
        DIMENSION  S(2), G(2,12), GP(2), EDW(2,12), FDW(2,12)
        CI=CMPLX(0.0,1.0)
        AA = GP(1) * S(1) + GP(2) * S(2)
        DO 10 J = 1, NG
        BB = AA - G(1,J) * S(1) - G(2,J) * S(2)
        TRL=CEXP(CI*BB)
        DO 10 I = 1, 2
10      FDW(I,J) = TRL * EDW(I,J)
        RETURN
        END
C-----------------------------------------------------------------------
C   SUBROUTINE TRANSP PERFORMS THE ORIGIN SHIFT (REGISTRY SHIFT) TO AN
C   AXIS OR PLANE OF SYMMETRY FOR SUBROUTINE MFOLT.
C    FDW= OUTPUT SHIFTED VALUES.
C    EDW= INPUT UNSHIFTED VALUES.
C    NG= NO. OF SYMMETRICAL BEAMS RELATED TO G (INCL. G).
C    G= BEAM G AND ITS SYMMETRICAL COUNTERPARTS.
C    GP= BEAM G(PRIME).
C    S= ORIGIN SHIFT (REGISTRY SHIFT).
C    POSS= POSITION VECTORS OF ATOMS IN THE SUBPLANES OF THE LAYER
C      TREATED BY MPERTI OR MTINV.
C    NLAY= NO. OF SUBPLANES IN LAYER.
C    AK2,AK3= PARALLEL COMPONENTS OF PRIMARY INCIDENT K-VECTOR.
C    INC= INCIDENCE DIRECTION (+-1 TOWARDS +-X).
        SUBROUTINE TRANSP(FDW,EDW,NG,G,GP,S,POSS,NLAY,AK2,AK3,INC)
        COMPLEX EDW(2,12),FDW(2,12),TRL,CI
```

```
       DIMENSION S(2),G(2,12),GP(2),POSS(NLAY,3)
       CI=(0.0,1.0)
       I1=1
       I2=NLAY
       IF (INC.EQ.1) GO TO 5
       I=I1
       I1=I2
       I2=I
5      CONTINUE
       AR=GP(1)*(POSS(I1,2)-S(1))+GP(2)*(POSS(I1,3)-S(2))
       AT=AR+AK2*(POSS(I1,2)-POSS(I2,2))+AK3*(POSS(I1,3)-POSS(I2,3))
       DO 10 J=1,NG
       BR=AR+G(1,J)*(S(1)-POSS(I1,2))+G(2,J)*(S(2)-POSS(I1,3))
       BT=AT+G(1,J)*(S(1)-POSS(I2,2))+G(2,J)*(S(2)-POSS(I2,3))
       FDW(1,J)=EDW(1,J)*CEXP(CI*BR)
10     FDW(2,J)=EDW(2,J)*CEXP(CI*BT)
       RETURN
       END
C-------------------------------------------------------------------------
C  SUBROUTINE TSCATF INTERPOLATES TABULATED PHASE SHIFTS AND PRODUCES
C  THE ATOMIC T-MATRIX ELEMENTS (OUTPUT IN AF AND TSF0). THESE ARE ALSO
C  CORRECTED FOR THERMAL VIBRATIONS (OUTPUT IN CAF AND TSF). AF AND CAF
C  ARE MEANT TO BE USED BY SUBROUTINE MSMF, TSF0 AND TSF BY MPERTI OR
C  MTINV.
C  IEL= CHEMICAL ELEMENT TO BE TREATED NOW, IDENTIFIED BY THE INPUT
C    SEQUENCE ORDER OF THE PHASE SHIFTS (IEL=1,2 OR 3).
C  L1= LMAX+1.
C  ES= LIST OF ENERGIES AT WHICH PHASE SHIFTS ARE TABULATED.
C  PHSS= TABULATED PHASE SHIFTS.
C  NPSI= NO. OF ENERGIES AT WHICH PHASE SHIFTS ARE GIVEN.
C  IT: NOT USED.
C  EB-V= CURRENT ENERGY (V CAN BE USED TO DESCRIBE LOCAL VARIATIONS
C    OF THE MUFFIN-TIN CONSTANT).
C  PPP= CLEBSCH-GORDON COEFFICIENTS FROM SUBROUTINE CPPP.
C  NN1= NN2+NN3-1.
C  NN2= NO. OF OUTPUT TEMPERATURE-CORRECTED PHASE SHIFTS DESIRED.
C  NN3= NO. OF INPUT PHASE SHIFTS.
C  DR0= FOURTH POWER OF RMS ZERO-TEMPERATURE VIBRATION AMPLITUDE.
C  DRPER= RMS VIBRATION AMPLITUDE PERPENDICULAR TO SURFACE.
C  DRPAR= RMS VIBRATION AMPLITUDE PARALLEL TO SURFACE.
C  TO= TEMPERATURE AT WHICH DRPER AND DRPAR HAVE BEEN COMPUTED.
C  T= CURRENT TEMPERATURE.
C  TSF0,TSF,AF,CAF: SEE ABOVE.
       SUBROUTINE TSCATF(IEL,L1,ES,PHSS,NPSI,IT,EB,V,PPP,NN1,NN2,NN3,
      1DR0,DRPER,DRPAR,TO,T,TSF0,TSF,AF,CAF)
       DIMENSION PHSS(NPSI,30),PHS(10),ES(NPSI),PPP(NN1,NN2,NN3)
       COMPLEX CI,DEL(16),CA,AF(L1),CAF(L1),TSF0(3,L1),TSF(3,L1)
700    FORMAT(42H TOO LOW ENERGY FOR AVAILABLE PHASE SHIFTS)
       CI=(0.0,1.0)
       E=EB-V
       IF (E.LT.ES(1)) GO TO 850
C  FIND SET OF PHASE SHIFTS APPROPRIATE TO DESIRED CHEMICAL ELEMENT
C  AND INTERPOLATE TO CURRENT ENERGY (OR EXTRAPOLATE TO ENERGIES
C  ABOVE THE RANGE GIVEN FOR THE PHASE SHIFTS)
       IO=(IEL-1)*L1
710    I = 1
720    IF ((E-ES(I))*(E-ES(I+1)))  750, 750, 730
730    I = I + 1
       IF (I-NPSI)  720, 740, 740
740    I = I - 1
750    FAC = (E - ES(I))/(ES(I + 1) - ES(I))
       DO 760 L = 1, L1
760    PHS(L) = PHSS(I,L + IO) + FAC * (PHSS(I + 1,L + IO) - PHSS(I,
      1L + IO))
C  COMPUTE TEMPERATURE-INDEPENDENT T-MATRIX ELEMENTS
       DO 790 L = 1, L1
       A = PHS(L)
```

```
      AF(L) = A * CI
      AF(L) = CEXP(AF(L))
      A = SIN(A)
      AF(L) = A * AF(L)
  790 TSFO(IEL,L)=AF(L)
C   AVERAGE ANY ANISOTROPY OF RMS VIBRATION AMPLITUDES
      DR=SQRT((DRPER*DRPER+2.0*DRPAR*DRPAR)/3.0)
C   COMPUTE TEMPERATURE-DEPENDENT PHASE SHIFTS (DEL)
      CALL  PSTEMP (PPP, NN1, NN2, NN3, DRO , DR, TO , T, E,PHS, DEL)
C   PRODUCE TEMPERATURE-DEPENDENT T-MATRIX ELEMENTS
      DO 840 L = 1, L1
      CA = DEL(L)
      CAF(L) = CA * CI
      CAF(L) = CEXP(CAF(L))
      CA = CSIN(CA)
      CAF(L) = CA * CAF(L)
  840 TSF(IEL,L)=CAF(L)
      RETURN
  850 WRITE(6,700)
      STOP
      END
C-----------------------------------------------------------------------
C   SUBROUTINE UNFOLD DETERMINES WHICH BEAMS ARE RELATED TO A GIVEN BEAM
C   BY A TYPE OF SYMMETRY SPECIFIED IN SYM.
C   IG= SERIAL NO. OF THE BEAM WHOSE SYMMETRICAL COUNTERPARTS ARE SOUGHT
C   G= OUTPUT LIST OF SYMMETRICAL BEAMS SOUGHT.
C   NG= OUTPUT NUMBER OF SYMMETRICAL BEAMS IN G.
C   LAY= INDEX FOR CHOICE OF SYMMETRY CODES SYM(LAY,I) (NORMALLY LAY=1
C      FOR OVERLAYER, LAY=2 FOR SUBSTRATE LAYERS).
C   PQ= LIST OF INPUT BEAMS G.
C   SYM= LIST OF SYMMETRY CODES FOR ALL BEAMS.
C   NT= TOTAL NO. OF BEAMS IN MAIN PROGRAM AT CURRENT ENERGY.
      SUBROUTINE  UNFOLD (IG, G, NG, LAY, PQ, SYM, NT)
      INTEGER  SYM
      DIMENSION  G(2,12), PQ(2,NT), SYM(2,NT)
      ACU=1.7320508
      DO 340 I = 1, 2
  340 G(I,1) = PQ(I,IG)
      LAB = SYM(LAY,IG)
      IF ((LAB .EQ. 10) .AND. (G(2,1) .LT. 0.0)) G(2,1) =  - G(2,1)
      GO TO (350,360,380,400,420,430,440,460,480,500,
     1511,512,513,514,515), LAB
  350 NG = 1
      GO TO 520
  360 NG = 2
      DO 370 I= 1, 2
  370 G(I,2) =  - G(I,1)
      GO TO 520
  380 NG = 2
      DO 390 I = 1, 2
  390 G(I,2) =  - G(3 - I,1)
      GO TO 520
  400 NG = 2
      DO 410 I = 1, 2
  410 G(I,2) = G(3 - I,1)
      GO TO 520
  420 NG = 2
      G(1,2) = G(1,1)
      G(2,2) =  - G(2,1)
      GO TO 520
  430 NG = 2
      G(1,2) =  - G(1,1)
      G(2,2) = G(2,1)
      GO TO 520
  440 NG = 4
      G(1,2) =  - G(2,1)
      G(2,2) = G(1,1)
```

```
      DO 450 I = 1, 2
      G(I,3) =  - G(I,1)
450 G(I,4) =  - G(I,2)
      GO TO 520
460 NG = 4
      DO 470 I = 1, 2
      G(I,2) = G(3 - I,1)
      G(I,3) =  - G(I,1)
470 G(I,4) =  - G(I,2)
      GO TO 520
480 NG = 4
      G(1,4) = G(1,1)
      G(2,4) =  - G(2,1)
      DO 490 I = 1, 2
      G(I,2) =  - G(I,4)
490 G(I,3) =  - G(I,1)
      GO TO 520
500 NG = 8
      G(1,8) = G(1,1)
      G(2,8) =  - G(2,1)
      DO 510 I = 1, 2
      G(I,2) = G(3 - I,1)
      G(I,3) = G(3 - I,8)
      G(I,4) =  - G(I,8)
      G(I,5) =  - G(I,1)
      G(I,6) =  - G(I,2)
510 G(I,7) =  - G(I,3)
      GO TO 520
511 NG=3
      G(1,12)=-ACU*G(2,1)
      G(2,12)=ACU*G(1,1)
      DO 521 I=1,2
      G(I,2)=0.5*(-G(I,1)+G(I,12))
521 G(I,3)=0.5*(-G(I,1)-G(I,12))
      GO TO 520
512 NG=6
      G(1,12)=-ACU*G(2,1)
      G(2,12)=ACU*G(1,1)
      DO 522 I=1,2
      G(I,3)=0.5*(-G(I,1)+G(I,12))
522 G(I,5)=0.5*(-G(I,1)-G(I,12))
      DO 532 I=2,6,2
      G(1,I)=G(1,7-I)
532 G(2,I)=-G(2,7-I)
      GO TO 520
513 NG=6
      G(1,12)=-ACU*G(2,1)
      G(2,12)=ACU*G(1,1)
      DO 523 I=1,2
      G(I,3)=0.5*(-G(I,1)+G(I,12))
523 G(I,5)=0.5*(-G(I,1)-G(I,12))
      G(1,2)=-G(1,1)
      G(2,2)=G(2,1)
      DO 533 I=4,6,2
      G(1,I)=-G(1,9-I)
533 G(2,I)=G(2,9-I)
      GO TO 520
514 NG=6
      G(1,12)=-ACU*G(2,1)
      G(2,12)=ACU*G(1,1)
      DO 524 I=1,2
      G(I,2)=0.5*(G(I,1)+G(I,12))
      G(I,3)=0.5*(-G(I,1)+G(I,12))
      G(I,4)=-G(I,1)
      G(I,5)=-G(I,2)
524 G(I,6)=-G(I,3)
      GO TO 520
```

```
  515 NG=12
      G(1,12)=-ACU*G(2,1)
      G(2,12)=ACU*G(1,1)
      DO 525 I=1,2
      G(I,3)=0.5*(G(I,1)+G(I,12))
      G(I,5)=0.5*(-G(I,1)+G(I,12))
      G(I,7)=-G(I,1)
      G(I,9)=-G(I,3)
  525 G(I,11)=-G(I,5)
      DO 535 I=2,12,2
      G(1,I)=G(1,13-I)
  535 G(2,I)=-G(1,13-I)
  520 RETURN
      END
C------------------------------------------------------------------------
C  SUBROUTINE XM PRODUCES THE INTRA-LAYER MULTIPLE-SCATTERING MATRIX X
C  FOR A SINGLE BRAVAIS-LATTICE LAYER. THE (L,M) SEQUENCE IS THE
C  'SYMMETRIZED' ONE AND X IS SPLIT INTO TWO PARTS CORRESPONDING TO
C  L+M= EVEN (XEV) AND L+M= ODD (XOD), AS A RESULT OF BLOCK-DIAGONALI-
C  ZATION.
C  FLM= LATTICE SUM FROM SUBROUTINES FMAT AND MSMF.
C  XEV,XOD= OUTPUT MATRIX X IN TWO PARTS.
C  LEV= (LMAX+1)*(LMAX+2)/2.
C  LOD= LMAX*(LMAX+1)/2.
C  AF: NOT USED.
C  CAF= TEMPERATURE-DEPENDENT ATOMIC SCATTERING T-MATRIX ELEMENTS.
C  LM= LMAX+1.
C  LX,LXI= PERMUTATIONS OF (L,M) SEQUENCE, FROM SUBROUTINE LXGENT.
C  LMMAX= (LMAX+1)**2.
C  KLM= (2*LMAX+1)*(2*LMAX+2)/2.
C  CLM= CLEBSCH-GORDON COEFFICIENTS, FROM SUBROUTINE CELMG.
C  NLM= DIMENSION OF CLM (SEE MAIN PROGRAM).
      SUBROUTINE  XM (FLM,XEV,XOD,LEV,LOD,AF,CAF,LM,LX,LXI,LMMAX,
     1 KLM,CLM,NLM)
      INTEGER  LX, LXI
      COMPLEX  XEV, XOD, FLM, AF, CZERO, ACC, CAF
      DIMENSION  CLM(NLM), XEV(LEV,LEV), XOD(LOD,LOD), FLM(KLM), AF(LM)
      DIMENSION  CAF(LM), LX(LMMAX), LXI(LMMAX)
      LMAX = LM-1
      CZERO = CMPLX(0.0,0.0)
      DO 440 I = 1, LEV
      DO 440 J = 1, LEV
  440 XEV(I,J) = CZERO
      DO 450 I = 1, LOD
      DO 450 J = 1, LOD
  450 XOD(I,J) = CZERO
      L2MAX = LMAX + LMAX
      NODD = LOD
      JSET = 1
      MM = LEV
      N = 1
C  FIRST XOD IS CREATED
  460 J = 1
      L = JSET
  470 M =  - L + JSET
      JL = L + 1
  480 K = 1
      LPP = JSET
  490 MPP =  - LPP + JSET
  500 MPA = IABS(MPP - M)
      LPA = IABS(LPP - L)
      IF (LPA-MPA) 520, 520, 510
  510 MPA = LPA
  520 MP1 = MPP - M + L2MAX + 1
      LP1 = L + LPP + 1
      ACC = CZERO
  530 JLM = (LP1 * LP1 + MP1 - L2MAX)/2
```

```
      ACC = ACC + CLM(N) * FLM(JLM)
      N = N + 1
      LP1 = LP1 - 2
      IF (LP1-1-MPA) 540, 530, 530
540 JX = LXI(J + MM)
      KX = LXI(K + MM)
      XEV(JX,KX) = ACC * CAF(JL)
      K = K + 1
      MPP = MPP + 2
      IF (LPP-MPP) 550, 500, 500
550 LPP = LPP + 1
      IF (LMAX-LPP) 560, 490, 490
560 J = J + 1
      M = M + 2
      IF (L-M) 570, 480, 480
570 L = L + 1
      IF (LMAX-L) 580, 470, 470
580 IF (JSET) 610, 610, 590
590 DO 600 J = 1, NODD
      DO 600 K = 1, NODD
600 XOD(J,K) = XEV(J,K)
      JSET = 0
      MM = 0
C  NOW RETURN TO CREATE XEV
      GO TO 460
610 RETURN
      END
C------------------------------------------------------------------
C  SUBROUTINE XMT PRODUCES THE INTRA-LAYER MULTIPLE-SCATTERING MATRIX
C  1-X FOR SUBROUTINE TAUMAT, OUTPUT IN THE TONG CONVENTION USING
C  INPUT IN THE PENDRY CONVENTION. XMT MUST BE CALLED TWICE, FIRST TO
C  PRODUCE THE VALUES FOR L+M= ODD, THEN FOR L+M= EVEN (IN THAT ORDER).
C  FOR QUANTITIES NOT EXPLAINED BELOW SEE SUBROUTINE TAUMAT.
C  IL=1 FOR L+M= ODD.
C  IL=2 FOR L+M= EVEN.
C  X= OUTPUT MATRIX 1-X, BLOCK-DIAGONALIZED, 2ND BLOCK LEFT-ADJUSTED.
C  LL= INPUT EITHER LOD OR LEV.
C  TSF= ATOMIC T-MATRIX ELEMENTS.
C  NTAU= NO. OF CHEMICAL ELEMENTS CONSIDERED.
C  IT= INDEX OF CURRENT CHEMICAL ELEMENT (.LE.NTAU).
C  N= RUNNING INDEX OF CLM: MAY NOT BE RESET BETWEEN THE TWO CALLS TO
C  XMT IN TAUMAT.
C DIMENSIONS SET FOR (NTAU,LM) = AT MOST (3,10)
      SUBROUTINE XMT(IL,FLMS,NL,X,LEV,LL,TSF,NTAU,IT,LM,LXI,LMMAX,
     1 KLM,CLM,NLM,N)
      COMPLEX TSF(3,10)
      COMPLEX X,FLMS,CZERO,ACC,CI,ST,SU,RU
      DIMENSION CLM(NLM),X(LEV,LL),FLMS(NL,KLM),LXI(LMMAX)
      LMAX = LM-1
      RU=(1.0,0.0)
      CZERO = CMPLX(0.0,0.0)
      CI=CMPLX(0.0,1.0)
      L2MAX = LMAX + LMAX
C  IF IL=1, CONSIDER L+M= ODD ONLY
C  IF IL=2, CONSIDER L+M= EVEN ONLY
      IF (IL-2) 455,450,450
450   JSET=0
      MM=0
      GO TO 457
455   JSET = 1
      MM = LEV
      N=1
457   CONTINUE
460 J = 1
      L = JSET
470 M = - L + JSET
      JL = L + 1
```

```
        ST=(-CI)**MOD(L,4)
  480 K = 1
        LPP = JSET
  490 MPP =   - LPP + JSET
        JLPP=LPP+1
        SU=((-CI)**MOD(LPP,4))/ST
  500 MPA = IABS(MPP - M)
        LPA = IABS(LPP - L)
        IF (LPA-MPA)  520, 520, 510
  510 MPA = LPA
  520 MP1 = MPP - M + L2MAX + 1
        LP1 = L + LPP + 1
        ACC = CZERO
  530 JLM = (LP1 * LP1 + MP1 - L2MAX)/2
        ACC=ACC+CLM(N)*FLMS(1,JLM)
        N = N + 1
        LP1 = LP1 - 2
        IF (LP1-1-MPA)   540, 530, 530
  540 JX = LXI(J + MM)
        KX = LXI(K + MM)
        X(KX,JX) = -ACC * TSF(IT,JLPP)*SU
        IF (J.EQ.K) X(KX,JX)=X(KX,JX)+RU
        K = K + 1
        MPP = MPP + 2
        IF (LPP-MPP)  550, 500, 500
  550 LPP = LPP + 1
        IF (LMAX-LPP)  560, 490, 490
  560 J = J + 1
        M = M + 2
        IF (L-M)  570, 480, 480
  570 L = L + 1
        IF (LMAX-L)  610, 470, 470
  610 RETURN
        END
C-----------------------------------------------------------------
C   SUBROUTINE ZGE PERFORMS GAUSSIAN ELIMINATION AS THE FIRST STEP IN THE
C   SOLUTION OF A SYSTEM OF LINEAR EQUATIONS. THIS IS USED TO MULTIPLY
C   THE INVERSE OF A MATRIX INTO A VECTOR, THE MULTIPLICATION BEING DONE
C   LATER BY SUBROUTINE ZSU.
C   A= INPUT (COMPLEX) MATRIX, WILL BE OVERWRITTEN BY A TRIANGULARIZED
C     MATRIX, TO BE TRANSMITTED TO SUBROUTINE ZSU.
C   INT= STORAGE FOR PERMUTATION OF MATRIX COLUMNS (AND ROWS), TO BE
C     TRANSMITTED TO SUBROUTINE ZSU.
C   NR= FIRST DIMENSION OF A (.GE.NC).
C   NC= ORDER OF A.
C   EMACH= MACHINE ACCURACY.
        SUBROUTINE  ZGE (A, INT, NR, NC, EMACH)
        COMPLEX  A, YR, DUM
        DIMENSION  A(NR,NC), INT(NC)
        N = NC
        DO 680 II = 2, N
        I = II - 1
        YR = A(I,I)
        IN = I
        DO 600 J = II, N
        IF (CABS(YR)-CABS(A(J,I)))  590, 600, 600
  590 YR = A(J,I)
        IN = J
  600 CONTINUE
        INT(I) = IN
        IF (IN-I)  610, 630, 610
  610 DO 620 J = I, N
        DUM = A(I,J)
        A(I,J) = A(IN,J)
  620 A(IN,J) = DUM
  630 IF (CABS(YR)-EMACH)  680, 680, 640
  640 DO 670 J = II, N
```

```
      IF (CABS(A(J,I))-EMACH)  670, 670, 650
  650 A(J,I) = A(J,I)/YR
      DO 660 K = II, N
  660 A(J,K) = A(J,K) - A(I,K) * A(J,I)
  670 CONTINUE
  680 CONTINUE
      RETURN
      END
C-------------------------------------------------------------------------
C  SUBROUTINE ZSU TERMINATES THE SOLUTION OF A SYSTEM OF LINEAR
C  EQUATIONS, INITIATED BY SUBROUTINE ZGE, BY BACK-SUBSTITUTING THE
C  CONSTANT VECTOR.
C   A= INPUT MATRIX, PREPARED BY SUBROUTINE ZGE.
C   INT= INPUT PERMUTATION FROM SUBROUTINE ZGE.
C   X= INPUT CONSTANT VECTOR AND OUTPUT RESULTING VECTOR.
C   NR= FIRST DIMENSION OF A (.GE.NC).
C   NC= ORDER OF A.
C   EMACH= MACHINE ACCURACY.
      SUBROUTINE  ZSU (A, INT, X, NR, NC, EMACH)
      COMPLEX A, X, DUM
      DIMENSION  A(NR,NC), X(NC), INT(NC)
      N = NC
      DO 730 II = 2, N
      I = II - 1
      IF (INT(I)-I)  690, 700, 690
  690 IN = INT(I)
      DUM = X(IN)
      X(IN) = X(I)
      X(I) = DUM
  700 DO 720 J = II, N
      IF (CABS(A(J,I))-EMACH)  720, 720, 710
  710 X(J) = X(J) - A(J,I) * X(I)
  720 CONTINUE
  730 CONTINUE
      DO 780 II = 1, N
      I = N - II + 1
      IJ = I + 1
      IF (I-N)  740, 760, 740
  740 DO 750 J = IJ, N
  750 X(I) = X(I) - A(I,J) * X(J)
  760 IF (CABS(A(I,I))-EMACH*1.0E-5)  770, 780, 780
  770 A(I,I) = EMACH * 1.0E - 5 * (1.0,1.0)
  780 X(I) = X(I)/A(I,I)
      RETURN
      END
```

12. Structural Results of LEED Crystallography

In this chapter, we summarize the surface structures obtained with LEED until early 1978. Further details may be found in the bibliography (Sect. 12.4), while a more systematic discussion is given elsewhere [16]. We include results obtained by kinematic (KLEED), Constant-Momentum-Transfer-Averaging (CMTA) and Quasi-Dynamical [17] (QDLEED) LEED theories, and identify those as such.

A conservative estimate of the accuracy of the quoted atomic positions is ±0.1Å in the direction perpendicular to the surface: this usually corresponds to greater absolute and relative accuracies in the bond lengths, which are the physically and chemically more interesting quantities. Below we quote mainly bond lengths (b.ℓ.) or their relative contractions/expansions with respect to their bulk values; where appropriate, the corresponding inter-layer spacings or their relative contractions/expansions are also quoted [between square brackets]. Expansions are given positive signs, contractions negative signs.

12.1 Non-Metals

Semiconductors

- Si (111) (1×1), impurity-stabilized: bulk structure; surface b.ℓ. contracted -2% [-15%];
- Si (100) (2×1): reconstruction of the bulk lattice, derived from the Schlier-Farnsworth dimer model, with atoms in the first five surface layers displaced from bulk positions ("optimized Appelbaum-Hamann model"; KLEED, QDLEED);
- Si (100) + (1×1) H: Si substrate has ideal bulk structure; position of H unknown;
- ZnO (0001), Zn termination: bulk structure; surface b.ℓ. contracted -3% [-25%];

- ZnO ($10\bar{1}0$): top Zn and O atoms pulled somewhat into surface;
- ZnSe (110): top Zn and Se atoms pulled into (respectively, pushed out of) surface;
- GaAs (110): top Ga and As atoms pulled into (respectively, pushed out of) surface; b.ℓ. of Ga and As surface back bonds contracted by -2.5% (resp., -3.6%); Ga-As bond in the surface is tilted at a projected angle of $\omega=27°$;
- GaAs (110) + p(1×1) As: bulk GaAs (110) substrate restored by As adsorption; As adatoms bonded as in bulk GaAs (but three neighbors missing);
- GaAs (100): bulk structure, As termination forced by molecular beam epitaxy; no change of interlayer spacings at the surface.

Ionic Compounds

- MgO (100), NiO (100): bulk structure, unrelaxed, no rumpling;
- CoO (111): polar face, O termination favored, Co-O b.ℓ. contracted -5% [-15%].

Layer Compounds

- MoS_2 (0001): cleaved between sandwiches; surface b.ℓ. contracted by -1.6% [-4.7%]; first Van der Waals spacing contracted by [-3%];
- $NbSe_2$ (0001): cleaved between sandwiches; surface b.ℓ. contracted by -0.2% [-0.6%]; first Van der Waals spacing contracted by [-1.4%];
- TiS_2 (0001): cleaved between sandwiches; surface b.ℓ. contracted by -1.7% [-5%]; first Van der Waals spacing contracted by [-5%];
- $TiSe_2$ (0001): cleaved between sandwiches; surface b.ℓ. expanded by +1.7% [+5%]; first Van der Waals spacing contracted by [-5%];
- Na_2O (111): termination between two Na layers; unrelaxed bulk structure.

12.2 Clean Metals

Results on reconstructed clean metal surfaces are available for one system:
- W (100) c(2×2): zigzag rows of touching top-layer W atoms; top layer spacing contracted [-6%].

The following metal surfaces are all "unreconstructed" (i.e., no super-lattices and no atomic rearrangements occur).

No registry changes (lateral shifts) of topmost atomic layers have been observed, although they are easily conceivable, on hcp (0001), fcc (111) and bcc (110): this includes the case of hcp Co (0001), a low-temperature phase, and fcc Co (111), a high-temperature phase. The list below shows the surfaces that have been investigated.

Bond length contractions in the back-bonds (i.e., in the bonds between top-layer and deeper-layer atoms) are mainly observed on rough (loose-packed) crystallographic faces. The observed bond length and interlayer spacing changes are shown in the following list between brackets:

- hcp (0001): Be (0%), Ti (-0.5% [-2%]), Co (0%), Zn (-0.5% [-2%]), Cd (0%);
- fcc (111): Aℓ (-1 to +1.5% [-3 to +5%]), Co (0%), Ni (0%), Cu (-1.2 to 0% [-4 to 0%]), Rh (0%), Ag (0%), Ir (-0.8% [-2.5%]), Pt (0%), Au (0%);
- bcc (110): Na (0%), Fe (0%), W(0%);
- fcc (100): Aℓ (0%), Co (-1.5% [-4%]), Ni (0%), Cu (0%), Rh (1% [3%]), Ag (0%), Xe (0%), Pt (0%), Au (0%);
- fcc (110): Aℓ (-3 to -4% [-9 to -15%]), Ni (-1.5% [-5%]), Cu (-3 to -4% [-10 to -12.5%]), Ag (-2 to -3% [-6 to -10%]);
- bcc (100): Fe (-0.7 to -1.5% [-1.4 to -4%]), Mo (-4% [-11%]), W (-2 to -4% [-5 to -11%]);
- bcc (111): Fe (-1.5% [-11%]);
- fcc (311): Cu (-1% [-5%]).

12.3 Adsorption on Metals

Atomic adsorption sites

All but two studies have yielded overlayers, rather than penetration (interstitially or substitutionally) into the surface: the exceptions are an interstitial underlayer and a substitutional penetration: see below. In most cases the number of nearest substrate neighbors is *maximized* by the adatom in its choice of an adsorption site, i.e., a hollow site is chosen. This is the case for the following surfaces:

fcc Aℓ (100) + c(2×2) Na	bcc Fe (100) + p(1×1) O
fcc Ag (111) + p(1×1) Au	bcc Fe (100) + c(2×2) S
fcc Ag (100) + c(2×2) Cℓ	bcc Mo (100) + c(2×2) N
fcc Ag (111) + (√3×√3)R30° I	bcc Mo (100) + p(2×1) O
fcc Ag (100) + c(2×2) Se	bcc Mo (100) + p(1×1) Si
fcc Co (100) + c(2×2) O	fcc Ni (111) + (2×2) 2H
fcc Cu (100) + c(2×2) N	fcc Ni (100) + c(2×2) Na
fcc Cu (100) + p(2×2) Te	fcc Ni (100) + c(2×2) O

```
fcc Ni (100) + p(2×2) O        fcc Ni (100) + p(2×2) Se
fcc Ni (111) + p(2×2) O        fcc Ni (100) + c(2×2) Te
fcc Ni (100) + c(2×2) S        fcc Ni (100) + p(2×2) Te
fcc Ni (100) + p(2×2) S        hcp Ti (0001) + p(1×1) Cd
fcc Ni (111) + p(2×2) S        bcc W (110) + p(2×1) O
fcc Ni (100) + c(2×2) Se
```

Exceptions:

- fcc Ni (110) + p(2×1) O: O in short-bridge site, only 2-fold coordinated;
- fcc Ni (110) + c(2×2) S: S in deepest site (i.e., site of missing Ni), but closest to one 2nd-layer Ni atom;
- fcc Ag (110) + p(2×1) O: O probably in long-bridge site.

On hcp (0001) and fcc (111) two non-equivalent hollow sites exist. Using lower-case letters to designate the registry of adatoms, one can symbolize the possible stacking sequences by bABAB... and cABAB... on hcp (0001) and by cABCA... and bABCA... on fcc (111). One finds, including the case of multilayers of adsorbates:

- Ti (0001) + p(1×1) Cd: cABAB...;
- Ti (0001) + [p(1×1) Cd]^2 (i.e., 2 monolayers): acABAB...;
- Ti (0001) + [p(1×1) Cd]^4 (i.e., 4 monolayers): acacABAB...;
- Ni (111) + (2×2) 2H: both cABCA... and bABCA... simultaneously (with essentially equal Ni-H bond lengths) in a graphite-layer-like arrangement;
- Ni (111) + p(2×2) O: unresolved;
- Ni (111) + p(2×2) S, Ag (111) + ($\sqrt{3} \times \sqrt{3}$) R30°I, Ag(111) + p(1×)Au: cABCA....

An *underlayer* is found for the system hcp Ti (0001) + p(1×1) N (symbolized by AcBAB...); in this case, a 3-layer site of (111) orientation of bulk TiN (NaCℓ structure) is formed, with a lateral lattice constant equal to that of the Ti substrate. With Cu (100) + c(2×2) O, CMTA LEED indicates a one-layer oxide formation by substitution of O for half the first-layer Cu atoms.

Atomic Adsorption Bond Lengths

Contractions of metal-metal bond lengths found on clean surfaces are cancelled by adsorption in all investigated cases. This is found for: bcc Fe (100) + c(2×2) S, bcc Mo (100) + c(2×2) N, bcc Mo (100) + p(2×1) O, bcc Mo (100) + p(1×1) Si, fcc Ni (110) + p(2×1) O, and fcc Ni (110) + c(2×2) S. In the following two cases, an expansion of the substrate bond lengths beyond the bulk value is observed:

- bcc Fe (100) + p(1×1) O: the Fe-Fe b.ℓ. expands to 3% larger than the bulk value (from a contraction of -1% in the clean surface);
- fcc Ni (100) + p(2×2) C: a 4% expansion of the Ni-Ni b.ℓ. is thought to occur, but further details of this structure are not known.

In another case, reconstruction of the substrate occurs:

- fcc Ni (110) + p(2×1) H: probably pairwise attraction of Ni atoms in adjacent close-packed rows; position of H unknown.

Adsorption by an incommensurate overlayer: the only system studied is the Xe overlayer on the Ag (111) surface. The Xe-Ag interlayer spacing was determined to be 3.55 Å, which presumably is an average value. Due to the variable registry the Xe-Ag bond length cannot be specified.

Table 12.1 compares adatom-metal bond lengths obtained at surfaces by LEED analyses with the corresponding bond lengths known in bulk compounds and molecules (found mainly in standard crystallographic and structure tables).

Table 12.1. Bond lengths and d-spacings (interlayer spacings; in square brackets) for surface adsorption, compared with known values in bulk compounds and molecules. Asterisks denote surfaces where the substrate geometry is changed by the adsorption. n.n. = nearest neighbor, n.n.n. = next-nearest neighbor

Atom Pair	Environment	Bond length [d-spacing] [Å]
Au-Ag	bulk	2.88
	Ag (111) + p(1×1) Au	2.88
Cd-Ti	sum of radii	3.01
	Ti (0001) + p(1×1) Cd	3.08 ± 0.03 [2.57 ± 0.05]
Cℓ-Ag	$Cs_2AgAuCℓ_6$	2.36
	sum of radii	2.43
	Ag (100) + c(2×) Cℓ	2.67 ± 0.06 [1.72 ± 0.1]
	AgCℓ bulk	2.77
H-Ni	NiH molecule	1.47
	Ni (111) + (2×2) 2H	1.84 ± 0.06
I-Ag	AgI molecule	2.54
	AgI bulk	2.80
	Ag (111) + ($\sqrt{3}×\sqrt{3}$)R30° I	2.80 [2.54]
	$[Ag_4I_6]^{2-}$	2.85
N-Cu	$...Cℓ_3CuNNCH_3...$ (molecule)	1.993
	Cu (100) + c(2×2) N	2.02 ± 0.05 [0.9 ± 0.1]
	$[Cu(NH_3)_2]^{2+}$	2.03
	$K_2Pb[Cu(NO_2)_6]$	2.11

Table 12.1 (cont'd)

N–Mo	sum of radii	2.11
	various molecules	2.29-2.33
	*Mo (100) + c(2×2) N	2.45 ± 0.05 [1.02 ± 0.1]
Na–Aℓ	sum of radii	2.82-3.00
	Aℓ (100) + c(2×2) Na	2.86-2.90 ± 0.08 [2.05-2.08 ± 0.1]
Na–Ni	Ni (100) + c(2×2) Na	2.84 ± 0.08 [2.23 ± 0.11]
	sum of radii	2.80-3.10
O–Co	Co_3O_4 bulk	1.915-1.946
	Co (100) + c(2×2) O	2.12 ± 0.05 [0.8 ± 0.01]
	CoO bulk	2.134
O–Fe	*Fe (100) + p(1×1) O n.n.	2.07 ± 0.06 [2.07 ± 0.06]
	*Fe (100) + p(1×1) O n.n.n.	2.09 ± 0.02 [0.53 ± 0.06]
	$FeCl_2H_8O_4$	2.09
	FeO bulk	2.15
O–Mo	U_2MoO_8	1.66-2.07
	$MoOPO_4$	1.65-1.98
	*Mo (100) + p(2×1) O n.n	2.275
	*Mo (100) + p(2×1) O n.n.n.	2.33
O–Ni	Ni (111) + p(2×2) O	1.88 ± 0.06 [1.20 ± 0.1]
	Ni $(C_5H_7O_2)_2$	1.90
	*Ni (110) + p(2×1) O	1.92 ± 0.04 [1.46 ± 0.05]
	Ni-chelate complexes	1.84-2.06
	Ni (100) + c(2×2) O	1.98 ± 0.05 [0.9 ± 0.1]
	Ni (100) + p(2×2) O	1.98 ± 0.05 [0.9 ± 0.1]
	NiO bulk	2.08
	$NiC_{10}H_{14}N_2O_8, H_2O$	2.03-2.18
O–W	$[WO_4]^{2-}$	1.75-1.81
	$BaWO_4$	1.95
	W (110) + p(2×1) O	2.08 ± 0.07 [1.25 ± 0.1]
	WOF_4	2.11
	sum of radii	2.12
S–Fe	$Fe_{0.35}ZrS_2$ bulk	1.99
	sum of radii	2.28
	*Fe (100) + c(2×2) S	2.30 ± 0.06 [1.09 ± 0.1]
	$Fe_{0.35}ZrS_2$ bulk	2.30
	Fe_7S_8 bulk	2.44
S–Ni	Ni (111) + p(2×2) S	2.02 ± 0.06 [1.40 ± 0.1]
	Ni-chelate complexes	2.10-2.23
	*Ni (110) + c(2×2) S n.n.	2.17 ± 0.1 [2.17 ± 0.1]
	γNiS bulk	2.18
	Ni (100) + c(2×2) S	2.19 ± 0.06 [1.30 ± 0.1]
	Ni (100) + p(2×2) S	2.19 ± 0.06 [1.30 ± 0.1]
	Ni_3S_2 bulk	2.28
	*Ni (1̄10) + c(2×2) S n.n.n.	2.35 ± 0.04 [0.93 ± 0.1]
	αNiS bulk	2.38
	NiS_2 bulk	2.34-2.42

Table 12.1 (cont'd)

Se-Ag	AgCrSe$_2$ bulk	2.46-2.80
	sum of radii	2.60-2.69
	Ag$_2$Se bulk	2.62-2.86
	Ag (100) + c(2×2) Se	2.80 ± 0.07 [1.91 ± 0.1]
Se-Ni	Ni (100) + c(2×2) Se	2.28 ± 0.06 [1.45 ± 0.1]
	Ni-chelate complex	2.32
	Ni (100) + p(2×2) Se	2.34 ± 0.07 [1.55 ± 0.1]
	Ni$_3$Se$_2$ bulk	2.36
	Ni$_{0.55}$Se$_{0.08}$Te$_{0.37}$ bulk	2.31-2.44
	Ni$_3$Se$_4$ bulk	2.47
	NiSe bulk	2.50
	NiRh$_2$Se$_4$ bulk	2.50-2.52
	NiSe$_2$ bulk	2.49-2.53
Si-Mo	*Mo (100) + p(1×1) Si	2.51 ± 0.05 [1.16 ± 0.1]
	sum of radii	2.53
Te-Cu	Cu (100) + p(2×2) Te	2.48 ± 0.10 [1.70 ± 0.15]
	Cu$_8$Cu$_{4-x}$Te$_6$ bulk	2.51-2.76
	Cu$_4$Te$_3$ bulk	2.65
	Cu$_2$Te bulk	2.67
	Cu$_{4-x}$Te$_2$ bulk	2.65-2.76
Te-Ni	Ni (100) + p(2×2) Te	2.52-2.58 ± 0.07 [1.80-1.90 ± 0.1]
	NiTe$_2$ bulk	2.58-2.59
	Ni (100) + c(2×2) Te	2.58 ± 0.07 [1.90 ± 0.1]
	NiTe bulk	2.64
	Ni$_{0.55}$Se$_{0.08}$Te$_{0.37}$ bulk	2.54-2.85

Coadsorption

Systems investigated: fcc Ni(100) + c/p(2×2)S + c/p(2×2)Na [i.e., 1/2 or
1/4 monolayers of S and Na coadsorbed on Ni(100)]. For independent S or Na
adsorption, see under atomic adsorption. For the case of coadsorption, Na
atoms adsorbed on S-covered Ni(100) choose those unoccupied hollow sites
of the substrate that provide the most S neighbors. The Ni-S bond length
is not noticeably affected by the presence of Na, while the Ni-Na bond
length is larger than in the absence of S (3.05 compared with 2.84Å) because of the space taken by the S atoms. The S-Na bond length of 2.76Å is
within the range found in bulk compounds of S and Na. No appreciable change
in these bond lengths is detected as the coverage is varied from quarter to
half-monolayer.

Molecular Adsorption

- Pt (111) + p(2×2) C$_2$H$_2$ in stable adsorption state: the acetylene lies
  parallel to the surface over the 3-fold hollow site (in cABCA... stacking

arrangement), with Pt-C bond lengths of 2.25 and 2.59Å [d-spacing 1.94 ± 0.1Å] (the C-C bond length is not well known, and assumed to be 1.20Å in the determination of the above-mentioned Pt-C bond lengths); the position of the H atoms is uncertain;

- Pt (111) + p(2×2) C_2H_2 in metastable adsorption state: the acetylene has the same orientation and average bond lengths as in the stable state, but the d-spacing is different [2.4Å] and the registry is different as well: in top view, the molecular axis passes 0.2Å away from the center of a top Pt atom;

- Ni (100) + c(2×2) CO: the CO molecule is undissociated and bonded by its carbon end to the top of one Ni atom; the Ni-C bond length is 1.8Å; the CO bond is tilted from the surface normal, with a Ni-C-O bond angle of 146 ± 10° if one assumes a C-O bond length of 1.15Å [the determined CO inter-layer spacing, i.e., vertical separation distance, is 0.95 ± 0.1Å];

- Ti (10001) + p(2×2) CO: dissociated CO with adatoms in separate but equivalent 3-fold hollow sites (whether cABCA... or bABCA... is not known), with possible additional underlayer formation.

Note added in proof

Combined LEED and EELS studies show that stable acetylene on Pt(111) actually stands about perpendicularly to the surface in threefold hollow sites, probably as ethylidyne ⯈ $C-CH_3$. The Pt-C bond lengths are 2.00 ± 0.05Å and the C-C bond length is 1.50 ± 0.05Å, the position of the hydrogen atoms remaining unknown [L.L. Kesmodel, L.H. Dubois and G.A. Somorjai: Chem. Phys. Lett. <u>56</u>, 267 (1978)].

12.4 Bibliography

Aℓ (111), (100), (110)

D.S. Boudreaux, V. Hoffstein: Phys. Rev. B3, 2447 (1971): (111), (110)
Groupe d'Etude des Surfaces: Surf. Sci. 62, 567 (1977): (100), (110)
D.W. Jepsen, P.M. Marcus, F. Jona: Phys. Rev. B6, 3684 (1972): (111), (100),
   (110); Phys. Rev. Lett. 26, 1365 (1971): (100)
G.E. Laramore, C.B. Duke: Phys. Rev. B5, 267 (1972): (111), (100), (110)
M.R. Martin, G.A. Somorjai: Phys. Rev. B7, 3607 (1973): (111), (110)
N. Masud, C.G. Kinniburgh, J.B. Pendry: J. Phys. C10, 1 (1977): (100), (110)
   by MEED
R.H. Tait, S.Y. Tong, T.N. Rhodin: Phys. Rev. Lett. 28, 553 (1972): (100),
   (110)
S.Y. Tong, T.N. Rhodin: Phys. Rev. Lett. 26, 711 (1971): (100)

Aℓ (100) + c(2×2) Na

B.A. Hutchins, T.N. Rhodin, J.E. Demuth: Surf. Sci. 54, 419 (1976)
M. Van Hove, S.Y. Tong, N Stoner: Surf. Sci. 54, 259 (1976)

Ag (111), (100), (110)

F. Forstmann: Jap. J. of Appl. Phys., Suppl. 2, Part 2, 657 (1974): (111)
D.W. Jepsen, P.M. Marcus, F. Jona: Phys. Rev. B8, 5523 (1973): (100)
M. Maglietta, E. Zanazzi, F. Jona, D.W. Jepsen, P.M. Marcus: J. Phys. C10,
   3287 (1977): (110)
W. Moritz: Doctoral Thesis (University of Munich, 1976): (100), (110)
T.C. Ngoc, M.G. Lagally, M.B. Webb: Surf. Sci. 35, 117 (1973): (111), by
   averaging
F. Soria, J.L. Sacédon, P.M. Echenique, D. Titterington: Surf. Sci. 68,
   448 (1977): (111)
E. Zanazzi, F. Jona, D.W. Jepsen, P.M. Marcus: J. Phys. C10, 375 (1977):
   (110)

Ag (111) + p(1×1) Au

F. Soria, J.L. Sacédon, P.M. Echenique, D. Titterington: Surf. Sci. 68,
   448 (1977)

Ag (100) + c(2×2) Cℓ

E. Zanazzi, F. Jona, D.W. Jepsen, P.M. Marcus: Phys. Rev. B14, 432 (1976)

Ag (111) + ($\sqrt{3}\times\sqrt{3}$)R30° I

F. Forstmann, W. Berndt, P. Büttner: Phys. Rev. Lett. 30, 17 (1973)

Ag (110) + (2×1) O

E. Zanazzi, M. Maglietta, U. Bardi, F. Jona, D.W. Jepsen, P.M. Marcus: Proc.
   7th IVC and 3rd ICSS (Vienna 1977) p. 2447

Ag (100) + c(2×2) Se

A. Ignatiev, F. Jona, D.W. Jepsen, P.M. Marcus: Surf. Sci. 40, 439 (1973)

Ag (111) + Incommensurate Xe

P.I. Cohen, J. Unguris, M.B. Webb: Surface Sci. 58, 429 (1976)
N. Stoner, M.A. Van Hove, S.Y. Tong, M.B. Webb: Phys. Rev. Lett. 40, 243
  (1978)
M.B. Webb, P.I. Cohen: CRC Solid State Sci. 6, 253 (1976)

Au (111), (100), (110)

R. Feder: Surf. Sci. 68, 229 (1977): (100)
R. Feder, N. Müller, D. Wolf: Z. f. Phys. B28, 265 (1977): (110)
F. Soria, J.L. Sacédon, P.M. Echenique, D. Titterington: Surf. Sci. 68,
  448 (1977): (111)

Be (0001)

J.A. Strozier, R.O. Jones: Phys. Rev. B3, 3228 (1971)

Cd (0001)

H.D. Shih, F. Jona, D.W. Jepsen, P.M. Marcus: Commun. on Physics 1, 25 (1976)

Co (0001), (111), (100)

B.W. Lee, R. Alsenz, A. Ignatiev, M.A. Van Hove: Bull. Am. Phys. Soc. 22,
  357 (1977): hcp (0001), fcc (111)
B.W. Lee, R. Alsenz, A. Ignatiev, M.A. Van Hove: Phys. Rev. B17, 1510 (1978):
  hcp (0001), fcc (111)
M. Maglietta, E. Zanazzi, F. Jona: Bull. Am. Phys. Soc. 22, 355 (1977):
  fcc (100)

CoO (111)

A. Ignatiev, B.W. Lee, M.A. Van Hove: Proc. 7th IVC and 3rd ICSS (Vienna
  1977) p. 1733

Co (100) + c(2×2) O

M. Maglietta, E. Zanazzi, U. Bardi, F. Jona, D.W. Jepsen, P.M. Marcus:
  Surf. Sci. 77, 101 (1978)

Cu (111), (100)

D.L. Adams, U. Landman: Phys. Rev. B15, 3775 (1977): (100), by Fourier-
  transform deconvolution
G. Capart: Surf. Sci. 26, 429 (1971): (100)
G.G. Kleiman, J.M. Burkstrand: Surf. Sci. 50, 493 (1975): (100), by averaging
G.E. Laramore: Phys. Rev. B9, 1204 (1974): (111), (100)
P.M. Marcus, D.W. Jepsen, F. Jona: Surf. Sci. 31, 180 (1972): (100)

J.B. Pendry: J. Phys. C4, 2514 (1971): (100)
J.B. Pendry: *Low Energy Electron Diffraction* (Academic Press, London 1974):
  (100)

Cu (100) + c(2×2) N

J.M. Burkstrand, G.G. Kleiman, G.G. Tibbetts, J.C. Tracy: J. Vac. Sci.
  Technol. 13, 291 (1976): by averaging
J.M. Burkstrand, S.Y. Tong, M.A. Van Hove: Unpublished

Cu (100) + c(2×2) O

L. McDonnell, D.P. Woodruff, K.A.R. Mitchell: Surf. Sci. 45, 1 (1974): by
  averaging

Cu (100) + p(2×2) Te

A. Salwén, J. Rundgren: Surf. Sci. 53, 523 (1975)

Fe (110), (100), (111)

R. Feder: Phys. Stat. Sol. 58, K137 (1973): (100), relativistic
R. Feder, G. Gafner: Surf. Sci. 57, 45 (1976): (110), relativistic
K.O. Legg, F. Jona, D.W. Jepsen, P.M. Marcus: J. Phys. C10, 937 (1977): (100)
H.D. Shih, F. Jona, D.W. Jepsen, P.M. Marcus: Bull. Am. Phys. Soc. 22, 357
  (1977): (111)

Fe (100) + p(1×1) O

K.O. Legg, F. Jona, D.W. Jepsen, P.M. Marcus: J. Phys. C8, L492 (1975)
K.O. Legg, F. Jona, D.W. Jepsen, P.M. Marcus: Phys. Rev. B16, 5271 (1977)

Fe (100) + c(2×2) S

R. Feder, H. Viefhaus: To be published, 1978
K.O. Legg, F. Jona, D.W. Jepsen, P.M. Marcus: Surf. Sci. 66, 25 (1977)

GaAs (110), (100)

C.B. Duke, A.R. Lubinsky, B.W. Lee, P. Mark: J. Vac. Sci. Technol. 13, 761
  (1976): (110)
A. Kahn, G. Cisneros, M. Bonn, P. Mark, C.B. Duke: Surf. Sci. 71, 387 (1978):
  (110), kinematical
A.R. Lubinsky, C.B. Duke, B.W. Lee, P. Mark: Phys. Rev. Lett. 36, 1058 (1976):
  (110)
P. Mark, G. Cisneros, M. Bonn, A. Kahn, C.B. Duke, A. Paton, A.R. Lubinsky:
  J. Vac. Sci. Technol. 14, 910 (1977): (110)
B.J. Mrstik, M.A. Van Hove, S.Y. Tong: Bull. Am. Phys. Soc. 23, 391 (1978):
  (100)
S.Y. Tong, A.R. Lubinsky, B.J. Mrstik, M.A. Van Hove: Phys. Rev. B17, 3303
  (1978): (110)
S.Y. Tong, M.A. Van Hove, B.J. Mrstik: Proc. 7th IVC and 3rd ICSS (Vienna
  1977) p. 2407: (100)

GaAs (110) + (1×1) As

B.J. Mrstik, M.A. Van Hove, S.Y. Tong: Bull. Am. Phys. Soc. <u>23</u>, 391 (1978)
S.Y. Tong, M.A. Van Hove, B.J. Mrstik: Proc. 7th IVC and 3rd ICSS (Vienna 1977) p. 2407

Ir (111)

C.-M. Chan, S.L. Cunningham, M.A. Van Hove, W.H. Weinberg, S.P. Withrow: Surf. Sci. <u>66</u>, 394 (1977)

LiF (100)

G.E. Laramore, A.C. Switendick: Phys. Rev. B<u>7</u>, 3615 (1973)

MgO (100)

C.G. Kinniburgh: J. Phys. C<u>8</u>, 2382 (1975)
C.G. Kinniburgh: J. Phys. C<u>9</u>, 2695 (1976)

Mo (100)

L.J. Clarke: Proc. 7th IVC and 3rd ICSS (Vienna 1977) p. A-2725
T.E. Felter, R.A. Barker, P.J. Estrup: Phys. Rev. Lett. <u>38</u>, 1138 (1977)
A. Ignatiev, F. Jona, H.D. Shih, D.W. Jepsen, P.M. Marcus: Phys. Rev. B<u>11</u>, 4787 (1975)

Mo (100) + C(2×2) N

A. Ignatiev, F. Jona, D.W. Jepsen, P.M. Marcus: Surf. Sci. <u>49</u>, 189 (1975)

Mo (100) + p(2×1) O

L.J. Clarke: Proc. 7th IVC and 3rd ICSS (Vienna 1977) p. A-2725

MoS_2 (0001)

B.J. Mrstik, R. Kaplan, T.L. Reinecke, M. Van Hove, S.Y. Tong: Phys. Rev. B<u>15</u>, 897 (1977)
B.J. Mrstik, R. Kaplan, T.L. Reinecke, M. Van Hove, S.Y. Tong: Il Nuovo Cimento <u>38</u>B, 387 (1977)
M.A. Van Hove, S.Y. Tong, M.H. Elconin: Surf. Sci. <u>64</u>, 85 (1977)

Mo (100) + p(1×1) Si

A. Ignatiev, F. Jona, D.W. Jepsen, P.M. Marcus: Phys. Rev. B<u>11</u>, 4780 (1975)

Na (110)

S. Andersson, J.B. Pendry, P.M. Echenique: Surf. Sci. <u>65</u>, 539 (1977)
P.M. Echenique: J. Phys. C<u>9</u>, 3193 (1976)

Na$_2$O (111)

S. Andersson, J.B. Pendry, P.M. Echenique: Surf. Sci. 65, 539 (1977)

NbSe$_2$ (0001)

B.J. Mrstik, R. Kaplan, T.L. Reinecke, M. Van Hove, S.Y. Tong: Phys. Rev
   B15, 897 (1977)
B.J. Mrstik, R. Kaplan, T.L. Reinecke, M. Van Hove, S.Y. Tong: Il Nuovo
   Cimento 38B, 387 (1977)
M.A. Van Hove, S.Y. Tong, M.H. Elconin: Surf. Sci. 64, 85 (1977)

Ni (111), (100), (110)

D.L. Adams, U. Landman: Phys. Rev. B15, 3775 (1977): (100), by Fourier-
   transform deconvolution
J.E. Demuth, P.M. Marcus, D.W. Jepsen: Phys. Rev. B11, 1460 (1975): (111),
   (100), (110)
R. Feder: Phys. Rev. B15, 1751 (1977): (111), spin-polarized
G.E. Laramore: Phys. Rev. B8, 515 (1973): (111), (100)
T.C. Ngoc, M.G. Lagally, M.B. Webb: Surf. Sci. 35, 117 (1973): (111),
   by averaging
R.H. Tait, S.Y. Tong, T.N. Rhodin: Phys. Rev. Lett. 38, 553 (1972): (100),
   (110)

Ni (100) + p(2×2) C

M.A. Van Hove, S.Y. Tong: Surf. Sci. 52, 673 (1975)

Ni (100) + c(2×2) CO

S. Andersson, J.B. Pendry: Surf. Sci. 71, 75 (1978)

Ni (110) + p(2×1) H

J.E. Demuth: J. of Colloid and Interface Science 58, 184 (1977)

Ni (111) + (2×2) 2H

M.A. Van Hove, G. Ertl, K. Christmann, R.J. Behm, W.H. Weinberg: Solid St.
   Commun., in press

Ni (100) + c(2×2) Na

S. Andersson, J.B. Pendry: J. Phys. C6, 601 (1973)
S. Andersson, J.B. Pendry: Solid St. Commun. 16, 563 (1975)
J.E. Demuth, D.W. Jepsen, P.M. Marcus: J. Phys. C8, L25 (1975)

NiO (100)

C.G. Kinniburgh, J.A. Walker: Surf. Sci. 63, 274 (1977)

Ni (100) + c/p(2×2) O

H.H. Brongersma, J.B. Theeten: Surf. Sci. $\underline{54}$, 519 (1976): c(2×2), by
ion scattering
J.E. Demuth, D.W. Jepsen, P.M. Marcus: Phys. Rev. Lett. $\underline{31}$, 540 (1973):
c(2×2)
M. Van Hove, S.Y. Tong: J. Vac. Sci. Technol. $\underline{12}$, 230 (1975): p(2×2)

Ni (110) + p(2×1) O

P.M. Marcus, J.E. Demuth, D.W. Jepsen: Surf. Sci. $\underline{53}$, 501 (1975)

Ni (100) + c/p(2×2) S

J.E. Demuth, D.W. Jepsen, P.M. Marcus: Phys. Rev. Lett. $\underline{31}$, 540 (1973):
c(2×2)
Groupe d'Etude des Surfaces: Surf. Sci. $\underline{48}$, 577 (1975): c(2×2)
H.D. Hagstrum, G.E. Becker: J. Vac. Sci. Technol. $\underline{14}$, 369 (1977): c and
p(2×2), by ion-neutralization spectroscopy
M. Van Hove, S.Y. Tong: J. Vac. Sci. Technol. $\underline{12}$, 230 (1975): p(2×2)

Ni (111) + p(2×2) S, Ni (110) + c(2×2) S

J.E. Demuth, D.W. Jepsen, P.M. Marcus: Phys. Rev. Lett. $\underline{32}$, 1182 (1974)
P.M. Marcus, J.E. Demuth, D.W. Jepsen: Surf. Sci. $\underline{53}$, 501 (1975): reviews
adsorption of chalcogens on nickel

Ni (100) + c/p(2×2) S + c/p(2×2) Na

S. Andersson, J.B. Pendry: J. Phys. C$\underline{9}$, 2721 (1976)

Ni (100) + c/p(2×2) Se/Te

J.E. Demuth, D.W. Jepsen, P.M. Marcus: Phys. Rev. Lett. $\underline{31}$, 540 (1973):
c(2×2) Se, Te
J.E. Demuth, D.W. Jepsen, P.M. Marcus: J. Phys. C$\underline{6}$, L307 (1973): c(2×2) Te
J.E. Demuth, P.M. Marcus, D.W. Jepsen: Phys. Rev. Lett. $\underline{32}$, 1182 (1974):
c(2×2) Te
M. Van Hove, S.Y. Tong: J. Vac. Sci. Technol. $\underline{12}$, 230 (1975): p(2×2) Se, Te

Pt (111), (100)

E. Bøgh, I. Stensgaard: Proc. 7th IVC and 3rd ICSS (Vienna 1977) p. A-2757:
(111), by ion channeling
R. Feder: Surf. Sci. $\underline{68}$, 229 (1977): (100)
L.L. Kesmodel, G.A. Somorjai: Phys. Rev. B$\underline{11}$, 630 (1975): (111)
L.L. Kesmodel, P.C. Stair, G.A. Somorjai: Surf. Sci. $\underline{64}$, 342 (1977): (111)
J.F. van der Veen, R.G. Smeenk, F.W. Saris: Proc. 7th IVC and 3rd ICSS
(Vienna 1977) p. 2515: (111), by ion channeling

Pt (111) + p(2×2) C_2H_2

L.L. Kesmodel, P.C. Stair, R.C. Baetzold, G.A. Somorjai: Phys. Rev. Lett.
$\underline{36}$, 1316 (1976)

L.L. Kesmodel, R.C. Baetzold, G.A. Somorjai: Surf. Sci. 66, 299 (1977)
W.J. Lo, Y.W. Chung, L.L. Kesmodel, P.C. Stair, G.A. Somorjai: Sol. St.
   Comm. 22, 335 (1977)

Rh (111), (100)

D.C. Frost, K.A.R. Mitchell, F.R. Shepherd, P.R. Watson: Proc. 7th IVC and
   3rd ISCC (Vienna 1977) p. A-2725: (111), (100)
K.A.R. Mitchell, F.R. Shepherd, P.R. Watson, D.C. Frost: Surf. Sci. 64,
   737 (1977): (100)

Si (111) (1×1), (7×7)

J.D. Levine, S.H. McFarlane, P. Mark: Phys. Rev. B16, 5415 (1977): (7×7),
   kinematic
P. Mark, J.D. Levine, S.H. McFarlane: Phys. Rev. Lett. 38, 1408 (1977):
   (7×7), kinematic
H.D. Shih, F. Jona, D.W. Jepsen, P.M. Marcus: Phys. Rev. Lett. 37, 1622 (1976):
   (1×1)

Si (100) p(2×1)

J.A. Appelbaum, D.R. Hamann: Surf. Sci., 74, 21 (1978)
A. Ignatiev, F. Jona, M. Debe, D.E. Johnson, S.J. White, D.P. Woodruff: J.
   Phys. C10, 1109 (1977): experiment
F. Jona, H.D. Shih, A. Ignatiev, D.W. Jepsen, P.M. Marcus: J. Phys. C10, L67
   (1977)
K.A.R. Mitchell, M.A. Van Hove: Surf. Sci. 75, 147L (1978)
T.D. Poppendieck, T.C. Ngoc, M.B. Webb: Surf. Sci. 75, 287 (1978)
S.Y. Tong, A.L. Maldonado: Surf. Sci., in press, 1978
S.Y. Tong, M.A. Van Hove: Unpublished

Si (100) + (1×1) H

S.J. White, D.P. Woodruff, B.W. Holland, R.S. Zimmer: Surf. Sci. 68, 457
   (1977)

Ti (0001)

H.D. Shih, F. Jona, D.W. Jepsen, P.M. Marcus: J. Phys. C9, 1405 (1976)

Ti (0001) + p(1×1) Cd multilayers

H.D. Shih, F. Jona, D.W. Jepsen, P.M. Marcus: Commun. On Physics 1, 25 (1976)
H.D. Shih, F. Jona, D.W. Jepsen, P.M. Marcus: Phys. Rev. B15, 5550, 5561
   (1977)

Ti (0001) + p(2×2) CO

H.D. Shih, F. Jona, D.W. Jepsen, P.M. Marcus: J. Vac. Sci. Technol. 15, 596
   (1978)

W (110), (100)

M.K. Debe, D.A. King, F.S. March: Surf. Sci. 68, 437 (1977): (100)

R. Feder: Phys. Stat. Solidi b62, 135 (1974): (110), (100)
R. Feder: Phys. Rev. Letters 36, 598 (1976): (100)
R. Feder: Surf. Sci. 63, 283 (1977): (100)
L.C. Feldman, R.L. Kauffman, P.J. Silverman, R.A. Zunr, J.H. Barrett: Phys.
    Rev. Lett. 39, 38 (1977): (100), by ion channeling
M. Kalisvaart, T.W. Riddle, F.B. Dunning, G.K. Walter: 37th Conf. on Phys.
    Electr. (Stanford 1977): (100)
J. Kirschner, R. Feder: Verh. Deutsch. Physik. Ges. 2, 557 (1978): (100)
M.G. Lagally, J.C. Buchholz, G.C. Wang: J. Vac. Sci. Technol. 12, 213 (1975):
    (110)
B.W. Lee, A. Ignatiev, S.Y. Tong, M. Van Hove: J. Vac. Sci. Technol. 14, 291
    (1977): (100)
M.A. Van Hove, S.Y. Tong: Surf. Sci. 54, 91 (1976): (110), (100)

W (100) c(2×2)

R.A. Barker, P.J. Estrup, F. Jona, P.M. Marcus: Sol. St. Comm. 25, 375 (1978)
M.K. Debe, D.A. King: Phys. Rev. Lett. 39, 708 (1977)
M.K. Debe, D.A. King: J. Phys. C10, L303 (1977)
T.E. Felter, R.A. Barker, P.J. Estrup: Phys. Rev. Lett. 38, 1138 (1977)

W (110) + p(2×1) O

M.G. Lagally, J.C. Buchholz, G.C. Wang: J. Vac. Sci. Technol. 12, 213 (1975):
    by averaging
M.A. Van Hove, S.Y. Tong: Phys. Rev. Lett. 35, 1092 (1975)
M.A. Van Hove, S.Y. Tong, M.H. Elconin: Surf. Sci. 64, 85 (1977)

Xe (100)

A. Ignatiev, J.B. Pendry, T.N. Rhodin: Phys. Rev. Lett. 26, 189 (1971)

Zn (0001)

W.N. Unertl, H.V. Thapliyal: J. Vac. Sci. Technol. 12, 263 (1975): by
    averaging

ZnO (0001), (10$\bar{1}$0)

C.B. Duke, A.R. Lubinsky: Surf. Sci. 50, 605 (1975): (0001)
C.B. Duke, A.R. Lubinsky, S.C. Chang, B.W. Lee, P. Mark: Phys. Rev. B15,
    4865 (1977): (1010)
C.B. Duke, A.R. Lubinsky, B.W. Lee, P. Mark: J. Vac. Sci. Technol. 13,
    761 (1976): (1010)
A.R. Lubinsky, C.B. Duke, S.C. Chang, B.W. Lee, P. Mark: J. Vac. Sci.
    Technol. 13, 189 (1976): (10$\bar{1}$0)

ZnSe (110)

C.B. Duke, A.R. Lubinsky, M. Bonn, G. Cisneros, P. Mark: J. Vac. Sci.
    Technol. 14, 294 (1977)
P. Mark, G. Cisneros, M. Bonn, A. Kahn, C.B. Duke, A. Paton, A.R.
    Lubinsky: J. Vac. Sci. Technol. 14, 910 (1977)

Appendix A. Symmetry Among Plane Waves

If an electron beam is incident on a crystal surface along an axis or a plane of symmetry of the surface structure, the diffracted plane waves will obviously have correspondingly symmetrical amplitudes. Standard group theory can be applied in this situation to reduce greatly the required computational effort, both in storage and in time. Here we shall consider the effects of symmetry on the diffracted plane waves travelling between atomic layers.

Two mutually symmetrical plane waves a_1 and a_2 can be combined to produce waves $b_s = (1/\sqrt{2})(a_1 + a_2)$ and $b_a = (1/\sqrt{2})(a_1 - a_2)$ that are respectively symmetrical and antisymmetrical about the axis or plane of symmetry relating a_1 and a_2: if the two waves a_1 and a_2 have equal amplitude due to the primary incident beam being parallel to the symmetry axis or plane, then the antisymmetrical combination will have zero amplitude. In the same symmetry situation, a layer-diffraction matrix

$$M = \begin{pmatrix} m_{11} & m_{12} \\ m_{21} & m_{22} \end{pmatrix} \quad \text{operating on the vector} \quad \begin{pmatrix} a_1 \\ a_2 \end{pmatrix}$$

has the symmetries $m_{11} = m_{22}$ and $m_{12} = m_{21}$ (if the origin of coordinates used in defining M lies on the symmetry axis or plane). Therefore M changes under the basis set transformation

$$\begin{pmatrix} a_1 \\ a_2 \end{pmatrix} \rightarrow \begin{pmatrix} b_s \\ b_a \end{pmatrix} = T \begin{pmatrix} a_1 \\ a_2 \end{pmatrix}, \quad \text{where } T = \frac{1}{\sqrt{2}} \begin{pmatrix} 1 & 1 \\ 1 & -1 \end{pmatrix},$$

into

$$M' = TMT^{-1} \begin{pmatrix} m_{11} + m_{21} & 0 \\ 0 & m_{11} - m_{21} \end{pmatrix}. \tag{A.1}$$

Due to the block-diagonal form of (A.1), the antisymmetrical wave b_a is independent of the symmetrical waves b_s and can be dropped from our calculations, because it is not excited by the wavefield incident on the crystal surface. Thus the matrix dimensions are reduced accordingly.

The basic symmetry exploitation described above extends to all types of symmetry encountered at crystal surfaces (mirror planes, 2, 3, 4, and 6 fold symmetry axes and combinations thereof). In general, matrix reduction is performed in the following way. Let us now describe a plane wave $a_{\underline{g}}$ together with the group of I-1 plane waves related to it by symmetry by the quantities $a_{\underline{g}i}$ ($i = 1,2, \ldots, I$). We obtain a symmetrical wavefunction by building

$$b_{\underline{g}} = I^{-1/2} \sum_{i=1}^{I} a_{\underline{g}i} ;$$

nonsymmetrical wavefunctions need not be considered, as they are not excited by the incident beam. We designate the diffraction matrix elements by

$$M_{\underline{g}'j,\underline{g}i}$$

($i = 1,2, \ldots, I$; $j = 1,2, \ldots, J$; I and J can be different). Here, J represents another group (or the same group) of plane waves related to each other by the same symmetry. The matrix element for diffraction from the symmetrical wavefunction $b_{\underline{g}}$ into the symmetrical wavefunction

$$b_{\underline{g}'} = J^{-1/2} \sum_{j=1}^{J} a_{\underline{g}'j}$$

becomes, generalizing the steps leading to (A.1),

$$M'_{\underline{g}'\underline{g}} = \sqrt{J/I} \sum_{i=1}^{I} M_{\underline{g}'1,\underline{g}i} . \qquad (A.2)$$

As will be apparent in Chaps. 5-7, the matrix element $M_{\underline{g}'j,\underline{g}i}$ depends on i only through the spherical harmonics

$$Y_{\ell-m} [\Omega(\underline{k}_{\underline{g}i})] ,$$

the absolute value $|\underline{k}_{\underline{g}i}|$ being independent of i by symmetry. Actually, the azimuth $\phi(\underline{g}i)$ of $\underline{k}_{\underline{g}i}$ depends on i, and $\phi(\underline{g}i)$ appears only in the factor $\exp[-im\phi(\underline{g}i)]$ contained in the spherical harmonics. So, if we write for convenience

$$M_{\underline{g}'j,\underline{g}i} = \sum_{\ell m} Z_{\ell m}(\underline{g}'j)Y_{\ell-m} [\Omega(\underline{k}_{\underline{g}i})] \qquad (A.3)$$

we obtain for the new matrix elements, using (A.2)

$$M'_{\underline{g}'\underline{g}} = \sum_{\ell m} Z_{\ell m}(\underline{g}'1)Y_{\ell-m} [\Omega(\underline{k}_{\underline{g}1})] \sqrt{J/I} \sum_{i=1}^{I} \exp[-im\phi(\underline{g}i) + im\phi(\underline{g}1)] . \qquad (A.4)$$

This result is seen to be only a small modification of (A.3) (involving the insertion of the sum of exponentials, that takes the particular symmetry situation into account), in exchange for a substantial reduction in the diffraction matrix dimensions.

Some comments about the use of symmetry should be made here. First, we must ensure that our origins of coordinates are properly positioned with respect to the symmetry axes or planes of the surface, in order to be able to perform the reduction illustrated in (A.4)(because, if the origin of co-ordinates is away from an axis or plane of symmetry, the amplitude of two symmetrical waves will have different phases, breaking the symmetry). In the case of Bravais-lattice layers, the diffraction amplitudes $M_{\underline{g}'\underline{g}}$ given in (A.3) assume an origin or coordinates situated at an atomic site of the layer under consideration. However, only in the simplest crystal surface structures will the symmetry axes or planes cut through the center of an atom in each layer. So the amplitudes $M_{\underline{g}'\underline{g}}$ will in general have to be referred to a new origin of coordinates positioned on a symmetry axis or plane of the surface. If $\underline{s}$ is the two-dimensional vector relating the new origin or coordinates to the old one (in any particular layer), $M_{\underline{g}'\underline{g}}$ will transform to

$$\tilde{M}_{\underline{g}'\underline{g}} = \exp[i(\underline{k}_{0//} + \underline{g}')\underline{s}]M_{\underline{g}'\underline{g}} \exp[-i(\underline{k}_{0//} + \underline{g})\underline{s}]$$

$$= M_{\underline{g}'\underline{g}} \exp[i(\underline{g}' - \underline{g})\underline{s}] . \qquad (A.5)$$

This new matrix can be reduced as described above, the exponential factor of (A.5) appearing then as a factor multiplying the exponentials in (A.4). As $\underline{s}$ is layer-dependent, layers with different values of $\underline{s}$ must be represented by different sets of symmetry-reduced matrices $\tilde{M}_{\underline{g}'\underline{g}}$, according to the "registry" of each layer. This need for registry-dependent matrices slightly lessens the total reduction in computer storage; however the computation time does not suffer noticeably, because the different sets of matrices can be efficiently generated simultaneously as by-products of one single calculation.

Appendix B. Lattice Sums Over Sublattices

In the case of a surface with a superlattice, it is useful to perform the
lattice sum of (37) over "sublattices" [2]. We shall describe the case where
every bulk layer has a Bravais lattice (more generality can be obtained with
our programs). Let us define sublattices: if the ratio of unit cell areas in
the superlattice and the bulk lattice is the integer P, the bulk layer can
be regarded as composed of P equal intermeshed sublattices; each of which,
except for registry, has a 2-dimensional structure identical to that of the
superlattice. Any bulk lattice vector $\underline{R}_j^{(B)}$ from an arbitrary origin can then
be decomposed into

$$\underline{R}_j^{(B)} = \underline{R}_j^{(C)} + \underline{R}_s^{(o)}, \text{ where each } \underline{R}_s^{(o)} \ (1 \leq s \leq P)$$

defines a vector from the origin to a lattice point of the sublattice to
which $\underline{R}_j^{(B)}$ belongs and $\underline{R}_j^{(C)}$ represents a vector within the sublattice. Similarly in reciprocal space, the plane waves can be classified by sublattices
(beam sets): the set of two-dimensional reciprocal-lattice vectors $\underline{g}_j^{(C)}$ of
the surface layer (representing all plane waves) can be decomposed into P
intermeshed subsets, each of which, except for a rigid shift, is equal to
the bulk-layer reciprocal set. Thus we can write

$$\underline{g}_j^{(C)} = \underline{g}_t^{(o)} + \underline{g}_j^{(B)}, \text{ where } \underline{g}_t^{(o)} \ (1 \leq t \leq P)$$

labels the subset and $\underline{g}_j^{(B)}$ is a reciprocal-lattice vector of the substrate
lattice. The following relations between lattices and reciprocal lattices
hold:

$$\exp[i\underline{g}_t^{(o)} \cdot \underline{R}_u^{(C)}] = \exp[i\underline{g}_j^{(B)} \cdot \underline{R}_s^{(o)}] = \exp[i\underline{g}_j^{(B)} \cdot \underline{R}_u^{(C)}] = 1 \ . \tag{B.1}$$

The layer diffraction matrices in the presence of a superlattice can now
efficiently be obtained from (34-37) in the following manner. For an incident wave corresponding to

$$g_{-j}^{(C)} = g_{-t}^{(0)} + g_{-j}^{(B)}$$

the wavevector $k_{0//}$ in (37) is replaced by $k_{0//} + g_{-t}^{(0)} + g_{-j}^{(B)}$. Using (B.1), (37) now can be written as

$$F_{\ell'm'} = \sum_{s=1}^{P} \exp\left(ig_{-t}^{(0)} \cdot R_{-s}^{(0)}\right) \sum_{u}' \exp[ik_{0//} \cdot (R_{-s}^{(0)}+R_{-u}^{(C)})] h_{\ell'}^{(1)}(k_0 |R_{-s}^{(0)}+R_{u}^{(C)}|)$$

$$\times (-1)^{m'} \exp[-im'\phi(R_{-s}^{(0)}+R_{-u}^{(C)})] \equiv \sum_{s=1}^{P} \exp\left(ig_{-t}^{(0)} \cdot R_{-s}^{(0)}\right) F_{\ell'm's} , \qquad (B.2)$$

where $\sum'$ excludes summation over $R_{-s}^{(0)} + R_{-u}^{(C)} = 0$. Each of the quantities $F_{\ell'm's}$ ($1 \le s \le P$) is a sum over one of the sublattices rather than over the full substrate lattice. Whereas for each different $g_{-t}^{(0)}$ ($1 \le t \le P$) direct application of (37) requires P summations over the full substrate lattice, using (B.2), one sums just once over all substrate lattice points (separated according to P sublattices) and then combines the quantities $F_{\ell'm's}$ P times to form the quantity $F_{\ell'm'}$ relevant to each $g_{-t}^{(0)}$, i.e., to each subset of beams that are independent in the bulk. This constitutes a valuable computational economy.

In addition, that $F_{\ell'm's}$ for which $R_{-s}^{(0)} = 0$ is just the lattice sum appropriate for a single Bravais-lattice layer with the superlattice periodicity (e.g., an overlayer) and also for a subplane of a composite layer.

Note: In the programs, the input quantities NL1 and NL2 help define the sublattices. Each of the lattice points $R_{-s}^{(0)} = I * ARA1 + J * ARA2$ with I = 0, 1,...,NL1-1 and J = 0,1,...,NL2-1 (where ARA1 and ARA2 are the substrate lattice basis vectors) is assumed to belong to a different sublattice: therefore NL1 and NL2 should be chosen by the user accordingly. (The product NL1 * NL2 should be equal to the number $\underline{P}$ of sublattices.) Thus, with a c(2×2) superlattice on fcc(100), NL1 = 2 and NL2 = 1 will produce the lattice points (0,0) and ARA1, which indeed belong to different sublattices.

Appendix C. A Line-Printer Plotting Program

A pen-plotter program to draw IV-curves is very installation-dependent and therefore not generally usable. On the other hand it is not entirely straightforward to produce a satisfactory plotting program for the particular purposes of a LEED analysis: such a program should include, among other features, the capability of domain-averaging, non-linear interpolation onto a fine grid (so that the time-consuming LEED programs can work on the coarsest and therefore most economical grid possible) and multiple-curve plotting to compare conveniently IV-curves for different geometries. For flexibility of use, it is convenient to not include such a plotting program in the LEED programs themselves.

For those reasons we list in this Appendix an independent program for line-printer plotting, with which one can plot calculated (and also digitized experimental) IV-curves, or produce by suitable extensions a pen-plotter program adapted to the user's particular hardware, or also an R-factor program.

The required form of the input is illustrated after the program with an example. The second line of this input, the lines with integers (KAV) describing the beam groupings and the last lines containing a negative energy are to be generated by the user, while all other lines are the punch output (unit 7) of the LEED programs.

```
C    MAIN PROGRAM FOR PRINTER-PLOTTING OF LEED IV-CURVES,
C    INCLUDING DOMAIN-AVERAGING AND 3RD-ORDER POLYNOMIAL
C    INTERPOLATION.
C      DIMENSIONS: 125.GE. NO. OF ENERGIES READ IN (NE) OR
C                         INTERPOLATED TO (NPTS).
C                  10 .GE. NO. OF GEOMETRIES (NS).
C                  19 .GE. NO. OF BEAMS READ IN (NB).
       DIMENSION WORX(125),WORY(10,125),ES(125),AT(10,19,100)
       DIMENSION PQ(2,19),PQAV(2,19),VALUES(10),KAV(19)
       DIMENSION XPL(125),YPL(10,125),ATAV(10,19,100),TEXT(20)
       INTEGER SYM(19)
C    READ AND WRITE DESCRIPTIVE TITLE (SUPPLIED BY LEED PROGRAM)
       READ(5,1)TEXT
       WRITE(6,11)TEXT
1      FORMAT(20A4)
11     FORMAT(1H ,20A4)
C    INCR IS STEP IN PLOTTING ENERGY GRID (EV)
       INCR=2
C    NS IS NO. OF GEOMETRIES TO BE READ IN (TO BE SUPPLIED
C    BY USER).
C    NSPL IS NO. OF GEOMETRIES ALLOWED PER GRAPH (TO BE SUPPLIED
C    BY USER)
       READ(5,2)NS,NSPL
C    NB IS NO. OF BEAMS TO BE READ IN (SUPPLIED BY THE LEED
C    PROGRAM)
       READ(5,2)NB
2      FORMAT(3I3)
C    READ IN CALCULATED ENERGIES AND INTENSITIES, AND INFORMATION
C    FOR DOMAIN-AVERAGING
       CALL RDPL(ES,AT,NS,NB,NE,VALUES,PQ,KAV,SYM)
C    PERFORM DOMAIN-AVERAGING
       CALL RINTAV(AT,NS,NB,NE,PQ,KAV,SYM,ATAV,NBAV,PQAV,ES)
C    LOOP OVER BEAMS
       DO 10 IB=1,NBAV
C    FOR PRESENT BEAM MOVE INTENSITIES FROM ATAV TO WORY
       CALL STRIP(ES,ATAV,NS,NB,NE,WORX,WORY,LENGTH,IB)
C    INTERPOLATE TO PLOTTING GRID AND FIND MAXIMUM INTENSITY
       CALL STFLIN(NPTS,XPL,YPL,WORX,WORY,LENGTH,INCR,NS,AM)
C    DO NOT PLOT FOR PRESENT BEAM IF ONLY ZERO INTENSITIES
C    OCCUR
       IF (AM.LT.1.0E-6) GO TO 10
       IS1=1
5      IS2=IS1+NSPL-1
       IS2=MINO(IS2,NS)
C    PLOT IV-CURVES FOR UP TO NSPL GEOMETRIES ON ONE GRAPH
       CALL GRAPHL(XPL,YPL,NS,NPTS,AM,IS1,IS2)
       WRITE(6,9)PQAV(1,IB),PQAV(2,IB)
9      FORMAT(8HOBEAM : ,2F8.3)
       IS1=IS2+1
C    LOOP BACK TO PLOT FOR REMAINING GEOMETRIES
       IF (IS1.LE.NS) GO TO 5
10     CONTINUE
       STOP
       END
C    SUBROUTINE BINSRX FINDS A REQUIRED INTERPOLATION INTERVAL
C    BY BINARY SEARCH (SUCCESSIVE HALVING OF INITIAL INTERVAL)
       SUBROUTINE BINSRX(IL, IH, X, WORX, LENGTH)
       DIMENSION WORX(LENGTH)
       I = (LENGTH + 1) / 2
       IHI = (I + 1) / 2
       IF(X.LT. WORX(1) .OR. X .GE. WORX(LENGTH)) GO TO 100
20     CONTINUE
       IF(I .LT. 1) GO TO 40
       IF(I .GE. LENGTH) GO TO 30
       IF(X .GE.WORX(I) .AND. X .LE. WORX(I + 1)) GO TO 50
       IF(X .LT. WORX(I)) GO TO 30
40     I = IHI + I
```

```
         IHI = (IHI + 1) / 2
         GO TO 20
30       I = I - IHI
         IHI =(IHI + 1) / 2
         GO TO 20
50       IL = I
         IH = I + 1
         GO TO 110
100      IF(X .LT. WORX(1)) GO TO 105
102      IL = LENGTH - 1
         IH = LENGTH
         GO TO 110
105      IL = 1
         IH = 2
110      RETURN
         END
C   SUBROUTINE GRAPHL PLOTS IS2-IS1+1 CURVES ON ONE GRAPH. THE
C   CURVES ARE DISTINGUISHED BY THE GEOMETRY NUMBER IS=1,2,...,NS,
C   (MODULO 10)
         SUBROUTINE GRAPHL(X,Y,NS,N,AM,IS1,IS2)
         DIMENSION X(N),Y(NS,N),GR(54,132),SYM(4),ANUM(10)
         DATA ANUM/1H1,1H2,1H3,1H4,1H5,1H6,1H7,1H8,1H9,1H0/
         DATA SYM/1H ,1H+,1H-,1HI/
         WRITE(6,5)
5        FORMAT(1H1)
C   PRODUCE GRID AND ENERGY SCALE
         DO 8 I=1,132
         DO 8 J=1,54
8        GR(J,I)=SYM(1)
         DO 12 I=12,132
         DO 12 J=1,51,10
12       GR(J,I)=SYM(3)
         DO 14 I=12,132,20
         DO 14 J=1,51
14       GR(J,I)=SYM(4)
         DO 20 I=62,102,10
20       GR(52,I)=ANUM(1)
         DO 30 I=112,132,10
30       GR(52,I)=ANUM(2)
         DO 40 I=22,122,50
40       GR(53,I)=ANUM(2)
         DO 50 I=32,132,50
50       GR(53,I)=ANUM(4)
         DO 60 I=42,92,50
60       GR(53,I)=ANUM(6)
         DO 70 I=52,102,50
70       GR(53,I)=ANUM(8)
         DO 80 I=62,112,50
80       GR(53,I)=ANUM(10)
         DO 90 I=12,132,10
90       GR(54,I)=ANUM(10)
C   POSITION PLOTTING POINTS IN GRAPH
         DO 110 J=IS1,IS2
         DO 110 I=1,N
         IF (Y(J,I).LT.0.0) Y(J,I)=0.0
         MY=51-IFIX(50.0*Y(J,I)/AM+0.5)
         MX=12+IFIX(X(I)*0.5+0.5)
         IS=MOD(J,10)
         IF (IS.EQ.0) IS=10
         GR(MY,MX)=ANUM(IS)
110      CONTINUE
         WRITE (6,115) AM
115      FORMAT(29H INTENSITY AT TOP OF FRAME = ,1E13.5)
         WRITE(6,120)((GR(I,J),J=1,132),I=1,54)
120      FORMAT(1H ,/,54(132A1,/))
         RETURN
         END
```

```
C     SUBROUTINE RDPL READS LEED INTENSITIES PUNCHED OUT BY LEED
C     SUBROUTINE RINT, FOR FURTHER TREATMENT IN AN INDEPENDENT PLOTTING,
C     LISTING, ETC. PROGRAM.
C     E= READ ENERGY GRID.
C     YDATA(K,J,I)= READ INTENSITIES: K INDEXES GEOMETRIES, J INDEXES
C     BEAMS, I INDEXES ENERGIES.
C     NS= INPUT NO. OF GEOMETRIES USED IN THE LEED COMPUTATION.
C     NB= INPUT NO. OF BEAMS FOR WHICH INTENSITIES WERE PUNCHED OUT.
C     NE= OUTPUT NO. OF ENERGIES ON GRID.
C     SPAC= NUMBERS SPECIFYING THE VARIOUS GEOMETRIES (E.G. INTERLAYER
C     SPACINGS).
C     PQ= READ BEAMS.
C     KAV= READ BEAM GROUPINGS (BEAMS WITH THE SAME VALUE OF KAV WILL BE
C     AVERAGED TOGETHER).
C     SYM= READ BEAM SYMMETRY CODES (USED TO DETERMINE THE AVERAGING
C     WEIGHTS).
C     NOTE: SUPPLY ABOUT 5 BLANK LINES AT END OF INPUT, AFTER A CARD WITH
C     NEGATIVE ENERGY SIGNIFYING THE END OF THE INPUT DECK.
      SUBROUTINE RDPL(E,YDATA,NS,NB,NE,SPAC,PQ,KAV,SYM)
      DIMENSION E(1),YDATA(NS,NB,1),SPAC(NS),LAB(30),PQ(2,NB),KAV(NB)
      DIMENSION SP(50),YD(50)
      INTEGER SYM(NB)
C     READ BEAM NAMES (PQ), IDENTIFIERS (LAB=NPU OF LEED PROGRAM;
C     NOT USED HERE) AND SYMMETRY PROPERTIES (SYM), AS SUPPLIED BY
C     THE LEED PROGRAM
      DO 10 J=1,NB
10    READ(5,20) LAB(J),(PQ(I,J),I=1,2),SYM(J)
20    FORMAT(I3,2F6.3,I3)
C     READ BEAM GROUPINGS FOR AVERAGING (TO BE SUPPLIED BY USER):
C     KAV=1 FOR THE FIRST GROUP, KAV=N FOR THE N-TH GROUP, ETC.
C     (KAV.LE.NB MUST ALWAYS HOLD)
      DO 25 J=1,NB
25    READ(5,26)KAV(J)
26    FORMAT(2I3)
      WRITE(6,27)
27    FORMAT(15H0BEAM GROUPINGS)
      DO 28 J=1,NB
28    WRITE(6,29)(PQ(I,J),I=1,2),KAV(J)
29    FORMAT(1H ,2F7.3,I5)
C     READ ENERGIES (E), GEOMETRICAL CHARACTERIZATION (SPAC) AND
C     INTENSITIES (YDATA), AS SUPPLIED BY THE LEED PROGRAM
      I=1
30    DO 50 J=1,NS
      READ (5,40) E(I),SP(J),(YD(K),K=1,NB)
40    FORMAT(F7.2,F7.4,4E14.5,/,/5(5E14.5,/))
C     CHECK FOR END OF INPUT
      IF (E(I).LT.0.0) GO TO 60
      SPAC(J)=SP(J)
      DO 35 K=1,NB
35    YDATA(J,K,I)=YD(K)
50    CONTINUE
      I=I+1
      GO TO 30
60    NE=I-1
      WRITE(6,65)
65    FORMAT(43H0SURFACE STRUCTURE NO. AND CHARACTERIZATION)
      DO 70 J=1,NS
70    WRITE(6,75)J,SPAC(J)
75    FORMAT(18X,I3,F15.4)
      RETURN
      END
C     SUBROUTINE RINTAV AVERAGES LEED INTENSITIES OVER DOMAINS, USING
C     WEIGHTS DETERMINED BY THE SYMMETRY PROPERTIES (SYM) USED IN THE
C     LEED PROGRAM (THIS IS NECESSARY FOR THE CASE WHEN SYMMETRY HAS
C     BEEN EXPLOITED, OTHERWISE ALL WEIGHTS ARE 1).
      SUBROUTINE RINTAV(AT,NS,NB,NE,PQ,KAV,SYM,ATAV,LAVM,PQAV,ES)
      DIMENSION AT(NS,NB,NE),KAV(NB),ATAV(NS,NB,NE),PQ(2,NB),
```

```
      1PQAV(2,NB),ES(NE)
       INTEGER SYM(NB)
       LAV=1
       LAVM=0
10     CONTINUE
       DO 20 IB=1,NB
C  FIND THE FIRST BEAM OF A GROUP TO BE AVERAGED TOGETHER
       IF (LAV.NE.KAV(IB)) GO TO 20
       LAVM=LAVM+1
       DO 11 IS=1,NS
       DO 11 IE=1,NE
11     ATAV(IS,LAV,IE)=0.0
C  USE NAME OF FIRST BEAM ENCOUNTERED IN THIS GROUP (FOR PRINTING)
       DO 12 K=1,2
12     PQAV(K,LAV)=PQ(K,IB)
       WT=0.0
       DO 15 JB=IB,NB
C  FIND THE OTHER BEAMS OF THE GROUP TO BE AVERAGED WITH THE FIRST
C  BEAM ENCOUNTERED ABOVE
       IF (LAV.NE.KAV(JB)) GO TO 15
       LAB=SYM(JB)
C  DETERMINE WEIGHTING FACTOR
       IF (LAB-10) 45,100,60
45     IF (LAB-2) 70,50,50
50     IF (LAB-7) 80,90,90
60     IF (LAB-12) 110,65,65
65     IF (LAB-15) 120,130,130
70     WGHT=1.0
       GO TO 200
80     WGHT=2.0
       GO TO 200
90     WGHT=4.0
       GO TO 200
100    WGHT=8.0
       GO TO 200
110    WGHT=3.0
       GO TO 200
120    WGHT=6.0
       GO TO 200
130    WGHT=12.0
200    CONTINUE
       DO 13 IS=1,NS
       DO 13 IE=1,NE
13     ATAV(IS,LAV,IE)=ATAV(IS,LAV,IE)+WGHT*AT(IS,JB,IE)
       WT=WT+WGHT
15     CONTINUE
       DO 17 IS=1,NS
       DO 17 IE=1,NE
17     ATAV(IS,LAV,IE)=ATAV(IS,LAV,IE)/WT
       GO TO 25
20     CONTINUE
25     LAV=LAV+1
       IF (LAV.NE.NB) GO TO 10
C  PRINT OUT ALL AVERAGED INTENSITIES
       DO 40 IS=1,NS
       WRITE(6,30)IS,(PQAV(1,IB),PQAV(2,IB),IB=1,LAVM)
30     FORMAT(1H ,///,19HOSURFACE STRUCTURE ,I3,//,45H BEAMS AND BEAM INTE
      1NSITIES AFTER AVERAGING :,/,22X,8(F7.3,F6.3),/,28X,8(F7.3,F6.3),/,
      222X,8(F7.3,F6.3))
       DO 32 IE=1,NE
32     WRITE(6,35)ES(IE),(ATAV(IS,IB,IE),IB=1,LAVM)
35     FORMAT(10H ENERGY = ,1F7.2,5H EV :,8E13.5,/,28X,8E13.5,/,22X,8E13.
      15)
40     CONTINUE
       RETURN
       END
C  SUBROUTINE STFLIN INTERPOLATES LEED INTENSITIES ONTO THE PLOTTING
```

```
C    GRID (0 TO 240 EV IN STEPS OF INCR EV)
     SUBROUTINE STFLIN(NPTS,X,Y,WORX,WORY,LENGTH,INCR,NS,AM)
     DIMENSION WORYT(200),X(1),Y(NS,1),WORX(LENGTH),WORY(NS,LENGTH)
     ITIL=0
     ITIH=0
     LMIN=INT(WORX(1)+0.000001)/INCR
     LMIN=MAXO(LMIN,0)
     XMIN=FLOAT(LMIN*INCR)
     LMAX=INT(WORX(LENGTH))/INCR
     LLIM=240/INCR
     LMAX=MINO(LMAX,LLIM)
     XMAX=FLOAT(LMAX*INCR)
C    NPTS IS NO. OF POINTS USED ON THE PLOTTING GRID
     NPTS=LMAX-LMIN+1
     XINCR=FLOAT(INCR)
     XVAL = XMIN - XINCR
     DO 2 I=1,NPTS
     X(I)=XVAL+XINCR
2    XVAL=X(I)
     DO 20 J=1,NS
     DO 5 I=1,LENGTH
5    WORYT(I)=WORY(J,I)
     DO 10 I = 1, NPTS
     XVAL = X(I)
C    INTERPOLATE
10   Y(J,I) = YVAL(XVAL, WORYT, WORX, LENGTH,ITIL,ITIH)
20   CONTINUE
C    FIND MAXIMUM INTENSITY
     AM=0.0
     DO 8 J=1,NS
     DO 8 I=1,NPTS
8    IF (Y(J,I).GT.AM) AM=Y(J,I)
     RETURN
     END
C    SUBROUTINE STFPTS FINDS, GIVEN THE INTERPOLATION INTERVAL, THE
C    FOUR NEAREST GRID POINTS AND THE CORRESPONDING ORDINATE VALUES
     SUBROUTINE STFPTS(IL, IH, WORX, WORY, LENGTH)
     DIMENSION WORX(LENGTH), WORY(LENGTH), TEMP(2, 4)
     COMMON / DATBLK / X0, X1, X2, X3, Y0, Y1, Y2, Y3
     I = IL - 1
     IF(IL .LE. 1) I = I + 1
     IF(IH .GE. LENGTH) I = I - 1
     DO 10 K = 1, 4
     N = K + I - 1
     TEMP(1,K) = WORX(N)
10   TEMP(2, K) = WORY(N)
     X0 = TEMP(1, 1)
     X1 = TEMP(1, 2)
     X2 = TEMP(1, 3)
     X3 = TEMP(1, 4)
     Y0 = TEMP(2, 1)
     Y1 = TEMP(2, 2)
     Y2 = TEMP(2, 3)
     Y3 = TEMP(2, 4)
     RETURN
     END
C    SUBROUTINE STRIP COPIES INTENSITIES FROM YDATA TO WORY FOR A
C    GIVEN BEAM
     SUBROUTINE STRIP(X,YDATA,NS,NB,NE,WORX,WORY,LENGTH,IB)
     DIMENSION X(NE), YDATA(NS,NB,NE),
    2 WORX(NE), WORY(NS,NE)
     JX = 0
     DO 10 J = 1, NE
C    DO NOT COPY LEADING ZERO INTENSITIES (OCCURING WHEN A BEAM
C    DOES NOT EMERGE INTO VACUUM)
     IF (YDATA(1,IB,J).LT.1.0E-10.AND.JX.EQ.0) GO TO 10
     JX = JX + 1
```

```
      DO 5 IS=1,NS
5     WORY(IS,JX) = YDATA(IS, IB, J)
      WORX(JX) = X(J)
10    CONTINUE
      LENGTH = JX
      RETURN
      END
C  SUBROUTINE XNTERP PERFORMS 3RD-ORDER POLYNOMIAL INTERPOLATION
      FUNCTION XNTERP(X)
      COMMON / DATBLK / X0, X1, X2, X3, Y0, Y1, Y2, Y3
      TERM = Y0
      FACT1 = X - X0
      FACT2 = (Y1 - Y0) / (X1 - X0)
      TERM = TERM + FACT1 * FACT2
      FACT1 = FACT1 * (X - X1)
      FACT2 = ((Y2 - Y1)/(X2 - X1) - FACT2) / (X2 - X0)
      TERM = TERM + FACT1 * FACT2
      FACT1 = FACT1 * (X - X2)
      TEMP = ((Y3 - Y2)/(X3 - X2) - (Y2 - Y1)/(X2 - X1))/(X3 - X1)
      FACT2 = (TEMP - FACT2) / (X3 - X0)
      TERM = TERM + FACT1 * FACT2
      XNTERP = TERM
      RETURN
      END
C  FUNCTION YVAL INTERPOLATES
      FUNCTION YVAL(X, WORY, WORX, LENGTH,ITIL,ITIH)
      DIMENSION WORY(LENGTH), WORX(LENGTH)
      COMMON /DATBLK/ X0,X1,X2,X3,Y0,Y1,Y2,Y3
C  FIND REQUIRED INTERPOLATION INTERVAL
      CALL BINSRX(IL, IH, X, WORX, LENGTH)
C  SKIP NEXT STEP IF SAME INTERVAL IS FOUND AS LAST TIME
      IF(IL .EQ. ITIL .AND. IH .EQ. ITIH) GO TO 5
C  FIND FOUR NEAREST GRID POINTS AND CORRESPONDING INTENSITIES
C  FOR 3RD-ORDER POLYNOMIAL INTERPOLATION
      CALL STFPTS(IL, IH, WORX, WORY, LENGTH)
C  DO ACTUAL 3RD-ORDER POLYNOMIAL INTERPOLATION
5     Y = XNTERP(X)
      ITIH = IH
      ITIL = IL
      YVAL = Y
      RETURN
      END
----------------------------------------------------------------
------ EXAMPLE INPUT FOLLOWS ----------------------------------
----------------------------------------------------------------
 NI(111)+P(2*1)H, RSP (OR BEEBY) + RFS, 1 MIRROR PLANE, 5 PH. SH.
  4  2
 19
  1 0.000 0.000   1
  2 1.000 0.000   6
  3-1.000 1.000   1
  4 0.000 1.000   6
  5 1.000-1.000   1
  6 1.000 1.000   6
  7-1.000 2.000   6
  8 2.000-1.000   6
  9 2.000 0.000   6
 10-2.000 2.000   1
 11 0.000 2.000   6
 12 2.000-2.000   1
 13 -.500  .500   1
 14  .500 -.500   1
 15  .500  .500   6
 16-1.500 1.500   1
 17 -.500 1.500   6
 20 1.500 -.500   6
 21 1.500-1.500   1
```

1
2
2
3
3
4
4
4
5
5
6
6
7
8
9
10
11
12
13

```
40.00 1.0000      .49899E-03      .10339E-01      .10399E-01      .13756E-02
     .14041E-02   0.              0.              0.              0.
  0.               0.              0.              .27782E-04      .30052E-03
     .25011E-04   0.              0.              0.              0.


40.00 2.0000      .31923E-02      .63228E-02      .64306E-02      .30378E-02
     .34254E-02   0.              0.              0.              0.
  0.               0.              0.              .45687E-03      .15910E-02
     .13155E-02   0.              0.              0.              0.


40.00 3.0000      .24681E-02      .83947E-02      .81963E-02      .29042E-02
     .27039E-02   0.              0.              0.              0.
  0.               0.              0.              .60061E-03      .52486E-04
     .35900E-03   0.              0.              0.              0.


40.00 4.0000      .65798E-02      .92897E-02      .10009E-01      .28738E-02
     .29412E-02   0.              0.              0.              0.
  0.               0.              0.              .23067E-02      .15145E-02
     .20024E-02   0.              0.              0.              0.
```

-1.00

References

1   C. Davisson, L. Germer: Nature $\underline{119}$, 558 (1927)

2   J.B. Pendry: *Low Energy Electron Diffraction* (Academic Press, London (1974)

3   J. Rundgren, A. Salwén: Comput. Phys. Commun. $\underline{9}$, 312 (1974); J. Phys. C$\underline{9}$, 3701 (1976)

4   G.A. Somorjai: *Principles of Surface Chemistry* (Prentice Hall, Englewood Cliffs, New Jersey, 1972)

5   G. Ertl, J. Küppers: *Low Energy Electrons and Surface Chemistry,* (Verlag Chemie, Weinheim 1974)

6   J.A. Strozier, Jr., D.W. Jepsen, F. Jona: In *Surface Physics of Materials,* ed. by J.M. Blakely (Academic Press, New York 1975)

7   F. Jona: Surf. Sci. $\underline{68}$, 204 (1977)

8   L.-H. Lee (Ed.): *Characterisation of Metal and Polymer Surfaces,* (Academic Press, New York 1977) Vol. 1, Part III

9   S.Y. Tong: Prog. Surf. Sci. $\underline{7}$, 1 (1975)

10 N. Stoner, M.A. Van Hove, S.Y. Tong, M.B. Webb: Phys. Rev. Lett. $\underline{40}$ 243 (1978)

11 V.L. Moruzzi, A.R. Williams, J.F. Janak: *Calculated Electronic Properties of Metals* (Pergamon Press, in press)

12 J.L. Beeby: J. Phys. C$\underline{1}$, 82 (1968)

13 S.Y. Tong, M.A. Van Hove: Phys. Rev. B$\underline{16}$, 1459 (1977)

14 R.S. Zimmer, B.W. Holland: J. Phys. C$\underline{8}$, 2395 (1975)

15 J.B. Pendry: J. Phys. C$\underline{4}$, 2514 (1971)

16 M.A. Van Hove: In *The Nature of the Surface Chemical Bond,* ed. by T.N. Rhodin, G. Ertl (North-Holland, Amsterdam 1978)

17 S.Y. Tong, M.A. Van Hove, B.J. Mrstik: Proc. 7th IVC and 3rd ICSS, (Vienna 1977) p. 2407

Subject Index

Electron Spectroscopy for Surface Analysis

Editor: H. Ibach

1977. 123 figures, 5 tables. XI, 255 pages
(Topics in Current Physics, Volume 4)
ISBN 3-540-08078-3

Contents:
H. Ibach: Introduction. – *D. Roy, J. D. Carette:* Design of Electron Spectrometers for Surface Analysis. – *J. Kirschner:* Electron-Excited Core Level Spectroscopies. – *M. Henzler:* Electron Diffraction and Surface Defect Structure. – *B. Feuerbacher, B. Fitton:* Photoemission Spectroscopy. – *H. Froitzheim:* Electron Energy Loss Spectroscopy.

Interactions on Metal Surfaces

Editor: R. Gomer

1975. 112 figures, XI, 310 pages
(Topics in Applied Physics, Volume 4)
ISBN 3-540-07094-X

Contents:
J. R. Smith: Theory of Electronic Properties of Surfaces. – *S. K. Lyo, R. Gomer:* Theory of Chemisorption. – *L. D. Schmidt:* Chemisorption: Aspects of the Experimental Situation. – *D. Menzel:* Desorption Phenomena- – *E. W. Plummer:* Photoemission and Field Emission Spectroscopy. – *E. Bauer:* Low Energy Electron Diffraction (LEED) and Auger Methods. – *M. Boudart:* Concepts in Heterogeneous Catalysis.

Monte Carlo Methods in Statistical Physics

Editor: K. Binder

1979. 91 figures, 10 tables. XV, 376 pages
(Topics in Current Physics, Volume 7)
ISBN 3-540-09018-5

Contents:
K. Binder: Introduction: Theory and „Technical" Aspects of Monte Carlo Simulations. – *D. Levesque, J. J. Weis, J. P. Hansen:* Simulation of Classical Fluids. – *D. P. Landau:* Phase Diagrams of Mixtures and Magnetic Systems. – *R. Müller-Krumbhaar:* Simulation of Small Systems. – *K. Binder, M. H. Kalos:* Monte Carlo Studies of Relaxation Phenomena: Kinetics of Phase Changes and Critical Slowing Down. – *H. Müller-Krumbhaar:* Monto Carlo Simulation of Crystal Growth. – *K. Binder, D. Stauffer:* Monto Carlo Studies of Systems with Disorders. – *D. P. Landau:* Applications in Surface Physics.

Solid Surface Physics

Editor: G. Höhler

1979. 101 figures, 13 tables.
Approx. 230 pages
(Springer Tracts in Modern Physics, Volume 85)
ISBN 3-540-09266-8

Contents:
J. Hölzl, F. K. Schulte: Work Function of Metals. – *H. Wagner:* Physical and Chemical Properties of Stepped Surfaces.

 **Springer-Verlag
Berlin Heidelberg NewYork**

Springer Series in
Solid-State Sciences

Editors:
M. Cardona, P. Fulde, H.-J. Queisser

This series is devoted to single- and multi-author graduate-level monographs and textbooks in the areas of solid-state physics, solid-state chemistry, and solid-state technology. Also covered are semiconductor physics and technology as well as surface physics. In addition, conference proceedings which delineate the directions for significant future research are considered for publication in the series.

Springer-Verlag
Berlin
Heidelberg
New York